Power System Analysis

Power System Analysis

A. AMBIKAPATHY
(M.tech)
Assistant Professor
Electronics and Instrumentation
Engineering
Galgotia College of Engineering and

KHANNA BOOK PUBLISHING CO. (P) LTD.
Publisher of Engineering and Computer Books
4C/4344, Ansari Road, Darya Ganj, New Delhi-110002
Phone : 011-23244447-48 **Mobile:** +91-9910909320
E-mail : contact@khannabooks.com
Website : www.khannabooks.com

Dear Readers,

To prevent the piracy, this book is secured with HIGH SECURITY HOLOGRAM on the front title cover. In case you don't find the hologram on the front cover title, please write us to at contact@khannabooks.com or whatsapp us at +91-99109 09320 and avail special gift voucher for yourself.

Specimen of Hologram on front Cover title:

Moreover, there is a SPECIAL DISCOUNT COUPON for you with EVERY HOLOGRAM.

How to avail this SPECIAL DISCOUNT:

Step 1: Scratch the hologram
Step 2: Under the scratch area, your "coupon code" is available
Step 3: Logon to www.khannabooks.com
Step 4: Use your "coupon code" in the shopping cart and get your copy at a special discount
Step 5: Enjoy your reading!

ISBN: 978-93-80016-81-8

Power System Analysis
by A. Ambikapathy

Reprint Edition: 2026

Published by:
Khanna Book Publishing Co. (P) Ltd.
CIN: U22110DL1998PTC095547

Visit us at: **www.khannabooks.com**
Write us at: **contact@khannabooks.com**

To view complete list of books, Please scan the QR Code: KPH

Printed in Bharat.

PREFACE

The text book on ''Power System Analysis'' has been written for undergraduate students of electrical engineering (EE) and electrical and electronics engineering (EEE) departments. The material which is presented in this book will be very useful to the students for understanding as well as from examination point of view. A number of solved and unsolved problems (With answers) have been included in each chapter. Short questions and answers are provided for every chapter so that the students will get benefit of it. The mathematical details and qualitative discussion has been followed throughout the book. The author will welcome criticisms of hte text and will acknoledge comments pertaining to the text and suggestions for the improvement of this book will be most welcome.

A. AMBIKAPATHY

Dedicated to

Parents

Husband

Mother-in-law

Brother and Sisters

Daughter

CONTENTS

Chapter-1

REPRESENTATION OF POWER SYSTEM COMPONENTS

1.1 INTRODUCTION TO POWER SYSTEM ANALYSIS

When the electric power is transmitted from generating station to consumer end the transmission lines are sutz safeguard the power system it is important to provide protecting equipments such as circuit breaker, isolator, lightning arrestor etc. The ratings of these protecting devices should be designed in a manner that the device will be able to protect the power system elements under fault conditions so, it is necessary to preestimate the value of fault current under various conditions. This subject "Power system analysis" helps the designer for estimating the ratings of protecting devices.

1.2 REPRESENTATION OF POWER SYSTEM COMPONENTS

Generally the Power system consists of grid, generating station, substation and distribution station.

The generating stations feeds energy to grid from which all substations taps energy. From the substations electrical energy is transmitted to distribution stations (or) distribution transformers from the distribution transformer, the energy is fed to various loads.

The components of Power system are generating station i.e. Alternators, Power transformers, Transmission lines, substations and loads. The various types of loads are synchronous motors, Induction motors, heating coils, lights etc.

The various components of power system are represented by their symbol. [Usually the power system and their interconnections are represented by single line diagram]. In single line diagram the components are represented by standard symbols and their interconnections are shown by single line, even though they are 3 phase circuits.

Let we see the representation of power system components.

1. Synchronous Machine (or) General Machine

Since the machines are having rotating part in it, the symbol which is used to represent the machine is round shape.

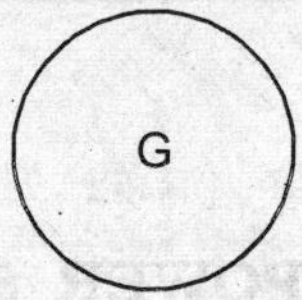

➡ Inside the circle instead of G anything can be changed in case of different machine is drawn.

2. Transformer

There are different types of transformers such as step up, step down, Two winding, three winding transformers. Let we see their representations

I. Two winding transformer

(i) Step-up-Transformer

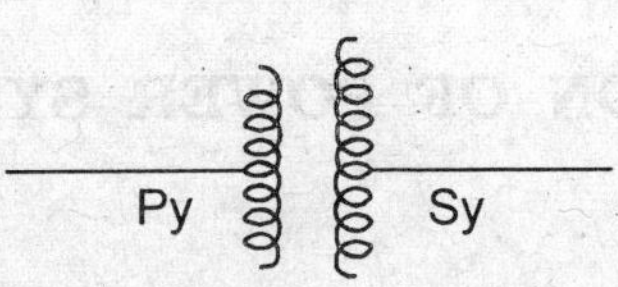

(ii) Step-down Transformer

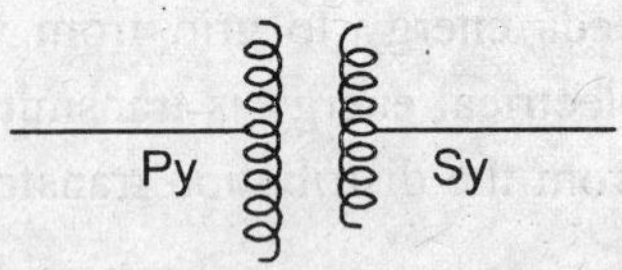

II. Three winding transformer

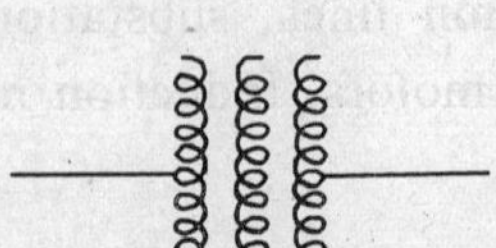

3. Transmission Line

The transmission lines are represented by a single straight line [for both single and 3 phase circuits only single line is drawn].

4. Now let we see some common symbols used in power system

(i) Fuse : 

(ii) Current transformer : 

(iii) Potential transformer: 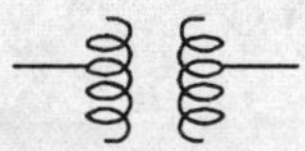

(iv) Circuit breaker (oil/gas) : 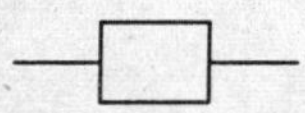

(v) Circuit breaker (Air): 

(vi) 3 phase, 3 wire delta connection : 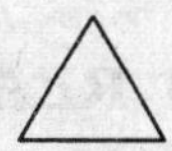

(vii) 3 phase star, neutral ungrounded: 

(viii) 3 phase star, neutral grounded : 

1.3 SINGLE LINE DIAGRAM

Generally in Power system analysis, a 3 phase system is always analysed on per phase basis by considering one phase of the three phase lines and neutral. Hence it is sufficient to show only one phase in the diagrammatic representation of power system. This is known as single line diagram.

In single line representation of power system, the components of the system are represented by standard symbols and the transmission lines are represented by straight lines. **"Hence a single line diagram is diagrammatic of power system in which the components are represented by their symbols and the interconnections between them are shown by straight lines".** Besides the symbols, the ratings and the impedance of the components are also marked on the single line diagram. The purpose of single line diagram is to make the representation in concise form.

For example, Let we see one typical single line diagram

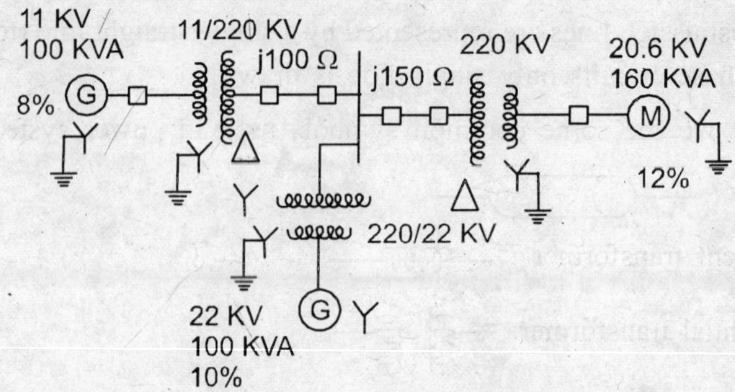

Figure 1.1 *A typical single line diagram*

1.4 IMPEDANCE AND REACTANCE DIAGRAM

The main purpose of this impedance and reactance diagram of power system is to analyse the performance of a system under load conditions (load flow studies) (or) to analyse the condition of the system under fault. The impedance (or) reactance diagram of a power system is the equivalent circuit of the power system in which the various components of the power system are represented by their approximate (or) simplified equivalent impedance (or) rectance.

Let we see about impedance diagram and reactance diagram in detail.

1.4.1 Impedance Diagram

The impedance diagram of a power system is used for load flow studies. This impedance diagram can be obtained from the single line diagram by replacing all the components of the power system by their single phase equivalent circuit. The following approximations are made while forming impedance diagram.

* If X_L <<<R, then the resistance can be omitted, because due to this only small deviation will be there in the result. i.e. only small error in calculations.
* Since the value of magnetising current is very low as compared to load current, the shunt branches in the equivalent circuit of the transformer can be neglected.
* The current limiting impedances connected between the generator neutral and ground are neglected because no current flows through neutral under balanced conditions.

1.4.2 Reactance Diagram

The reactance diagram of a power system is used for fault calculations. This is also same as impedance diagram. This is also obtained from the single line diagram by replacing all the components of the power system by their single phase equivalent reactance. The following approximations are made while forming reactance diagram.

* The capacitance of the transmission lines are neglected.
* Since the value of magnetising current is very low as compared to load current, the shunt branches in the equivalent circuit of the transformer can be neglected.
* The neutral to ground impedance of the generator is neglected for symmetrical faults.
* All static loads are neglected.
* The resistance in the equivalent circuits of various components of the system are omitted.
* Since, the current contributed by an induction motor dies out very quickly after the induction motor is short circuited, the induction motors are neglected in computing fault current a few cycles after the fault occurs.

The reactance diagram can be obtained from impedance diagram if we omit all static loads, resistances, short branches of transformer and capacitance of transmission lines in the impedance diagram.

The impedance and reactance diagram obtained from the above approximations are also called +ve sequence impedance diagram and +ve sequence reactance diagram.

1.4.3 Procedure to Form Reactance Diagram From Single Line Diagram :

Step 1: Select base power rating **ie. select base kilovolt ampere (or) mega volt-ampere (KVA_b (or) MVA_b).** This power rating will be same for all sections of power system.

Step 2: Select a base kilovolt (KV_b) for one section of power systems.

➡ The base power rating (KVA_b (or) MVA_b) will be same throughout the power system. But, the voltage rating get differs at various stages. So, the KV_b rating will not be same throughout the power system. The KV_b will get change where transformer is presented. The new base kilovolt should be found using transformer voltage ratio.

Let we see the formula for finding new base kilovolt (KV_b)

$$KV_b \text{ on LT section} = KV_b \text{ on HT section} \times \frac{\text{LT voltage rating}}{\text{HT voltage rating}} \quad ...(1.1)$$

$$KV_b \text{ on HT section} = KV_b \text{ on LT section} \times \frac{\text{HT voltage rating}}{\text{LT voltage rating}} \quad ...(1.2)$$

Step 3: By using base power and base voltage ratings find the reactance value of every components of single line diagram.

> ➡ The impedance of the components of power system are expressed either in ohms (or) in per unit which is calculated using the component rating as base values. In reactance diagram, the resistances are neglected and the reactances of all the components are expressed on a common base.
>
> [i.e. Starting form one end of power system the reactances of each component should be converted to per unit reactances on the selected new base.

(*i*) When the specified reactance of the component is in ohms then

$$\text{Per unit reactance} = \frac{\text{Actual reactacne in } \Omega}{\text{Base impedance}} \quad ...(1.3)$$

(*ii*) When the specified reactance of the component is in P.o. on the component rating as base values, then consider the component rating as old base values and selected base values as new bases. Now the P.u. reactance on new base can be calculated as

$$X_{P.u.(new)} = X_{pu(old)} \times \left[\frac{KV_{b(old)}}{KV_{b(new)}}\right]^2 \times \left[\frac{MVA_{b(new)}}{MVA_{b(old)}}\right] \quad ...(1.4)$$

1.5 PER UNIT SYSTEM

The components of power system may operate at different voltage and power levels. It will be convenient to analyse the power system if the voltage, Power, current and impedance ratings of power system are expressed with reference to a common value called "base value."

Hence for analysis purpose a base value is chosen for voltage, power, current and impedance. Then all the voltage, power, current and impedance ratings of the components are expressed as a per cent (or) per unit of the base value.

* **The per unit value of any quantity is defined as the ratio of the actual value of the quantity to the base value expressed as a decimal. The ratio in percent is 100 times the value in per unit.**

The base value is an arbitrary chosen value of the quantity.

$$\text{Per unit value} = \frac{\text{Actual value}}{\text{Base value}} \quad ...(1.5)$$

$$\% \text{ per unit value} = \left[\frac{\text{Actual value}}{\text{Base value}} \times 100\right]\% \quad ...(1.6)$$

1.5.1 Single Phase System

Now we will derive the formula for per unit impedance.

Let, $KVA_b \rightarrow$ Base power
$KV_b \rightarrow$ Base voltage
$I_b \rightarrow$ Base current in Ampere
$Z_b \rightarrow$ Base impedance in ohms.

By ohm's Law,

$$\text{Base current,} \quad I_b = \frac{KVA_b}{KV_b} \text{ in amp} \quad ...(1.7)$$

$$\text{Base impedance,} \quad Z_b = \frac{KV_b \times 1000}{I_b} \text{ in } \Omega \quad ...(1.8)$$

substitute (1.7) in equation (1.8)

$$Z_b = \frac{KV_b \times 1000}{\frac{KVAb}{KV_b}} = \frac{(KV_b)^2}{\frac{KVA_b}{1000}} = \frac{KV_b^{\ 2}}{MVA_b}$$

$$\text{i.e.,} \quad Z_b = \frac{KV_b^{\ 2}}{MVA_b} \quad ...(1.9)$$

$$\text{Per unit impedance, } Z_{pu} = \frac{\text{Actual impedance, } \Omega}{\text{Base impedance, } \Omega}$$

1.5.2 Three Phase System

In 3 phase system let we see the relation between line to neutral voltage and base line to neutral voltage then the relation between line to line voltage and base line to neutral voltage.

(under balanced condition)

$$\frac{V_{LN}}{V_{b,LN}} = \frac{V_{LL}}{V_{b,LL}} \quad ...(1.10)$$

Similarly for 3 phase power rating

$$\frac{3 \text{ phase } KVA}{3 \text{ phase base } KVA} = \frac{KVA/\text{phase}}{\text{Base } KVA/\text{phase}} \quad ...(1.11)$$

From the equation (1.10) and (1.11) in three phase systems, the line value of voltage and 3 phase KVA are directly used for per unit calculations.

Base impedance (Z_b) and base current (I_b) of three phase system can be computed directly from 3 phase value of base KVA and line value of base KV.

Let $KV_b \rightarrow$ Line to Line KV

$KVA_b \rightarrow$ 3 phase base KVA

$I_b \rightarrow$ Line value of base current

We know, $$KVA_b = KV_b \times I_b \times \sqrt{3} \quad ...(1.12)$$

$$[KVA = \sqrt{3}\ V_L I_L \times 10^{-3} = \sqrt{3}\ (KV_L)\ (I_L)]$$

From equation (1.12) we get,

Base current, $$I_b = \frac{KVA_b}{\sqrt{3} \times KV_b} \quad ...(1.13)$$

In a balanced power system the phase voltage is $\frac{1}{\sqrt{3}}$ times $\left[ie, V_{ph} = \frac{V_L}{\sqrt{3}}\right]$ the line voltage. Hence the base impedance per phase is given by

Base impedance/phase, $$Z_b = \frac{KV_b/\sqrt{3} \times 1000}{I_b}$$

$$Z_b = \frac{KV_b \times 1000}{\sqrt{3} I_b} \quad ...(1.14)$$

Substitute equation (1.13) in equation (1.14) we get,

$$Z_b = \frac{KV_b \times 1000}{\sqrt{3} \times \dfrac{KVA_b}{\sqrt{3} \times KV_b}}$$

$$= \frac{(KV_b)^2}{\dfrac{KVA_b}{1000}} = \frac{(KV_b)^2}{MVA_b}$$

$$Z_b = \frac{(KV_b)^2}{MVA_b} \quad ...(1.15)$$

Here, the equation (1.9) and (1.15) are similar, but in 3 phase system, the KV_b is a line value and MVA_b is a 3 phase MVA.

1.5.3 Changing the Base of Per Unit Quantities on Common Base

From single line diagram we can see that, the impedance of a device (or) component is usually specified in per unit on the base of name plate rating. As we know that, the power system is interconnection of various devices, so it will be convenient to analyse if all the impedances are converted to common base. So Let we see how to convert per unit impedances from one base to another base.

Let, $Z \rightarrow$ Actual impedance Ω

$Z_b \rightarrow$ Base impedance Ω

$$\text{Per unit impedance } Z_{Pu} = \frac{Z}{Z_b} = \frac{Z}{\frac{(KV_b)^2}{MVA_b}}$$

$$Z_{pu} = \frac{Z \times MVA_b}{(KV_b)^2} \quad ...(1.16)$$

Using the equation (1.16) we can derive an expression to convert the pu impedance expressed in one base value (old base) to another base (new base)

Let the old base values are

$KV_{b,\text{old}}$

$MVA_{b,\text{old}}$

Let the new base values are

$KV_{b,\text{new}}$

$MVA_{b,\text{new}}$

Now we can rewrite the equation (1.16) in terms of old base values, to find old per unit impedance.

$$Z_{pu,\text{old}} = \frac{Z \times MVA_{b,\text{old}}}{\left(KV_{b,\text{old}}\right)^2} \quad ...(1.17)$$

From equation (1.17)

$$Z_{(actual\ \text{impedance})} = \frac{(KV_{b,\text{old}})^2}{MVA_{b,\text{old}}} Z_{pu,\text{old}} \quad ...(1.18)$$

Again the equation (1.16) can be rewritten in terms of new base values,

$$Z_{pu,\text{new}} = \frac{Z \times MVA_{b,\text{new}}}{(KV_{b,\text{new}})^2} \qquad ...(1.19)$$

Substitute equation (1.18) in equation (1.19) we get

$$Z_{pu,\text{new}} = Z_{p.u,\text{old}} \frac{\left(KV_{b,\text{old}}\right)^2}{MVA_{b,\text{old}}} \times \frac{MVA_{b,\text{new}}}{\left(KV_{b,\text{new}}\right)^2}$$

$$Z_{p.u,\text{new}} = Z_{p.u,\text{old}} \times \left[\frac{KV_{b,\text{old}}}{KV_{b,\text{new}}}\right]^2 \times \left[\frac{MVA_{b,\text{new}}}{MVA_{b,\text{old}}}\right] \qquad ...(1.20)$$

The equation (1.20) is used to convert the per unit impedance on one base value to another base.(new base)

1.5.4 Advantages of Per Unit Computations

1. Type of connection of transformer (Y or Δ) does not affect the per unit impedance of the equivalent circuit
2. The per unit impedances of machines of the same type and widely different rating usually lie within a narrow range, although the ohmic Values differ widely for machines of different ratings.
3. Impedance of a machine is specified in percent (or) per unit on the name plate rating.
4. The per unit impedance of circuit element connected by transformers expressed or a proper base will be if it is referred to either side of a transformer.

1.6 EQUIVALENT CIRCUITS OF COMPONENTS OF POWER SYSTEM

To analyse the power system under various stages like load flow analysis, fault level calculations, the equivalent circuit of each component of power system is very important. The various components of power system are generator, transformer, transmission line, induction motor, synchronous motor, resistive and reactive loads. Now we will see the equivalent circuit of each component.

(*i*) **Equivalent circuit of generator**

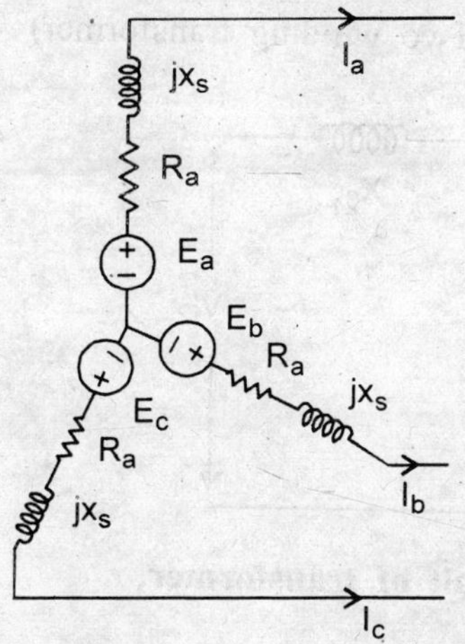

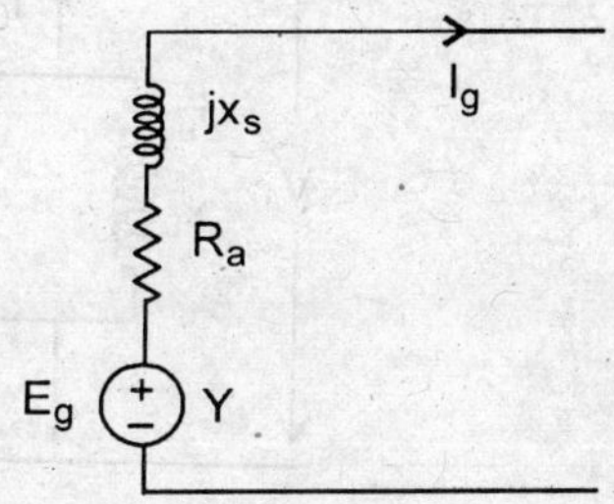

Figure 1.2 *3 Phase Equivalent Circuit of Generator*

Figure 1.3 *Single Phase Equivalent Circuit of Generator*

The equivalent circuit of single and three phase generators are shown in Fig. (1.2) and Fig. (1.3). It consists of a source which represents induced emf per phase, series armature reactance and leakage reactance (jXs) and a series armature resistance (Ra).

(*ii*) **Equivalent circuit of synchronous motor**

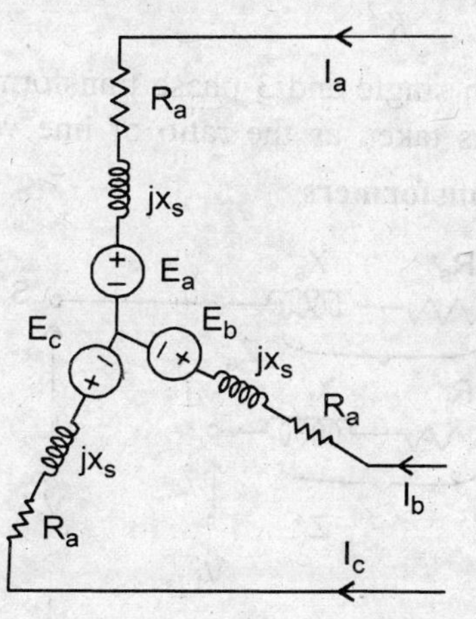

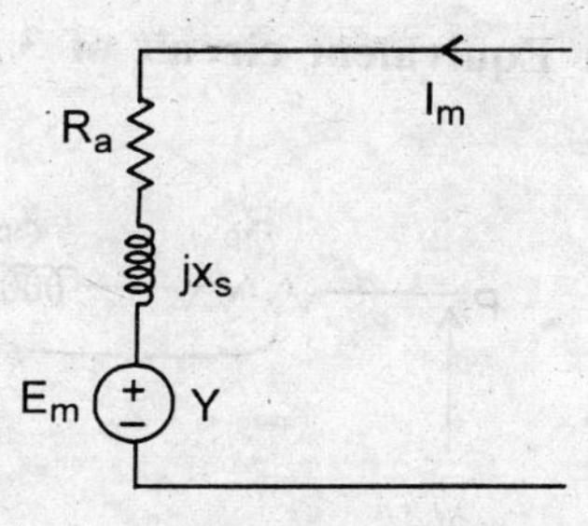

Figure 1.4 *Equivalent Circuit of 3 Phase Synchronous Motor*

Figure 1.5 *Equivalent Circuit of 1 ϕ Synchronous Motor*

The equivalent circuit of synchronous motor is similar to synchronous generator except current direction and I_g should be replaced by I_m

(*iii*) **Equivalent circuit of transformer:** (Two winding transformer)

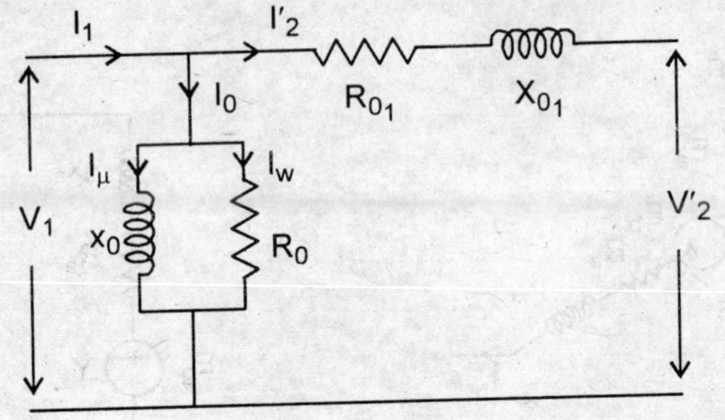

Fig. (1.6): Equivalent circuit of transformer.

The equivalent circuit of a single phase, two winding transformer referred to primary side is shown in Fig. (1.6).

In Fig. (1.6),

$$K = \frac{E_2}{E_1} = \frac{N_2}{N_1} \approx \frac{V_2}{V_1} = \frac{I_1}{I_2}$$

$$R_{0_1} = R_1 + R_2^1 = R_1 + \frac{R_2}{K^2}$$

$$X_{0_1} = X_1 + X_2^1 = X_1 + \frac{X_2}{K^2}$$

The above Fig. (1.6) is common for both single and 3 phase transformers. But in 3 phase transformers, the turns ratio, K is taken as the ratio of line voltages.

(*iv*) **Equivalent circuit of 3 winding transformers**

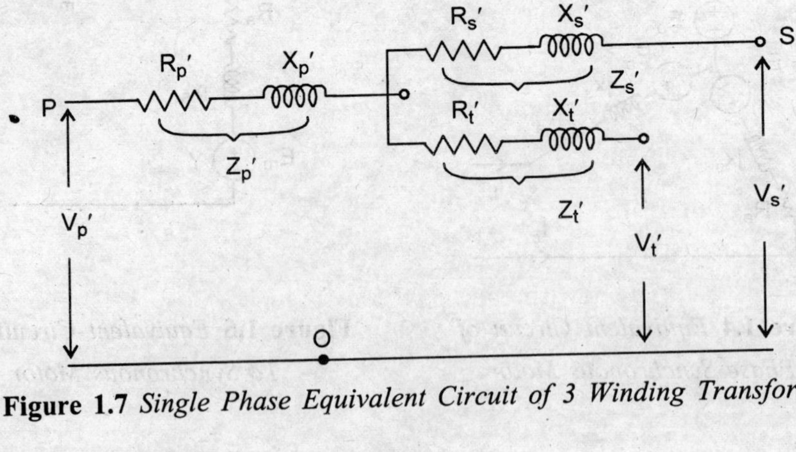

Figure 1.7 *Single Phase Equivalent Circuit of 3 Winding Transformer*

The three winding transformers has one additional winding called tertiary winding in addition with primary and secondary windings Among 3 windings, the two windings are connected is star and one winding in delta (or) two windings in delta and one winding in star. The advantages of having tertiary winding is

* Tertiary winding can be used to measure voltage of an HV testing transformer.
* It can interconnect 3 transmission line at different voltages
* To Supply voltage for the substation auxiliary devices can be provided. [auxiliary devices may work at different voltage level as compared to primary and secondary winding]
* A delta connected tertiary winding reduces the impedance offered to the zero sequence currents there by allowing a large earth fault current to flow to proper protection equipments.
* The tertiary winding can be used for injecting reactive power into the system for voltage control by connecting series capacitors (or) synchronous condensers.
* This tertiary winding limits voltage unbalance when the load is unbalanced.
* It permits the third harmonic current to flow there by reducing 3rd harmonic voltages.

From above reasons the tertiary winding is also called as stabilization winding.

In 3 winding transformer, the tertiary winding alone and primary and secondary winding together can have same *KVA* rating. The impedance of each winding of a 3 winding transformer may be given in *p.u.* Which can be calculated by using their own winding rating as bases. But while representing in reactance (or) impedance diagram it is necessary to convert the *p.u.* reactances to a common base.

The equivalent circuit of 3 winding transformer can be represented by the single phase equivalent circuit which is shown in Fig. (1.7).

In Fig. (1.7), the points P, T. & S represents primary, tertiary and secondary winding terminals and the point O is the common terminal.

Z_P → Primary winding impedance

Z_S^1 → Secondary winding impedance referred to primary

Z_t^1 → Tertiary winding impedance referred to primary

Z_{ps} → Leakage impedance measured in primary with secondary short circuited and tertiary open.

Z_{pt} → Leakage impedance measured in primary with tertiary short circuited and secondary open.

Z^1_{st} → Leakage impedance measured in secondary winding with tertiary short circuited and primary open and then referred to primary.

The Leakage impedance measured by short circuit test are referred to winding impedance as follows,

$$Z_{ps} = Z_p + Z_s^1 \quad ...(1.21)$$

$$Z_{pt} = Z_p + Z_t^1 \quad ...(1.22)$$

$$Z_{st} = Z_s^1 + Z_t^1 \quad ...(1.23)$$

The equation (1.21), (1.22) and (1.23) can be expressed in the form of matrix,

$$\begin{bmatrix} 1 & 1 & 0 \\ 1 & 0 & 1 \\ 0 & 1 & 1 \end{bmatrix} \begin{bmatrix} Z_p \\ Z_s^1 \\ Z_t^1 \end{bmatrix} = \begin{bmatrix} Z_{ps} \\ Z_{pt} \\ Z'_{st} \end{bmatrix} \quad ...(1.24)$$

From equation (1.24)

$$\Delta = \begin{bmatrix} 1 & 1 & 0 \\ 1 & 0 & 1 \\ 0 & 1 & 1 \end{bmatrix} = -2$$

$$\Delta_1 = \begin{bmatrix} Z_{ps} & 1 & 0 \\ Z_{pt} & 0 & 1 \\ Z'_{st} & 1 & 1 \end{bmatrix}$$

$$= Z_{ps}\,(-1)\,(-1)\,(Z_{pt} - Z'_{st}) = -Z_{ps} - Z_{pt} + Z'_{st}$$

$$\Delta_2 = \begin{bmatrix} 1 & Z_{ps} & 0 \\ 1 & Z_{pt} & 1 \\ 0 & Z'st & 1 \end{bmatrix}$$

$$= 1 \times (Z_{pt} - Z'_{st}) - Z_{ps}(1) = -Z_{ps} - Z'_{st} + Z_{pt}$$

$$\Delta_3 = \begin{bmatrix} 1 & 1 & Z_{ps} \\ 1 & 0 & Z_{pt} \\ 0 & 1 & Z'st \end{bmatrix}$$

$$= 1 \times (-Z_{pt}) - 1\,(Z'_{st}) + Z_{ps}\,(1) = -Z_{pt} - Z'_{st} + Z_{ps}$$

$$Z_p = \frac{\Delta_1}{\Delta} = \frac{1}{-2}[-Z_{ps} - Z_{pt} + Z'_{st}] = \frac{1}{2}[+Z_{ps} + Z_{pt} - Z'_{st}] \quad ...(1.25)$$

$$Z'_s = \frac{\Delta_2}{\Delta} = \frac{1}{-2}[-Z_{ps} + Z_{pt} - Z'_{st}]$$

$$= \frac{1}{2}[+Z_{ps} - Z_{pt} + Z'_{st}] \quad ...(1.26)$$

$$Z'_t = \frac{\Delta_3}{\Delta} = \frac{1}{-2}[-Z_{pt} - Z'_{st} + Z_{ps}]$$

$$= \frac{1}{2}[+Z_{pt} + Z'_{st} - Z_{ps}] \quad ...(1.27)$$

The equations (1.25), (1.26) and (1.27) can be used to calculate the impedances of the 3 winding transformer using short circuit test data (referred to primary)

(v) Equivalent circuit of Induction on motor

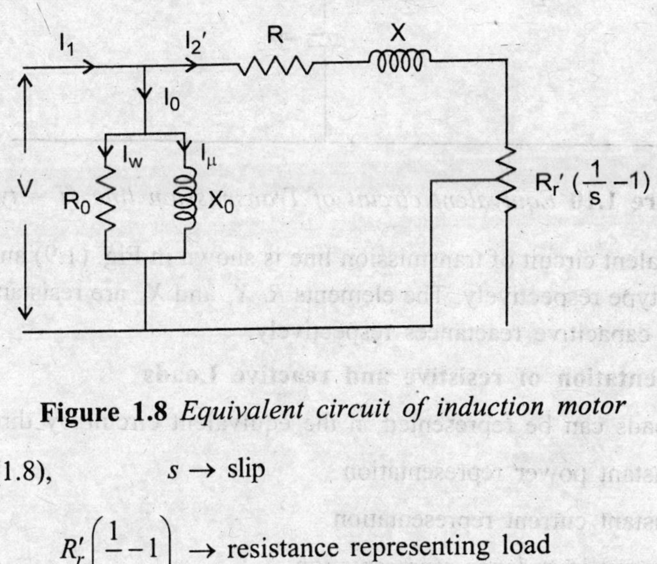

Figure 1.8 *Equivalent circuit of induction motor*

In Fig. (1.8), $s \rightarrow$ slip

$R'_r\left(\frac{1}{s}-1\right) \rightarrow$ resistance representing load

$R = R_s + R'_r \rightarrow$ Equivalent resistance referred to stator

$X = X_r + X'_r \rightarrow$ Equivalent resistance referred to stator

$R_s \rightarrow$ Resistance of stator

$X_s \rightarrow$ Reactance of stator

$R_r \rightarrow$ Resistance of rotor

$X_r \rightarrow$ Reactance of rotor

The equivalent circuit of an induction motor is shown in Fig. (1.8). It is similar to the equivalent circuit of an two winding transformer.

(*vi*) Equivalent circuit of Transmission line

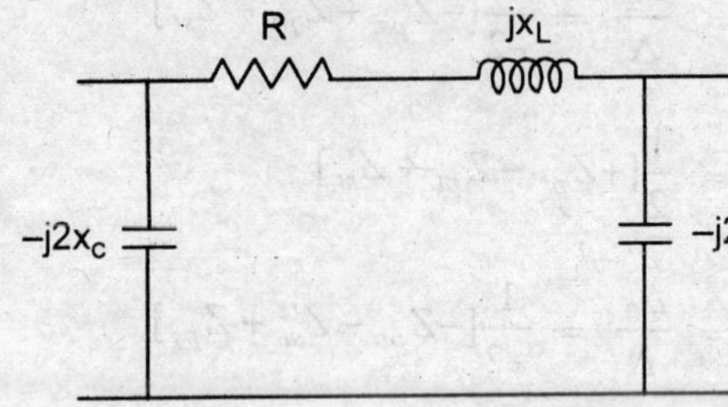

Figure 1.9 *Equivalent circuit of Transmission line (Π – type)*

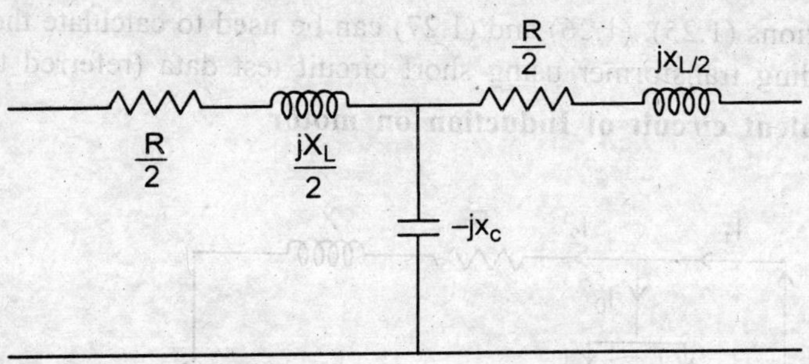

Figure 1.10 *Equivalent circuit of Transmission line (T – type)*

The equivalent circuit of transmission line is shown in Fig. (1.9) and Fig. (1.10), Π-type and T-type respectively. The elements R, X_L and X_C are resistance, inductive reactance and capacitive reactances respectively.

(*vii*) Representation of resistive and reactive Loads

The Loads can be represented in the equivalent circuit by three ways.

1. Constant power representation
2. Constant current representation
3. Constant impedance representation

In constant power representation, the active power (MW) and reactive power (MVA_R) are constant. In constant current representation the magnitude of load current is constant. In constant impedance representation the magnitude of impedance is constant.

(*i*) Single phase load

Let, $P \rightarrow$ Active power

$Q \rightarrow$ Reactive power

$S \rightarrow$ Complex power

The complex power, $S = VI^* = P + jQ$

$$(VI^*)^* = (P + jQ)^*$$

$$V^* - I = P - jQ$$

$$\therefore \quad I = \frac{P - jQ}{V^*} \qquad \text{...(1.28)}$$

Let, $$V = |V|\underline{|\delta} \qquad (1.29)$$

$$V^* = |V|\underline{|-\delta} \qquad (1.30)$$

$$P - jQ = \sqrt{P^2 + Q^2}\angle \tan^{-1}\frac{Q}{P} = \sqrt{P^2 + Q^2}\angle -\theta \qquad \text{...(1.31)}$$

Where, $$\theta = \tan^{-1}\frac{Q}{p} \qquad \text{...(1.32)}$$

From equations (1.28), (1.29), (1.30), (1.31) and (1.32)

$$I = \frac{\sqrt{P^2 + Q^2}\angle -\theta}{|V|\underline{|-\delta}} = \frac{\sqrt{P^2 + Q^2}}{|V|}\angle \delta - \theta \qquad \text{...(1.33)}$$

$$I = |I| \ \underline{|\delta - \theta} \qquad \text{...(1.34)}$$

Where, $$|I| = \frac{\sqrt{P^2 + Q^2}}{|V|} \qquad \text{...(1.35)}$$

In constant impedance representation the load is represented by its impedance (or) admittance. The impedance of the load can be calculated from the specified voltage, active and reactive powers of the load as shown below

Load Impedance, $$Z = \frac{V}{I} \qquad \text{...(1.36)}$$

On substituting I from equation (1.28) in equation (1.36) we get,

Load impedance, $$Z = \frac{V}{\frac{(P - jQ)}{V^*}}$$

$$\text{Load impedance} = \frac{VV^*}{P - jQ} = \frac{|V|^2}{P - jQ} \qquad \text{...(1.37)}$$

Load admittance, $$Y = \frac{1}{Z} = \frac{P - jQ}{|V|^2} \qquad \text{...(1.38)}$$

Three phase Load

(*i*) Balanced star connected Load

Let, $P \rightarrow 3\phi$ active power in W

$Q \rightarrow 3\phi$ reactive power in VAR

$V \rightarrow$ phase Voltage of load

$V_L \rightarrow$ Line Voltage of load

$I \rightarrow$ phase Current of load

$I_L \rightarrow$ Line Current of load

3ϕ Complex power, $S = 3VI^* = P + jQ$

$$(3VI^*)^* = (P + jQ)^*$$

$$3V^*I = P - jQ$$

$$\therefore \quad I = \frac{P - jQ}{3V^*}$$

Let,
$$\left.\begin{aligned} V &= |v|\angle\delta \\ V^* &= |v|\angle-\delta \end{aligned}\right\} \qquad ...(1.40)$$

In star connected load,

$$|V| = \frac{|V_L|}{\sqrt{3}} \text{ and}$$

$$I = I_L$$

$$\therefore \quad V^* = \frac{|V_L|}{\sqrt{3}} \angle -\delta \qquad ...(1.41)$$

Let
$$P - jQ = \sqrt{P^2 + Q^2} \angle \tan^{-1}\frac{Q}{P}$$

$$p - jQ = \sqrt{P^2 + Q^2} \angle -\theta \qquad ...(1.42)$$

Using the equations and (1.41) and (1.42) the equation (1.39) can be written as,

$$I = \frac{P - jQ}{3V^*} = \frac{\sqrt{P^2 + Q^2} \angle -\theta}{3\frac{|V_L|}{\sqrt{3}} \angle -\delta}$$

$$= \frac{\sqrt{P^2 + Q^2}}{3|V_L|} \angle \delta - \theta$$

$$\therefore \quad I = I_L = \frac{\sqrt{P^2 + Q^2}}{3|V_L|} \angle \delta - \theta \qquad ...(1.43)$$

$$\therefore \qquad |I| = |I_L| = \frac{\sqrt{P^2+Q^2}}{3|V_L|} \qquad ...(1.44)$$

Load impedance per phase, $Z = \frac{V}{I}$ ...(1.45)

On substituting for I from equation (1.39) in equation (1.45) we get,

$$Z = \frac{V}{\frac{(P-jQ)}{3V^*}} = \frac{3VV^*}{P-jQ} = \frac{3|V|^2}{P-jQ}$$

$$= \frac{3\left(|V_L|\sqrt{3}\right)^2}{P-jQ} = \frac{|V_L|^2}{P-jQ} \qquad ...(1.46)$$

$\therefore$ Load admittance per phase, $Y = \frac{1}{Z} = \frac{P-jQ}{|V_L|^2}$ ...(1.47)

***(ii)* Balanced delta connected load**

For balanced delta connected load upto equation (1.40) the steps are same as star connected load.

Now,

In delta connected load, $V = V_L$

and $|I| = \frac{|I_L|}{\sqrt{3}}$ ...(1.48)

Let, $P - jQ = \sqrt{P^2+Q^2}\angle -\tan^{-1}\frac{Q}{P}$

$$= \sqrt{P^2+Q^2}\angle -\theta \qquad ...(1.49)$$

where $\theta = \tan^{-1}\frac{Q}{P}$

using equations (1.48) and (1.49) the equation (1.39) can be written as

$$I = \frac{\sqrt{P^2+Q^2}\angle -\theta}{3|V|\angle -\delta} = \frac{\sqrt{P^2+Q^2}}{3|V|}\angle(\delta-\theta)$$

$$= \frac{\sqrt{P^2+Q^2}}{3|V_L|}\angle(\delta-\theta)$$

$$= |I|\angle(\delta-\theta)$$

where, $$|I| = \frac{\sqrt{P^2+Q^2}}{3|V_L|}$$

$$\therefore \quad |I_L| = \sqrt{3}\ |I| = \sqrt{3}\left[\frac{\sqrt{P^2+Q^2}}{3|V_L|}\right] = \frac{\sqrt{P^2+Q^2}}{\sqrt{3}|V_L|} \quad ...(1.50)$$

Load impedance/phase, $$Z = \frac{V}{I} \quad ...(1.51)$$

on substituting for I from equation (1.39) in equation (1.51) we get,

$$Z = \frac{V}{\frac{(P-jQ)}{3V^*}}$$

$$= \frac{3VV^*}{P-jQ} = \frac{3|V_L|^2}{P-jQ}$$

Load impedance/phazse, $$Z = \frac{3|V_L|^2}{P-jQ} \quad ...(1.52)$$

Load admittance/phase, $$Y = \frac{1}{Z} = \frac{P-jQ}{3|V_L|^2} \quad ...(1.53)$$

1.7 Example for drawing impedance reactance diagram

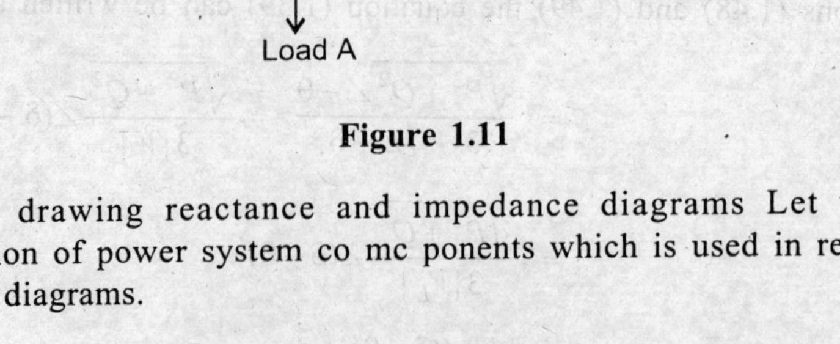

Figure 1.11

Before drawing reactance and impedance diagrams Let we see the representation of power system co mc ponents which is used in reactance and impedance diagrams.

Table 1 *Equivalent circuit of power system components to draw reactance diagram*

Component	Equivalent Circuit
Three phase generator	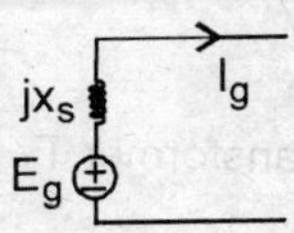 $I_g \rightarrow$ generated current $E_g \rightarrow$ generated emf $jX_S \rightarrow$ Reactance
synchronous motor	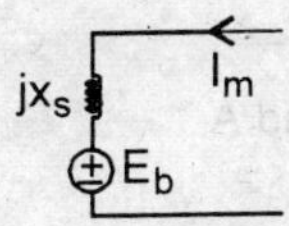 $I_m \rightarrow$ motor current $jX_S \rightarrow$ Reactance $E_b \rightarrow$ Back emf
Transformer	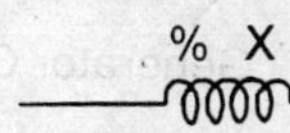
Transmission line	

Table 2 *Equivalent circuit of power system components to draw impedance diagram*

Component	Equivalent Circuit
Three phase generator	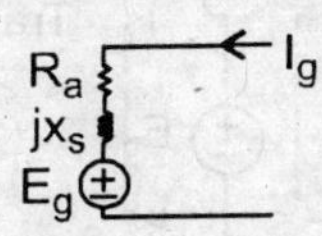
generator synchronous motor	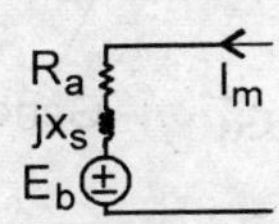
Transformer	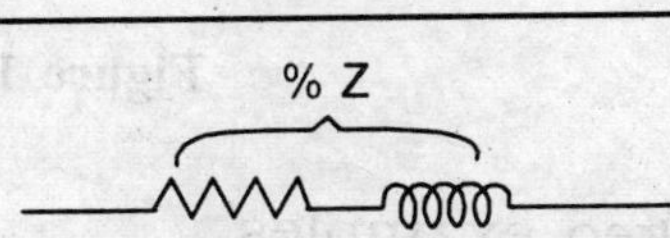
Transmission line	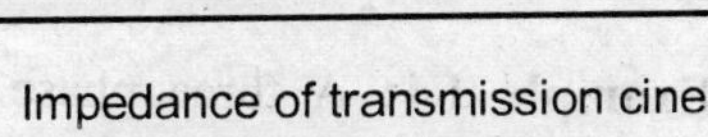

Let we draw impedance diagram for Fig. (1.11)

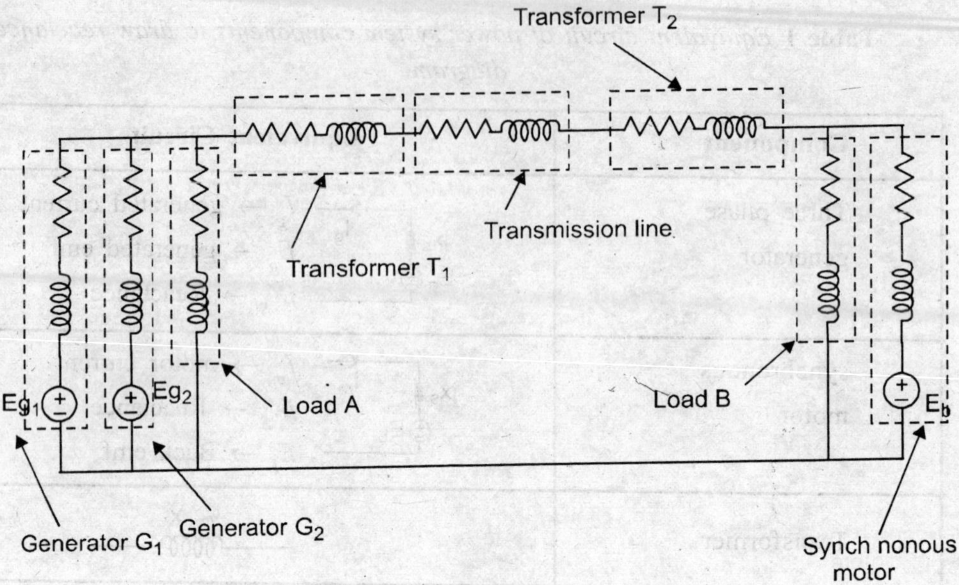

Figure 1.12 *Impedance diagram*

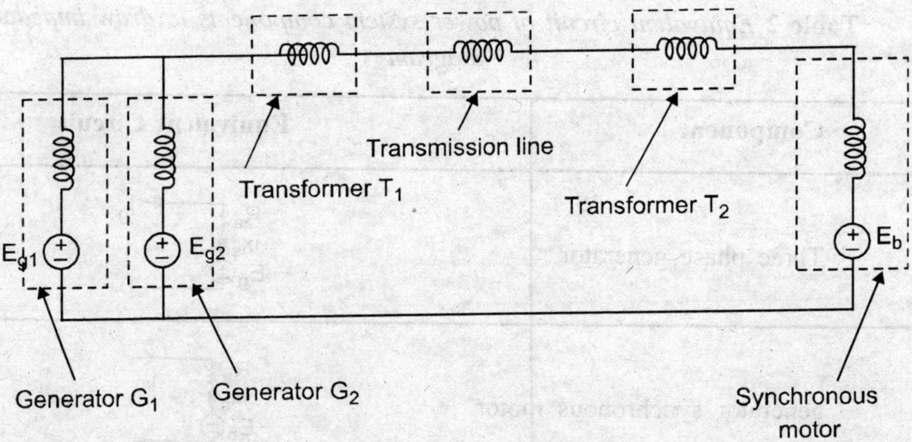

Figure 1.13 *Reactance diagram*

Worked examples

Example 1.1: A three phase generator with rating 1200KVA, 33KV has its armature resistance and synchronous reactance as 30Ω/phase and 80Ω/phase.

Calcutate p.u impedance of the generator.

Given data: $P = 1200\ KVA$

$V = 33\ KV$

$R_a = 30\Omega/\text{phase}$

$X_s = 80\Omega/\text{phase}$

To find: $Z_{p.u}$ of generator

Solution: Choose generator ratings as base KV and base KVA (*i.e* Base Voltage and base power)

$\therefore$ Base kilovolt, $KV_b = 33\ KV$

Base kilovolt ampere, $KVA_b = 1200\ KVA$

Base impedance/phase, $$Z_b = \frac{(KV_b)^2}{MVA_b}$$

$$Z_b = \frac{(33)^2}{\frac{1200}{1000}} = 907.5\Omega.$$

Actual impedance/phase, $Z = 30 + j80\Omega/\text{phase}$

$\therefore$ Per unit impedance, $$Z_{p.u} = \frac{\text{Actual impedance}}{\text{Base impedance}}$$

$$Z_{p.u} = \frac{Z}{Z_b} = \frac{30 + j80}{907.5}$$

$$= .033 + j.088\ pu$$

$$Z_{p.u} = .033 + j.088\ pu$$

Result: The per unit impedance of the generator is $Z_{p.u} = .033 + j.088\ pu$

Example 1.2: A three phase, Δ-Y transformer with rating KVA, 11 KV/400 V has its primary and secondary Leakage reactance as 10 Ω/phase and 0.04Ω/phase respectively. calculate the pu. reactance of transformer.

Given data: $P = 120\ KVA$

$V_{LT} = 11\ KV$

$V_{HT} = 400\ V$

$X_1 = 10\ \Omega/\text{phase}$

$X_2 = .04\Omega/\text{phase}$

To find: $X_{p.u}$ of transformer.

Solution: This $X_{p.u}$ can be calculated by two ways *i.e*) with respect to primary side and with respect to secondary side.

Case (*i*): With respect to primary side *i.e. HT* side

The high Voltage side ratings are chosen as base values.

$$KV_b = 11\ KV$$
$$KVA_b = 120\ KVA$$
$$Z_b = \frac{(KV_b)^2}{MVA_b} = \frac{(11)^2}{\left(\frac{120}{1000}\right)}$$
$$= 1008.33\Omega$$

Transformer Voltage ratio, $K = \frac{400}{11000} = .0364$

$$\left.\begin{array}{r}\text{Total reactance referred to}\\ \text{Primamry, } X_{0_1}\end{array}\right\} = X_1 + X_2'$$
$$= X_1 + \frac{X_2}{K^2} = 10 + \frac{.04}{(.0364)^2}$$
$$X_{0_1} = 40.30\ \Omega/\text{phase}$$
$$\therefore \quad X_{p.u} = \frac{X_{0_1}}{Z_b} = \frac{40.30}{1008.33} = .0399\ \text{pu}$$

Case (*ii*) The Low voltage side ratings are chosen as base values.

$$KV_b = \frac{400}{1000} = .4KV$$
$$KVA_b = 120\ KVA$$
$$Z_b = \frac{(KV_b)^2}{MVA_b} = \frac{(.4)^2}{\left(\frac{120}{1000}\right)} = 1.33\Omega$$
$$Z_b = 1.33\ \Omega$$

Transformer voltage ratio, $K = \frac{400}{11000} = .0364$

$$\left.\begin{array}{r}\text{Total Leakage reactance}\\ \text{referred to secondary}\end{array}\right\} X_{02} = X_1' + X_2$$
$$X_{02} = K^2\ X_1 + X_2$$
$$(0.0364)^2 \times 10 + 0.04$$
$$X_{02} = .053\ \Omega/\text{phase}$$

$P.u$ reactance/phase, $X_{P.u} = \dfrac{\text{Total Leakage reactance}}{\text{Base impedance}}$

$$= \frac{X_{0_2}}{Z_b} = \frac{.053}{1.333}$$

$$X_{pu} = .039$$

➡ The value of per unit reactance of both the cases are same.

Resut: The $p.u$ reactance of the transformer is $X_{pu} = .039$

Example 1.3: A 3ϕ Y-Δ transformer is constructed using 3 identical single phase transformers of rating 260 KVA, 127 KV/11KV transformer. The impedance of primary and secondary are $22 + j50\Omega$ and $0.2 + j0.3\Omega$ respectively. calculate the $p.u$ impedance of the transformer.

Given data:

$$P = 200\ KVA$$
$$V_1 = 127\ KV$$
$$V_2 = 11\ KV$$
$$Z_1 = 22 + j50\Omega$$
$$Z_2 = 0.2 + j0.3\Omega$$

To find: Per unit impedance (Z_{pu})

Solution: The three phase transformer is formed by 3 numbers of identical single phase transformers. Hence the KVA rating of three phase transformer is 3 times that of single phase transformer

∴ KVA rating of 3ϕ transformer

$$= 3 \times 260 = 780\ KVA$$

Line voltage rating of Y-Δ transformer

$$= 127 \times \sqrt{3}KV/11KV$$
$$= 220\ KV/11\ KV$$

(∴ only Primary is connected in star and delta secondary is connected in delta)

Case (*i*) Choose High voltage side ratings as base values

∴

$$KV_b = 220\ KV$$
$$KVA_b = 780\ KVA$$

$$Z_b = \frac{(KV_b)^2}{MVA_b} = \frac{(220)^2}{\left(\dfrac{780}{1000}\right)} = 62051.2\Omega$$

Transformer voltage ratio, $K = \dfrac{11}{220} = .05$

Total impedance referred to primary $\Big\}\ Z_{0_1} = Z_1 + Z_2' = Z_1 + \dfrac{Z_2}{K^2}$

$$Z_{0_1} = 22 + j50 + \frac{0.2 + j0.3}{(.05)^2}$$

$$Z_{0_1} = 22 + j50 + 80 + j120$$

$$Z_{0_1} = 102 + j170 \ \Omega/\text{phase}$$

$$Z_{p.u} = \frac{\text{Total impedance } (Z_{0_1})}{\text{Base impedance } (Z_b)}$$

$$= \frac{102 + j170}{62051.2}$$

$$Z_{p.u} = .00164 + j.0027 \ pu.$$

Case (*ii*) The Low voltage side ratings are choosen as base values

$$KV_b = 11 \ KV$$

$$KVA_b = 780 \ KVA$$

$$Z_b = \frac{(KV_b)^2}{MVA_b} = \frac{(11)^2}{\left(\frac{780}{1000}\right)} = 155.12$$

Total impedace referred to secondary $\Big\}\ Z_{0_2} = Z'_1 + Z_2$

$$Z_{0_2} = (.05)^2 \ (22 + j50) + (.2 + j0.3)$$

$$Z_{0_2} = .055 + j.125 + .2 + j0.3$$

$$Z_{0_2} = .255 + j.425$$

p.u reactance/phase, $Z_{pu} = \dfrac{\text{Total impedance } Z_{0_2}}{\text{Base impedance}}$

$$Z_{pu} = \frac{Z_{0_2}}{Z_b} = \frac{.255 + j.425}{155.122}$$

$$Z_{pu} = .00164 + j.0027 \ pu.$$

Result: The pu. impedance of the transformer is

$$Z_{pu} = .00164 + j0.0027 \ pu.$$

Example 1.4: A 100 *KW*, 3ϕ, star connected (*Y*) load is connected with **300** *KVA* transformer with 11 *KV*/400*V* through a feeder. The Length of the **feeder is** 1km and the impedance of the feeder is 0.2 + *j*0.3Ω/Km. If the load power **factor** is 0.8, calculate the *p.u* impedance of the load and feeder.

Given data: $P_L = 100 \ KW$ (Load power)

Power rating of transformer = 300 *KVA*

$$V_1 = 11 \ KV$$

$$V_2 = 400 \ V$$

Length of the feeder = 1km

Z of feeder = 0.2 + *j*0.3Ω/km

$$\cos\phi_L = 0.8$$

To find: (1) Z_{pu} of load

(2) Z_{pu} of feeder

Solution: Let us choose secondary winding rating of a transformer as **base** value.

$$\therefore \quad KV_b = 400 \text{ V} = 0.4 \ KV$$

$$KVA_b = 300 \ KVA$$

$$Z_b/\text{phase} = \frac{(KV_b)^2}{MVA_b} = \frac{(.4)^2}{\left(\frac{300}{1000}\right)} = .533\Omega$$

$$\left.\begin{matrix}\text{Actual impedance}\\ \text{of feeder}\end{matrix}\right\}_{Z_{fed}} = (0.2 + j0.3) \times 1$$

$$= .2 + j0.3\Omega/\text{phase}$$

$$\left.\begin{matrix}p.u \text{ impedance of}\\ \text{feeder}\end{matrix}\right\}_{Z_{pu,fed}} = \frac{0.2 + j0.3}{.533} = .375 + j.562$$

$$\left[\therefore Z_{pu,fed} = \frac{\text{Actual impedance}}{\text{Base impedance}} = \frac{Z_{fed}}{Z_b}\right]$$

From load power, power factor which is given, we can calculate,

$$P_L = 100 \ KW \ \& \ \cos\phi_L = .8$$

$$\therefore \quad \sin\phi_L = \sin(\cos^{-1} .8) = 0.6$$

Reactive power, $Q = \frac{P}{\cos\phi} \times \sin\phi$

$$= \frac{100}{.8} \times .6$$

$$Q = 75\ KVAR$$

Load impedance/phase, $Z_L = \frac{|V_L|^2}{P - jQ}$

$$= \frac{(400)^2}{(100 - j75) \times 10^3} = \frac{160}{100 - j75}$$

$$= \frac{160}{125\angle -36.86°} = 1.28\angle -36.86°$$

$$Z_L = 1.28\angle +36.86\ \Omega/\text{phase}$$

P.u value of load impedance $Z_{L,pu} = \frac{Z_L}{Z_b} = \frac{1.28\angle 36.86}{.533}$

$$Z_{L,pu} = 2.40\angle 36.86\ pu$$

Result:

$$Z_{pu,\ fed} = .375 + j.562\ pu$$

$$Z_{L,\ fed} = 2.40\angle 36.86\ pu$$

Example 1.5: The 3 phase ratings of a three winding transformer are

Primary : Star connected, 110 *KV*, 20 *MVA*

Secondary : Star connected, 13.2 *KV*, 15 *MVA*

Tertiary : Delta connected, 2.1 *KV*, .5 *MVA*

Three short circuit test performed on this transformer yielded the following results

(*i*) Primary excited, Secondary shorted : 2290 V, 52.5A

(*ii*) Primary excited, Tertiary shorted: 1785 V, 52.5 A

(*iii*) Secondary excited, Tertiary shorted : 148 V, 328 A.

Find the p.u impedances of the star connected single phase equivalent circuit for a base of 20 *MVA*, 110 *KV* in the Primary circuit. Neglect resistances.

Given data:

$$V_p = 110\ KV$$

$$V_s = 13.2\ KV$$

$$V_t = 2.1\ KV$$

$$P_p = 20\ MVA$$
$$P_s = 15\ MVA$$
$$P_t = .5\ MVA$$
$$MVA_b = 20\ MVA$$
$$KV_b = 110\ KV$$

and test results are given.

To find: Per unit impedance

Solution:

➡ Test (*i*) and (*ii*) are performed on primary winding, hence the *p.u* impedances of Z_{ps} and Z_{pt} can be obtained directly using the primary winding ratings as base values.

The test (*iii*) is performed on secondary winding, hence the *p.u* impedance Z_{st} is obtained using secondary winding rating as bases and then it can be converted to primary winding base

Finding *p.u* values of Z_{ps} and Z_{pt}

KV_b of primary, $KV_{b,py} = 110\ KV$

MVA_b of primary, $MVA_{b,py} = 20\ MVA$

$$Z_{b,py} = \frac{\left(KV_{b,py}\right)^2}{MVA_{b,py}} = \frac{(110)^2}{20}$$

$$= 605\ \Omega/\text{phase}$$

The readings of test (*i*) can be used to Calculate the value of Z_{ps} in Ω/phase

$$Z_{ps} \text{ is } \Omega/\text{phase} = \frac{\frac{2290}{\sqrt{3}}}{52.5} = 25.1835\Omega/\text{phase}$$

(∵ primary is star connected)

$$P.u \text{ value of } Z_{ps} = \frac{Z_p \text{ in } \Omega/\text{phase}}{Z_{b,py}} = \frac{25.1835}{605} = .0416\ p.u$$

The readings of test (*ii*) can be used to calculate the value of Z_{pt} in Ω/phase

$$Z_{pt} \text{ in } \Omega/\text{phase} = \frac{\frac{1785}{\sqrt{3}}}{52.5} = 19.6299\Omega/\text{phase}$$

(∴ primary is star connected)

$$p.u \text{ value of } Z_{pt} = \frac{Z_{pt} \text{in } \Omega/\text{phase}}{Z_{b,py}}$$

$$= \frac{19.6299}{605} = .0324\ p.u$$

Finding $p.u$ value of Z'_{st}

$$KV_b \text{ of secondary, } KV_{b,\text{sec}} = 13.2\ KV$$

$$MVA_b \text{ of secondary, } MVA_{b,\text{sec}} = 15\ MVA$$

$$Z_{b,\text{sec}} = \frac{(KV_{b,\text{sec}})^2}{MVA_{b,\text{sec}}} = \frac{(13.2)^2}{15}$$

$$= 11.616\Omega/\text{phase}$$

The readings of test (*iii*) can be used to calculate the value of Z_{st} in Ω/phase.

$$Z_{st} \text{ in } \Omega/\text{phase} = \frac{\frac{148}{\sqrt{3}}}{328} = .2605\Omega/\text{phase}$$

(∴ secondary is star connected)

$$\left.\begin{array}{r} p.u \text{ value of } Z_{st} \text{on the} \\ \text{base values of secondary circuit}\end{array}\right\} Z_{st} = \frac{Z_{st} \text{in } \Omega/\text{phase}}{Z_{b,\text{sec}}}$$

$$= \frac{.2605}{11.616} = .0224\ p.u$$

The $p.u$ value of Z_{st} on the secondary circuit base values can be converted to primary circuit base using the following formula.

$$Z_{pu,\text{new}} = Z_{pu,\text{old}} \times \left[\frac{KV_{b,\text{old}}}{KV_{b,\text{new}}}\right]^2 \times \left[\frac{MVA_{b,\text{new}}}{MVA_{b,\text{old}}}\right]$$

Here new refers to primary and old refers to secondary

$$\left.\begin{array}{r} p.u \text{ value of } Z_{st} \text{ on the base} \\ \text{value of primary circuit}\end{array}\right\} Z'_{st} = .0224 \times \left[\frac{110}{110}\right]^2 \times \left[\frac{20}{15}\right]$$

$$= .0299\ p.u$$

➡ $KV_{b,\text{sec}}$ when refened to primary side will be $\left(\frac{110}{13.2}\right) \times 13.2 = 110\ KV$

Finding Z_p, Z'_s and Z'_t

$$Z_p = \frac{1}{2}[Z_{ps} + Z_{pt} - Z_{st}{}^1]$$

$$= \frac{1}{2}[.0416 - .0299 - 0.324] = 0.0221\ pu$$

$$Z'_s = \frac{1}{2}[Z_{ps} + Z'_{st} - Z_{pt}] = \frac{1}{2}[.0416 + .0299 - .0324]$$

$$= .0196\ P.u$$

$$Z'_t = \left[Z_{pt} + Z'_{st} - Z_{Ps}\right] = Z'_t = \frac{1}{2}[.0324 + .0299 - 0.416] = 0.1035\ pu.$$

Result:

$$Z_p = j0.0221$$

$$Z^1_s = j0.0196$$

$$Z^1_t = j0.01035$$

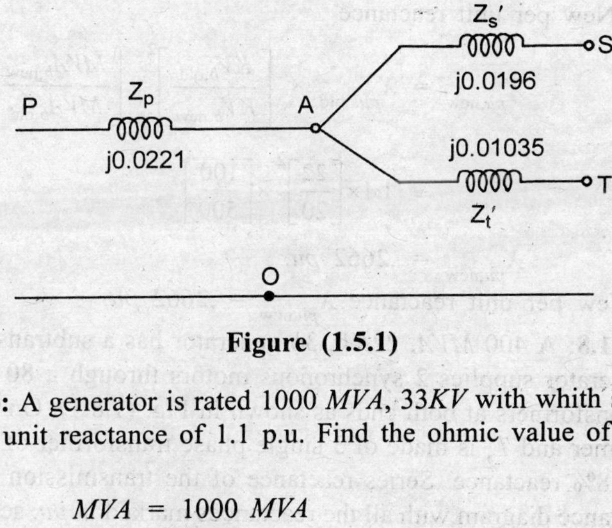

Figure (1.5.1)

Example 1.6: A generator is rated 1000 *MVA*, 33*KV* with whith star connected which has a per unit reactance of 1.1 p.u. Find the ohmic value of the reactance of winding.

Given data:

$$MVA = 1000\ MVA$$

$$KV = 33\ KV$$

$$X_{pu} = 1.1\ pu$$

To find Actual reactance

Solution: Let choose generator ratings as base value.

$$KV_b = 33\ KV$$

$$MVA_b = 1000\ MVA$$

$$Z_b = \frac{(KV_b)^2}{MVA_b} = \frac{(33)^2}{1000} = 1.089\ \Omega$$

$$X_{pu} = \frac{\text{Actual reactance}}{\text{Base impedance}} = \frac{X}{Z_b}$$

$$X_{pu} = \frac{X}{1.089}$$

$$X = X_{pu} \times 1.089 = 1.1 \times 1.089$$

$$X = 1.19\ \Omega/\text{phase}$$

Result: Actual reactance, $X = 1.19\Omega$/phase

Example 1.7: If the generator bases are specified as 100 *MVA*, 20 *KV*. Then find the *pu*. value of reactance of generator winding on the specified base. The generator values are given as 22*KV*, 500 *MVA* and $X_{pu} = 1.1\ pu$

Given data: $X_{pu} = 1.1\ pu$

New base: $KV_{b,new} = 20\ KV$

$MVA_{b,\ new} = 100\ MVA$

Old base: $KV_{b,old} = 22\ KV$

$MVA_{b,old} = 500\ MVA,$

To find: New per unit reactance

Solution:

$$X_{pu,new} = X_{pu.\ old} \times \left[\frac{KV_{b,old}}{KV_{b,new}}\right]^2 \times \left[\frac{MVA_{b,new}}{MVA_{b,old}}\right]$$

$$= 1.1\times\left[\frac{22}{20}\right]^2 \times \left[\frac{100}{500}\right]$$

$$X_{pu,new} = .2662\ pu.$$

Result: New per unit reactance $X_{pu,new} = .2662\ pu.$

Example 1.8: A 400 *MVA*, 22*KV*, 3ϕ generator has a subtransient reactance of 15%. The generator supplies 2 synchronous motors through a 80 km transmission line having transformers at both ends as shown in Fig. (1.8.1). On this, T_1 is a three phase transformer and T_2 is made of 3 single phase transformer of rating 250 *MVA*, 127/13.2 *KV*, 8% reactance. Series reactance of the transmission line is 0.8Ω/km. Draw the reactance diagram with all the reactances marked in *pu*. select the generator rating as base values

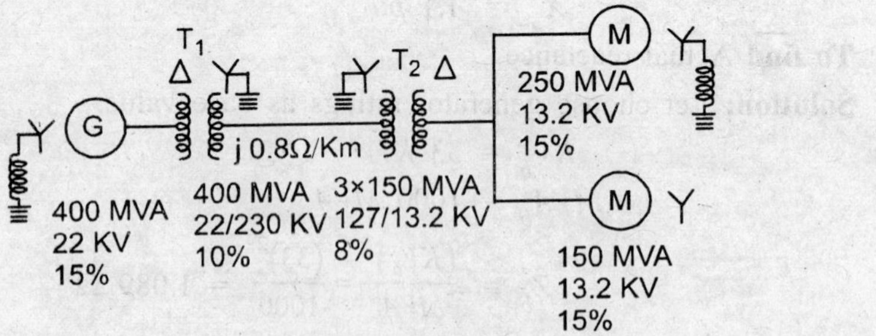

Figure (1.8.1)

Given data:

(*i*) Generator Ratings

Power = 400 *MVA*

Voltage = 22 *KV*

P.u. reactance = 15%

(*ii*) Transformer (T_1) Ratings

Power = 400 *MVA*

Voltage = 22/230 *KV*

P.u. reactance = 10%

(*iii*) Transmission line resistance value

$j0.8\Omega$/km

(*iv*) Transformer (T_2) Ratings

Power = 3 × 150 MVA = 450 MVA

[∵ 3 single phase transform connected]

Voltage = 127/13.2 KV

P.u. reactance = 8%

(*v*) Synchronous motor (M_1) rating:

Power = 250 MVA

Voltage = 13.2 KV

P.u. reactance = 15%

(*vi*) Synchronous motor (M_2) ratings:

Power = 150 MVA

Voltage = 13.2 KV

P.u. reactance = 15%

(*vii*) Length of the transmission line = 80 km.

To find: reactances of components of single line diagram which is given in Fig. (1.8.1) and to draw reactance diagram.

Solution: Base Mega Volt ampere, $MVA_{b,new}$ = 400 MVA

Base Kilo volt, $KV_{b,new}$ = 22 KV

Now Let we find p.u Reactance of Generator G

➡ Since the generator rating and the base values are same, the generator p.u reactance does not change.

P.u reactance of generator, $X_{p.u,g}$ = 15% = .15 p.u

P.u Reactance of Transformer T_1

$$\left.\begin{array}{r}\text{The new } P.u \text{ reactance of} \\ \text{Tranformer } T_1\end{array}\right\} = X_{pu,\text{old}}\left[\frac{KV_{b,\text{old}}}{KV_{b,\text{new}}}\right]^2 \times \left[\frac{MVA_{b,\text{new}}}{MVA_{b,\text{old}}}\right]$$

$$= 0.1 \times \left[\frac{22}{22}\right]^2 \times \left[\frac{400}{400}\right]$$

$$= 0.1 \text{ p.u}$$

$$X_{pu,T_1} = 0.1 \text{ p.u.}$$

> ➡ Since the transformer rating and the base values are same, as the generator case, the transformer P.u. reactance also does not change.

P.u. Reactance of transmission line

Reactance of transmission line = 0.8Ω/km

Total reactance of transmission line = 0.8 × 80 = 64 Ω

$$\left.\begin{array}{r}\text{Base } KV \text{ on } HT \text{ side} \\ \text{of transformer } T_1\end{array}\right\} = \text{Base } KV \text{ on } LT \text{ side} \times \frac{HT \text{ Voltage rating}}{LT \text{ Voltage rating}}$$

$$= 22 \times \frac{230}{22}$$

$$KV_b = 230 \text{ KV}$$

Now, New base values are,

$$\boxed{\begin{array}{l} KV_b = 230 \text{ KV} \\ MVA_b = 400 \text{ MVA}\end{array}}$$

Base impedance,
$$Z_b = \frac{(KV_b)^2}{MVA_b} = \frac{(230)^2}{400}$$

$$Z_b = 132.25 \ \Omega$$

$$\left.\begin{array}{r}\text{per unit reactance of} \\ \text{transmission line}\end{array}\right\} = \frac{\text{Actual reactance}}{\text{Base impedance}}$$

$$= \frac{64}{132.25}$$

$$\boxed{X_{pu,TL} = .483 \, p.u}$$

P.u. reactance of transformer T_2

The transfer T_2 is formed by 3 identical single phase transformers with voltage rating 127/13.2 KV. In this the high voltage side is star connected and low Voltage side is delta connected.

∴ Voltage ratio of line voltage of 3 – phase transformer bank

$$= \frac{\sqrt{3}\times 127}{13.2} = \frac{220}{13.2} KV$$

Base KV on LT side of Transformer T_2

$$= \text{Base KV on HT side} \times \frac{LT \text{ Voltage rating}}{HT \text{ Voltage rating}}$$

$$= 230 \times \frac{13.2}{220}$$

$$KV_b = 13.8 \text{ KV}$$

Now, New base ⇒ $\boxed{\begin{aligned} KV_b &= 13.8KV \\ MVA_b &= 400MVA \end{aligned}}$

The new P.u reactance of transformer T_2

$$= X_{pu,\text{old}} \left[\frac{KV_{b,\text{old}}}{KV_{b,\text{new}}}\right]^2 \times \left[\frac{MVA_{b,\text{new}}}{MVA_{b,\text{old}}}\right]$$

$$= 0.8\left[\frac{13.2}{13.8}\right]^2 \times \left[\frac{400}{450}\right] = .731[.88]$$

$$\boxed{X_{pu,T_2} = .643pu}$$

P.u reactance of synchronous motor M_1

P.u reactance of M_1 on new base $= X_{pu,\text{old}} \left[\frac{KV_{b,\text{old}}}{KV_{b,\text{new}}}\right]^2 \times \left[\frac{MVA_{b,\text{new}}}{MVA_{b,\text{old}}}\right]$

$$= .15\times\left[\frac{13.2}{13.8}\right]^2 \times \left[\frac{400}{250}\right]$$

$$\boxed{X_{pu,m_1} = .219_{pu}}$$

P.u reactance of synchronous motor M_2

P.u reactance of M_2 on new base

$$= X_{pu,\text{old}} \left[\frac{KV_{b,\text{old}}}{KV_{b,\text{new}}}\right]^2 \left[\frac{MVA_{b,\text{new}}}{MVA_{b,\text{old}}}\right]$$

$$= .15\left[\frac{13.2}{13.8}\right]^2 \left[\frac{400}{150}\right]$$

$$\boxed{X_{pu,m_2} = .365p.u}$$

Now draw the reactance diagram of Fig. (1.8.1) and mark the corresponding p.u reactance values on reactance diagram

Reactance diagram

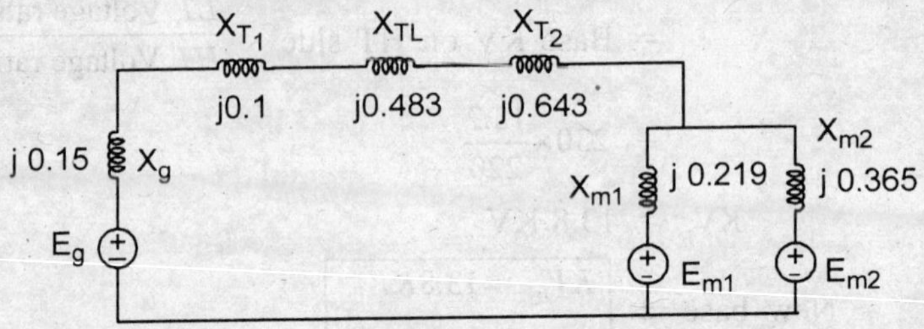

Figure (1.8.2)

> ➡ The Base Kilovolt get changed wherever transformer is presented.But base MVA will not change

1.7.1 Guidelines to Draw Reactance Diagram from Single Line Diagram

The guidelines to draw reactance diagram from single line diagram has following points. To understand them, Let we take one example of Fig. (1.8.1).

* First choose Base KV (KV_b) and Base MVA (MVA_b)

Hint to choose base value

1. In some problems, base value will be specified for example in example 1.8, base values are specified. ie. generator rating is base value
2. When the base values are not specified, base values can be chosen from given single line diagram (i.e.) any one of the component rating can be choosed as base value (or) it can be assumed value also.
   * Find new P.u. reactance of all the components referred to base value which is chosen.
   * The base MVA (MVA_b) will be same through out the single line diagram.
   * The base KV (KV_b) get changed whenever transformer is presented. So find new base KV (KV_b) with respect to transformer voltage ratio.

Now we will see the explanation of example (1.8)

1. Choose generator rating as base values
2. Find new p.u. reactance of generator.

3. Find new P.u reactance of transformer (T_1) either referred to primary (or) secondary. [in this problem new p.u reactance for transformer is calculated referred to primary]
4. Find new p.u reactance of transmission line. [Since, the transmission line is connected with secondary of transformer T_1, its voltage value is not same as primary side. So find new base kilovolt with respect to voltage ratio of transformer.]
5. Find new per unit reactance of transformer T_2 either referred to primary (or) secondary.

   [In this problem, $X_{pu,new}$ for transformer is calculated referred to primary].
6. Find new per unit reactance of synchronous motor (M_1). [Since Motor's M_1 and M_2 are connected with secondary of transformer, find new base kilo volt with respect to voltage ratio of transformer.]
7. Find new per unit reactance of synchronous motor (M_2).
8. By combining all P.u reactances, draw the reactance diagram.

Example 1.9: A 200 *MVA*, 22*KV* generator has a synchronous reactance of .2 *P.u* and it is connected to a transmission line through a transformer rated 250 *MVA*, 220/20 *KV* (*Y*/Δ) with *X* = .2 *P.u.*

(*a*) Calculate P.u reactances by taking generator rating as base values.

(*b*) Calculate the *P.u* reactance by taking transformer rating as base values.

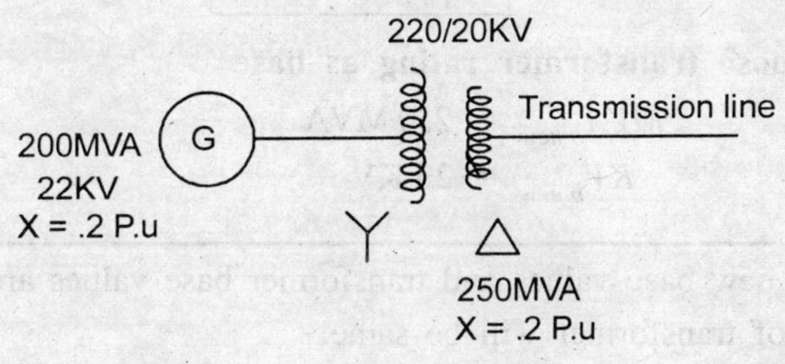

Figure 1.17

Given data:

(*i*) Generator rating	Base power	= 200 *MVA*
	Base *KV*	= 22 *KV*
	P.u. reactance, *X*	= .2 *P.u*
(*ii*) Transformer rating	Base power	= 250 *MVA*
	Base *KV*	= 220/20 KV
	P.u. reactanced, *X*	= .2 *P.u*

Solution:

(*a*) Let choose generate rating as base

$$\therefore \quad MVA_{b,\ new} = 200\ MVA$$

$$KV_{b,\ new} = 22\ KV$$

$$\text{New P.u reactance of generator} = X_{P.u,old}\left[\frac{KV_{b,old}}{KV_{b,new}}\right]^2\left[\frac{MVA_{b,new}}{MVA_{b,old}}\right]$$

$$= .2\left[\frac{22}{22}\right]^2\left[\frac{200}{200}\right]$$

$$\boxed{X_{pu,g} = .2P.u}$$

➡ Since, the new base values and generator base values are same, the *P.u* reactance value of generator will be same.

$$\left.\begin{array}{l}\text{New P.u reactanceof}\\ \text{transformer}\end{array}\right\} = X_{P.u,old}\left[\frac{KV_{b,old}}{KV_{b,new}}\right]^2\left[\frac{MVA_{b,new}}{MVA_{b,old}}\right]$$

$$= .2\left[\frac{20}{22}\right]^2\left[\frac{200}{250}\right]$$

$$\boxed{X_{Pu,T} = .132P.u}$$

(*b*) Let choose transformer rating as base

$$MVA_{b,new} = 250\ \text{MVA}$$

$$KV_{b,new} = 20\ \text{KV}$$

➡ Since, the new base values and transformer base values are same the *P.u* reactance of transformer will be same.

$$\therefore \quad Z_{p.u,\ T} = .0\ P.u$$

New *P.u* reactance of generator,

$$X_{p.u,\ g} = X_{P.u,old}\left[\frac{KV_{b,old}}{KV_{b,new}}\right]^2\left[\frac{MVA_{b,new}}{MVA_{b,old}}\right]$$

$$= .2\left[\frac{22}{20}\right]^2\left[\frac{250}{200}\right]$$

$$\boxed{X_{P.u,g} = .3025P.u}$$

Result:

(*a*) $X_{Pu,\,g} = .2\ Pu$

$X_{Pu,\,T} = .132\ Pu$

(*b*) $X_{Pu,\,g} = .3025\ Pu$

$X_{pu,\,T} = .2\ Pu$

Example 1.10: The single line diagram of an unloaded power system is shown in Fig. (1.10.1). The ratings of the components of single line diagram as follows.

G_1 = 25 MVA, 13.2 KV, $X'' = 25\%$

G_2 = 35 MVA, 20 KV, $X'' = 25\%$

G_3 = 35 MVA, 22 KV, $X'' = 25\%$

T_1 = 30 MVA, 230/ 13.2 KV, $X = 15\%$

T_2 = 3 single phase units each rated at 15 MVA, 130/20 K

X = 15%

T_3 = 40 MVA, 230/23 KV, $X = 15\%$

Draw the reactance diagram using *a* base of 100 MVA and 13.2 KV on the generator G_1

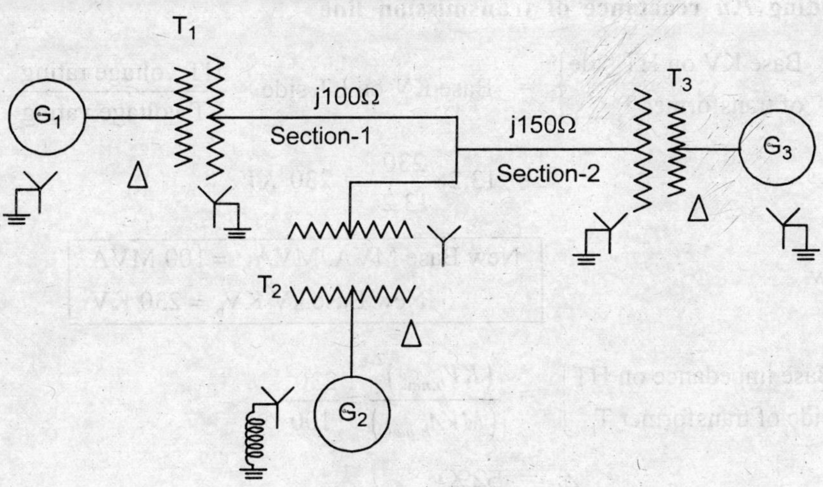

Figure (1.10.1)

Solution:

Base Values:

Base MVA, MVA_b = 100 MVA
Base KV, KV_b = 13.2 KV

Finding *P.u* reactance of generator G_1

$$\left.\begin{array}{l}\text{New P.u reactance}\\ \text{of genevator } G_1\end{array}\right\} = X_{\text{Pu,old}}\left[\frac{KV_{b,old}}{KV_{b,new}}\right]^2\left[\frac{MVA_{b,new}}{MVA_{b.old}}\right]$$

$$= .25\left[\frac{13.2}{13.2}\right]^2\left[\frac{100}{25}\right]$$

$$\boxed{X_{P.u,g} = 1\,p.u}$$

Finding *P.u* reactance of transformer T_1

$$\left.\begin{array}{l}\text{Now P.u reactance}\\ \text{at transformer } T_1\end{array}\right\} = X_{\text{Pu, old}}\left[\frac{KV_{b,old}}{KV_{b,new}}\right]^2\left[\frac{MVA_{b,new}}{MVA_{b,old}}\right]$$

$$= .15\left[\frac{13.2}{13.2}\right]^2\left[\frac{100}{30}\right]$$

$$\boxed{X_{pu,T_1} = .0495\,P.u}$$

Finding *P.u* reactance of transmission line

$$\left.\begin{array}{l}\text{Base KV on HT side}\\ \text{of transformer } T_1\end{array}\right\} = \text{Base KV on LT side} \times \frac{\text{HT voltage rating}}{\text{LT voltage vating}}$$

$$= 13.2\times\frac{230}{13.2} = 230\ KV$$

Now,

$$\boxed{\begin{array}{r}\text{New Base MVA, MVA}_b = 100\text{ MVA}\\ \text{New Base kV KV}_b = 230\text{ KV}\end{array}}$$

$$\left.\begin{array}{l}\text{Base impedance on HT}\\ \text{Side of transformer } T_1\end{array}\right\} = \frac{\left(KV_{b,new}\right)^2}{\left(MVA_{b,new}\right)} = \frac{230^2}{100}$$

$$Z_b = 529\Omega$$

$$\left.\begin{array}{l}\text{P.u reactance of section 1}\\ \text{of transmission line}\end{array}\right\} = \frac{\text{Actual reactance, }\Omega}{\text{Base impedance, }\Omega}$$

$$= \frac{100}{529} = .189\text{ P.u}$$

$$\boxed{X_{pu,TL_1} = .189\,P.u}$$

$$\left.\begin{array}{l}\text{P.u reactance of Section 2}\\ \text{of transmission line}\end{array}\right\} = \frac{\text{Actual reactance, }\Omega}{\text{Base impedance, }\Omega}$$

$$= \frac{150}{529} = .283\ P.u$$

$$\boxed{X_{Pu,TL_2} = .283 P.u}$$

Finding *P.u* reactance of transformer T_2

The transformer T_2 is a connection of three single phase transformers with voltage rating 130/20KV. In this the *HT* side is star connected and *LT* side is Δ connected.

$$\therefore \left.\begin{array}{l}\text{Voltage ratio of line voltage}\\ \text{of 3-phase transformer}\end{array}\right\} = \frac{\sqrt{3}\times 130}{20} = \frac{225}{20}\text{KV}$$

$$\left.\begin{array}{l}\text{Base KV on LT side}\\ \text{of transfor T}_2\end{array}\right\} = \text{Base KV on HT side}\times\frac{\text{LT voltage rating}}{\text{HT voltage rating}}$$

$$= 230\times\frac{20}{30} = 20\text{KV}$$

Now $$\boxed{\begin{array}{l}\text{New Base MVA, MVA}_b = 100\text{MVA}\\ \text{New Base KV, KV}_b = 20\text{KV}\end{array}}$$

$$\left.\begin{array}{l}\text{New P.u reactance}\\ \text{of transformer T}_2\end{array}\right\} = X_{Pu,old}\left[\frac{KV_{b,old}}{KV_{b,new}}\right]^2\left[\frac{MVA_{b,new}}{MVA_{b,old}}\right]$$

$$= .15\left[\frac{20}{20}\right]^2\left[\frac{100}{3\times 15}\right]$$

$$\boxed{X_{Pu,}T_2 = .33Pu}$$

Finding *P.u* reactance of generate G_2

$$\left.\begin{array}{l}\text{New P.u reactance}\\ \text{of generator G}_2\end{array}\right\} = X_{Pu,old}\left[\frac{KV_{b,old}}{KV_{b,new}}\right]^2\left[\frac{MVA_{b,new}}{MVA_{b,old}}\right]$$

$$= .25\left[\frac{20}{20}\right]^2\left[\frac{100}{35}\right]$$

$$\boxed{X_{Pu,g_2} = .714P.u}$$

Finding *P.u* reactance of Transformer T_3

$$\left.\begin{array}{l}\text{Base KV on LT side}\\ \text{of transformer } T_3\end{array}\right\} = \text{Base KV on HT side}\times\frac{\text{LT voltage rating}}{\text{HT voltage rating}}$$

$$= 230\times\frac{23}{230} = 23 \text{ KV}$$

Now,

$$\boxed{\begin{array}{r}\text{New base MVA, } MVA_b = 100\text{MVA}\\ \text{New base KV, } KV_b = 23 \text{ KV}\end{array}}$$

$$\left.\begin{array}{l}\text{New P.u reactance}\\ \text{of transformer } T_3\end{array}\right\} = X_{Pu,old}\left[\frac{KV_{b,old}}{KV_{b,new}}\right]^2\left[\frac{MVA_{b,new}}{MVA_{b,old}}\right]$$

$$= .15\left[\frac{23}{23}\right]^2\left[\frac{100}{40}\right]$$

$$\boxed{X_{pu,T_3} = .375 \ P.u}$$

Finding *P.u* reactance of generator G_3

$$\left.\begin{array}{l}\text{New P.u reactance of}\\ \text{generator } G_3\end{array}\right\} = X_{Pu,old}\left[\frac{KV_{b,old}}{KV_{b,new}}\right]^2\left[\frac{MVA_{b,new}}{MVA_{b,old}}\right]$$

$$= .25\left[\frac{22}{23}\right]^2\left[\frac{100}{35}\right]$$

$$\boxed{Z_{pu,g_3} = .653 P.u}$$

Reactance diagram of single line diagram which is shown in Fig. (1.10.1)

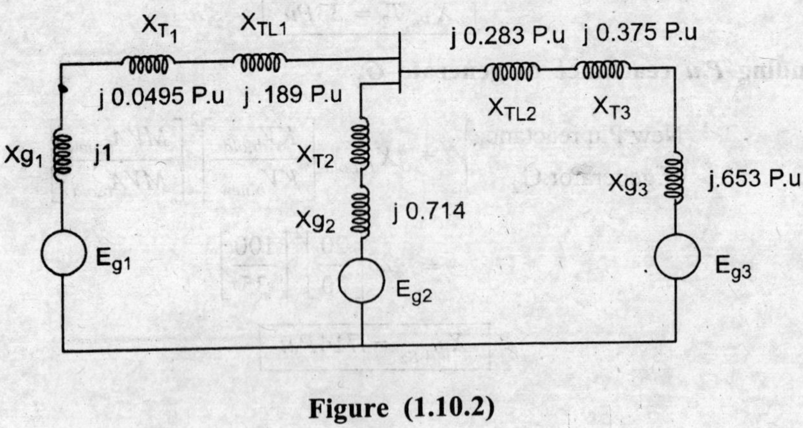

Figure (1.10.2)

Result: The Fig (1.10.2) is the reactance diagram of Fig(1.10.1)

Example 1.11: Draw the reactance diagram for the power system shown in Fig (1.11.1) The ratings of generator , motor and transformers are given below. Neglect resistance and use a base of 50 *MVA*, 138 *KV* in the 40 Ω line.

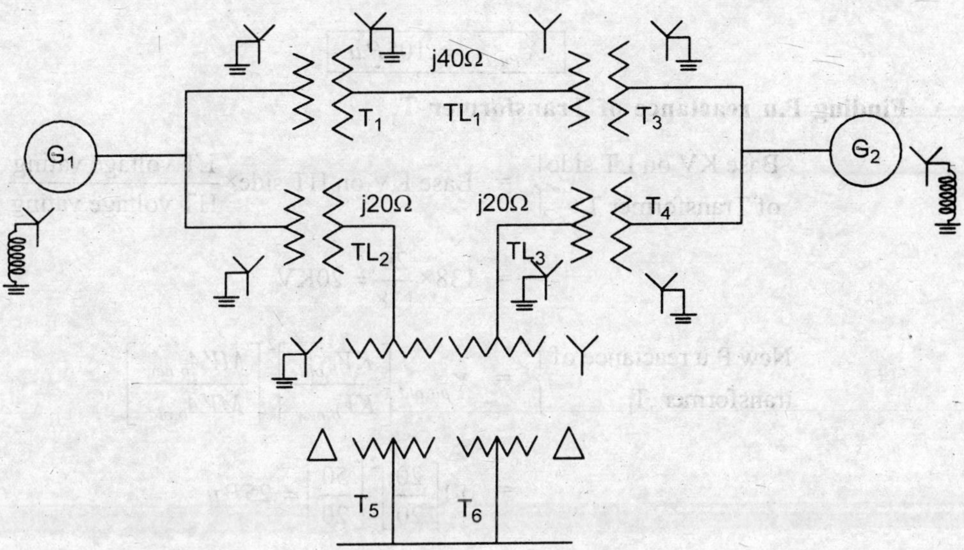

Figure (1.11.1)

Generator G_1: 20 *MVA*, 18*KV*, $X'' = 20\%$

Generator G_2: 20 *MVA*, 18 *KV*, $X'' = 20\%$

Synchronous motor : 30 MVA, 13.8 KV, $X'' = 20\%$

3 Phase, *Y-Y* transformer: 20 MVA, 138/20 KV, $X = 10\%$

3 Phase, *Y-Δ* transformer : 15 MVA, 138/13.8 KV, $X = 10\%$

Solution:

$$\text{Base MVA, } MVA_{b,new} = 50 \text{ MVA}$$
$$\text{Base KV, } KV_{b,new} = 138 \text{ KV}$$

Finding *P.u* reactance of *j* 40Ω transmission line:

$$\text{Base impedance, } Z_b = \frac{(KV_b)^2}{MVA_b}$$

$$Z_b = \frac{138^2}{50} = 380.88\ \Omega$$

$$\left.\begin{array}{l}\text{P.u reactance of} \\ 40\ \Omega \text{ tranmission line}\end{array}\right\} = \frac{\text{Actual reactance, }\Omega}{\text{Base impedance, }\Omega}$$

$$= \frac{40}{380.88}$$

$$\boxed{X_{Pu,TL1} = .105P.u}$$

Finding P.u reactance of Transformer T_1

$$\left.\begin{array}{l}\text{Base KV on LT side} \\ \text{of Transformer } T_1\end{array}\right\} = \text{Base KV on HT side} \times \frac{\text{LT voltage vating}}{\text{HT voltage vating}}$$

$$= 138 \times \frac{20}{138} = 20\text{KV}$$

$$\left.\begin{array}{l}\text{New P.u reactance of} \\ \text{transformer },T_1\end{array}\right\} = X_{pu,old}\left[\frac{KV_{b,old}}{KV_{b,new}}\right]^2\left[\frac{MVA_{b,new}}{MVA_{b,old}}\right]$$

$$= 0.1\left[\frac{20}{20}\right]^2\left[\frac{50}{20}\right] = .25P.u$$

$$\boxed{X_{pu,T_1} = .25P.u}$$

Finding *P.u* reactance of generator G_1

$$\left.\begin{array}{l}\text{New P.u reactance} \\ \text{of generator } G_1\end{array}\right\} = X_{Pu,odl}\left[\frac{KV_{b,old}}{KV_{b,new}}\right]^2\left[\frac{MVA_{b,new}}{MVA_{b,old}}\right]$$

$$= .2\left[\frac{18}{20}\right]^2 \times \left[\frac{50}{20}\right]$$

$$\boxed{X_{Pu,g_1} = .405P.u}$$

Finding *P.u* reactance of transformer T_2

$$\left.\begin{array}{l}\text{New P.u reactance of} \\ \text{Transformer } T_2\end{array}\right\} = X_{Pu,old}\left[\frac{KV_{b,old}}{KV_{b,new}}\right]^2\left[\frac{MVA_{b,new}}{MVA_{b,old}}\right]$$

$$= 0.1\left[\frac{20}{20}\right]^2 \times \left[\frac{50}{20}\right] = .25P.u$$

$$\boxed{X_{pu,T_2} = .25P.u}$$

Finding *P.u* reactance of *j*20Ω Transmission line

$$\left.\begin{array}{l}\text{Base KV onHT side}\\ \text{of transformer, } T_2\end{array}\right\} = \text{Base KV on LT side} \times \frac{\textbf{HT voltage vating}}{\textbf{LT voltage vating}}$$

$$= 20\times\frac{138}{20} = 138 \text{ KV}$$

Now,

$$\boxed{\begin{array}{l}\text{New Base MVA, } MVA_{b,new} = 50\text{MVA}\\ \text{New Base KV } KV_{b,new} = 138 \text{ KV}\end{array}}$$

Base impedance,
$$Z_b = \frac{KV_b^2}{MVA_b} = \frac{138^2}{50} = 380.88\ \Omega$$

$$\left.\begin{array}{l}\text{P.u reactance of } 20\Omega\\ \text{Transmission line}\end{array}\right\} = \frac{\text{Actual reactance, } \Omega}{\text{Base impedance, } \Omega} = \frac{20}{380.88}$$

$$\boxed{X_{pu} = .0525 P.u}$$

Since, the transmission lines TL_2 and TL_3 are having same values of reactances and base $KV's$, their P.u reactance will be same.

$$\therefore \quad X_{pu'}\ TL_2 = .0525\ P.u$$
$$X_{pu,}\ TL_3 = .0525\ P.u$$

Finding P.u reactance of transformer T_5

$$\left.\begin{array}{l}\text{New P.u reactance of}\\ \text{Transformer, } T_5\end{array}\right\} = X_{Pu,old}\left[\frac{KV_{b,old}}{KV_{b,new}}\right]^2\left[\frac{MVA_{b,new}}{MVA_{b,old}}\right]$$

$$= 0.1\times\left[\frac{138}{138}\right]^2\left[\frac{50}{15}\right] = .333 \text{ P.u}$$

$$\boxed{X_{Pu,T_5} = .333 P.u}$$

Finding P.u reactance of synchronous motor (M)

$$\left.\begin{array}{l}\text{Base KV on LT side}\\ \text{of Transformer, } T_2\end{array}\right\} = \text{Base KV on 1 + HT side} \times \frac{\text{LT Voltage rating}}{\text{HT voltage rating}}$$

$$= 138\times\frac{13.8}{138} = 13.8 \text{ KV}$$

$$\boxed{\begin{array}{l}\text{New Base MVA, } MVA_{b,new} = 50\text{MVA}\\ \text{New base KV, } KV_{b,new} = 13.8 \text{ KV}\end{array}}$$

$$\left.\begin{array}{l}\text{New P.u reactance}\\ \text{of synchronous motor}\end{array}\right\} = X_{\text{Pu,old}}\left[\frac{KV_{b,old}}{KV_{b,new}}\right]^2\left[\frac{MVA_{b,new}}{MVA_{b,old}}\right]$$

$$= 0.2\left[\frac{13.8}{13.8}\right]^2\left[\frac{50}{30}\right] = .333\ P.u$$

$$\boxed{X_{pu,m} = .333 P.u}$$

Finding P.u reactances of T_6, T_4, T_3 and G_2

- The transformer T_6 is identical (or) same as of transformer T_5. Hence P.u reactance of transformer T_5 and T_6 are same.

$$\boxed{\therefore X_{pu,Tb} = .333 P.u}$$

- The transformers T_1, T_2, T_3 and T_4 are identical (or) all are same rating. Hence their P.u reactances are same.

$$\therefore \quad \boxed{\begin{array}{l} X_{pu,T_1} = .25 P.u \\ X_{pu,T_2} = .25 P.u \\ X_{pu,T_3} = .25 P.u \\ X_{pu,T_4} = .25 P.u \end{array}}$$

- The generator G_2 is identical (or) same as G_1. Hence their P.u reactances are same.

$$\boxed{\therefore X_{pu,g_2} = .405 P.u}$$

The reactance diagram of Single line diagram (1.11.1) is

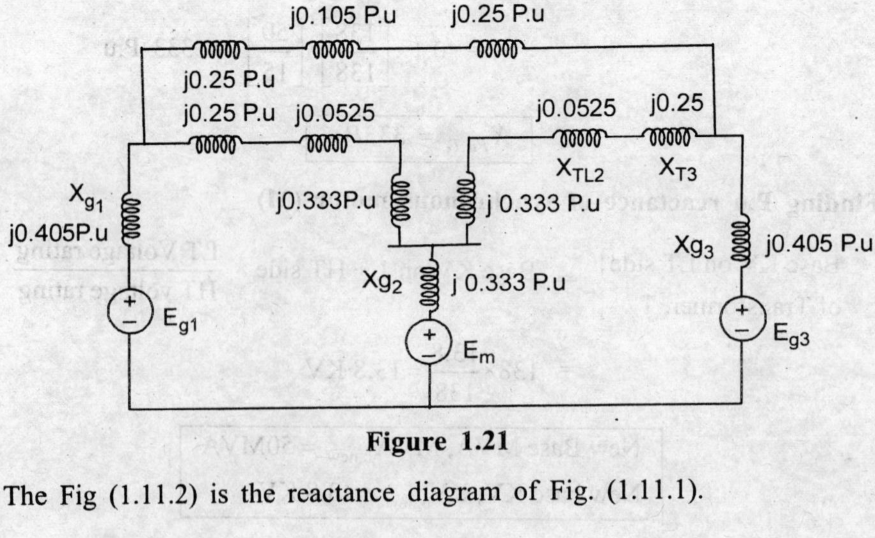

Figure 1.21

The Fig (1.11.2) is the reactance diagram of Fig. (1.11.1).

Example 1.12: A 15 *MVA*, 8.5 *KV*, 3 phase generator has a subtransient reactance of 20% It is connected through a Δ-*Y* transformer to a high voltage transmission line having a total series assistance of 100 Ω. The load end the line has *Y-Y* step down transformer. Both transformer. Banks are composed of single phase transformers connected for 3-phase operation. Each of three transformers composing three phase bank is rated 6667 KVA, 10/100 KV, with a reactance of 10%. The load represented as impedance-is drawing 10 MVA at 12.5 KV and 0.8 P.f lagging. Draw the single line diagram of the power network. Choose a base of 10 MVA, 12.5 KV in the load circuit and determine the reactance diagram. Determine also the voltage at the terminals of the generator

Solution: The single line diagram of the power shown in Fig. (1.12.1)

Base values: Base megavolt ampre,

$$MVA_{b,new} = 10 \text{ MVA}$$

Base kilovolt $\quad KV_{b,\ new} = 12.5 \text{ KV}$

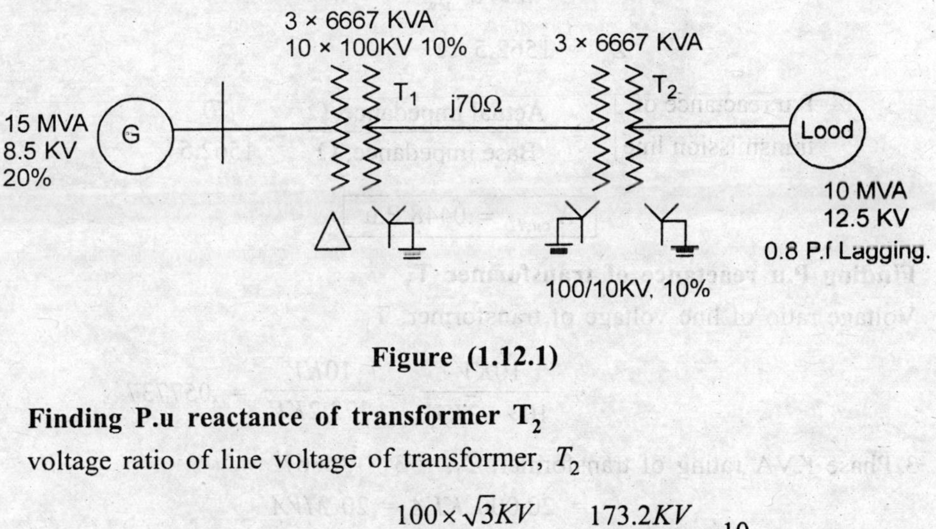

Figure (1.12.1)

Finding P.u reactance of transformer T_2

voltage ratio of line voltage of transformer, T_2

$$= \frac{100 \times \sqrt{3}KV}{10 \times \sqrt{3}KV} = \frac{173.2KV}{17.32KV} = 10$$

3 Phase KVA rating of transformer T_2 = 3 × 6667

$$= 20{,}000 \text{ KVA} = 20 \text{ MVA}$$

$\therefore\ KV_{b,old} = 17.32$ KV (on LT Side)

$$MVA_{b,old} = 20 \text{ MVA}$$

$$\left.\begin{array}{l}\text{New P.u reactance}\\ \text{of transformer}\end{array}\right\} = X_{\text{Pu,old}}\left[\frac{KV_{b,old}}{KV_{b,new}}\right]^2\left[\frac{MVA_{b,new}}{MVA_{b,old}}\right]$$

$$= 0.1\left[\frac{17.32}{12.5}\right]^2\left[\frac{10}{20}\right]$$

$$\boxed{X_{pu,T_2} = .096 P.u}$$

Finding P.u reactance of transmission line

$$\left.\begin{array}{l}\text{Base KV on HT side}\\ \text{of transformer } T_2\end{array}\right\} = \text{Base KV on LT side} \times \frac{\text{HT voltage rating}}{\text{LT voltage rating}}$$

$$= 12.5 \times \frac{173.2}{17.32} = 125 \text{ KV.}$$

New Base:

$$\boxed{\begin{array}{l}KV_{b,new} = 125KV\\ MVA_{b,new} = 10MVA\end{array}}$$

Base impedance, $$Z_b = \frac{(KV_{b,\,new})^2}{MVA_{b,\,new}} = \frac{(125)^2}{10}$$

$$Z_b = 1562.5\ \Omega$$

$$\left.\begin{array}{l}\text{P.u reactance of}\\ \text{transmission line}\end{array}\right\} = \frac{\text{Actual impedance, }\Omega}{\text{Base impedance, }\Omega} = \frac{70}{1562.5}$$

$$\boxed{X_{pu,TL} = .0448 \text{ P.u}}$$

Finding P.u reactance of transformer T_1

Voltage ratio of line voltage of transformer, T_1

$$= \frac{10KV}{10 \times \sqrt{3}KV} = \frac{10KV}{173.2KV} = .057737$$

3 Phase KVA rating of transformer, T_1, = 3 × 36667

$$= 20{,}000\ KVA = 20\ MVA$$

$$\therefore \quad KV_{b,\text{old}} = 173.2\ KV \text{ (on HT Side)}$$

$$MVA_{b,\text{ old}} = 20\ MVA$$

$$\left.\begin{array}{l}\text{New P.u reactance}\\ \text{of transformer } T_1\end{array}\right\} = X_{Pu,} old \times \left(\frac{KV_{b,old}}{KV_{b,new}}\right)^2 \times \left(\frac{MVA_{b,new}}{MVA_{b,old}}\right)$$

$$= 0.1 \times \left(\frac{173.2}{125}\right)^2 \times \left(\frac{10}{20}\right) = 0.096\ P.u$$

Finding P.u reactance of generator

$$\left.\begin{array}{l}\text{Base KV on LT side}\\ \text{of transformer } T_1\end{array}\right\} = \text{Base KV on H T Side} \times \left[\frac{\text{LT Voltage rating}}{\text{HT Voltage rating}}\right]$$

$$= 125 \times \frac{10}{173.2} = 7.217\text{KV}$$

Now, $KV_{b,new} = 7.217\ KV$

$$\left.\begin{array}{l}\text{New P.u realtance}\\ \text{of generator}\end{array}\right\} = X_{Pu,old} \times \left(\frac{KV_{b,old}}{KV_{b,new}}\right)^2 \times \left(\frac{MVA_{b,new}}{MVA_{b,old}}\right)$$

$$= 0.2 \times \left(\frac{8.5}{7.217}\right)^2 \times \left(\frac{10}{15}\right) = 0.185\ P.u$$

Load: This can be represented as Constant current Load

$$P.f \text{ of load} = 0.8 \text{ lag}$$

$$\therefore \quad P.f \text{ angle} = -\text{Cos}^{-1}0.8 = --36.87°$$

$$\text{Complex load power} = 10\angle -36.87° \text{ MVA}$$

$$P.u \text{ value of load (Power)} = \frac{\text{Actual load MVA}}{\text{Base value of MVA}}$$

$$= \frac{10\angle -36.87°}{10} = 1\angle -36.81° \text{ P.u}$$

Reactance diagram: The reactance diagram of single line diagram shown in Fig (1.12.2) is,

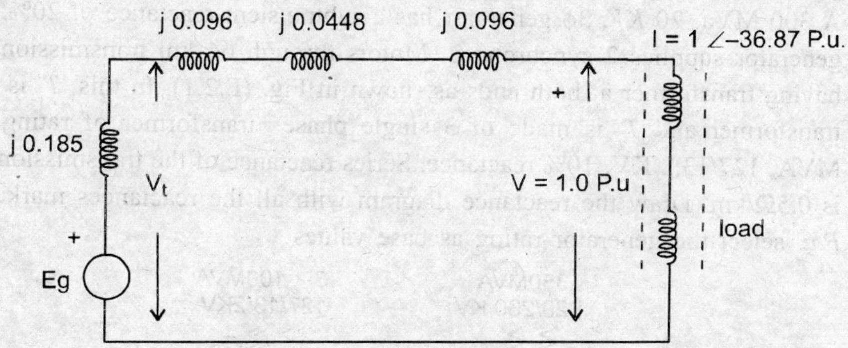

Figure (1.12.2)

The find the terminal voltage of the generator:

With reference to above fig (1.12.2)

The terminal voltage of the generator,

$$Vt = \text{V} + \text{I} \ (\text{j } 0.096 + \text{j } 0.0448 + j\ 0.096)$$

$$= 1.0 + 1\ \angle -36.87° \ (j\ 0.2368)$$

$= 1.0 + 1\angle{-36.87°} \times 0.2368\angle 90°$

$= 1.0 + 0.2368\angle 53.13°$

$= 1.0 + 0.1421 + j0.1894$

$= 1.1421 + j\ 0.1894$

$= 1.1577 \angle 9.6°\ P.u$

$$\left.\begin{array}{l}\text{Actual value of}\\ \text{generator terminal}\\ \text{voltage}\end{array}\right\} = \left[\begin{array}{l}P.u\text{ value of}\\ \text{voltage}\end{array}\right] \times \left[\begin{array}{l}\text{Base KV on } LT\\ \text{side on transformer } T_1\end{array}\right]$$

$= 1.1577 \angle 9.4° \times 7.217 = 8.355\angle 9.4°$ KV

EXERCISE

1. (*a*) A generator is rated 500 *MVA*, 22 *KV*. Its *Y*-connected winding has a reactance of 1.1P.u. Find the onmic value of the reactance of winding

(*b*) If the generator is working in a circuit for which the bases are specified as 100 *MVA*, 20 *KV*, Then find the *P.u* value of reactance of generator winding on the specified base.

Ans. (*a*) $X = 1.0648\ \Omega$/Phase

(*b*) $X_{pu,}$ new $= .2662\ P.u$

2. A 300 MVa, 20 *KV*, 3φ generator has a subtransient reactance of 20%. The generator supplies 2 synchronous. Motors through 64 km transmission line having transformer at both ends as shown in Fig. (E.2.1). In this, T_1 is a 3φ transformer and T_2 is made of 3 single phase transformer of rating 100 MVA, 127/13.2 KV, 10% reactance. Series reactance of the transmission line is 0.5Ω/km. Draw the reactance diagram with all the reactances marked in *P.u.* select the generator rating as base values

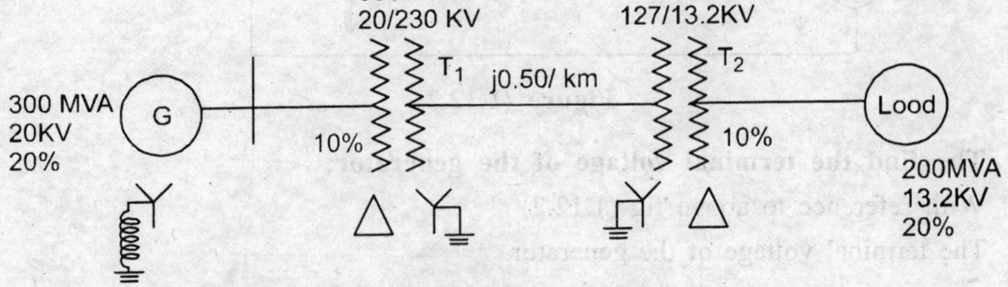

Fig (E. 2. 1)

Ans.

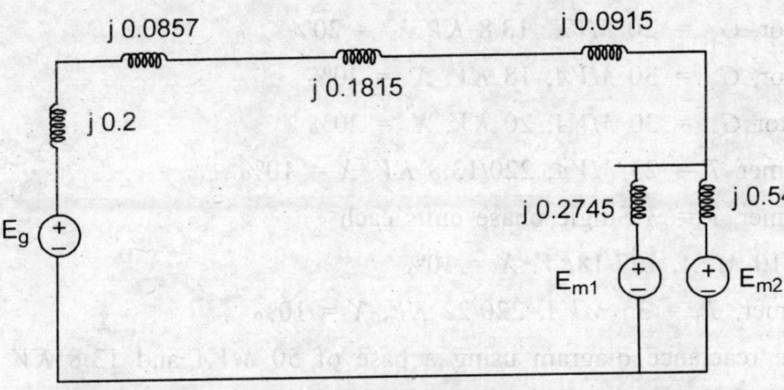

Fig (E. 2. 2)

3. A 120*MVA*, 19.5 *KV* generator has a synchronous reactance of 0.15 *P.u* and it is connected to a transmission line through a transformer rated 150 MVA, 230/18 KV (*Y*/Δ) with X = 0.1 *P.u*

(a) Calculate the P.*u* reactance by taking generator raking as base values

(b) Calculate the *P.u* reactance by taking transformer rating as base values

(c) Calculate the *P.u* reactances for a base value of 100 MVA and 220 KV HT side of transformer

Ans: (*a*) $X_{Pu,\ new}$ = .15 *P.u* (generator) $X_{Pu,\ new}$ = .0682 *P.u* (Transformer)

(*b*) $X_{Pu,\ new}$ = .22 *P.u* (generator) $X_{Pu,new}$ = .1 *P.u* (transformer)

(*c*) $X_{Pu,\ new}$ = .1603 *P.u* (Generator) $X_{Pu,\ new}$ (transformer) = .0727 *P.u*

4. The single line diagram of an unloaded power system is shown in Fig. (E. 1. 4. 1) The The generator and transformers are rated as follows

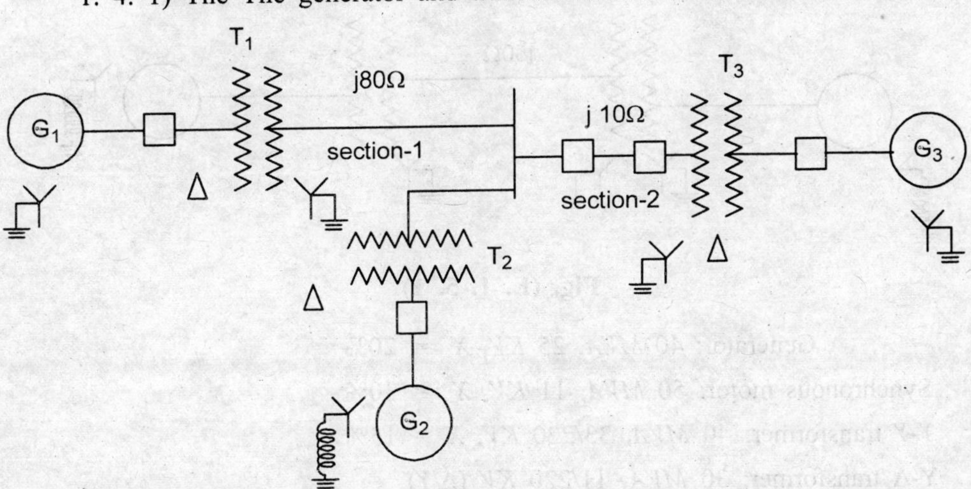

Fig (E. 4. 1)

Generator, G_1 = 20 *MVA*, 13.8 *KV* $X'' = 20\%$

Generator, G_2 = 30 *MVA*, 18 *KV*, $X'' = 20\%$

Generator G_3 = 30 *MVA*, 20 *KV*, $X'' = 20\%$

Transformer T_1= 25 *MVA*, 220/13.8 *KV*, $X = 10\%$

Transformer T_2= 3 Single phase units each rated at 10 *MVA*, 127/18*KV*, $X = 10\%$

Transformer, T_3 = 35 *MVA*, 220/22 *KV*, $X = 10\%$

Draw the reactance diagram using a base of 50 *MVA* and 13.8 *KV* on the generator G_1.

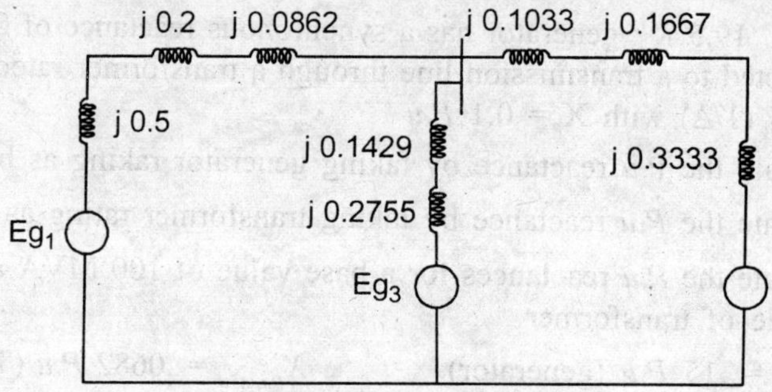

Fig. (E. 1. 4. 2)

5. Draw the reactance diagram for the power system shown in the Fig. (E. 1. 5. 1) Neglect resistance and use a base of 100 *MVA*, 220 *KV* in 50Ω line. the ratings of the generator, motor and transformer are given below

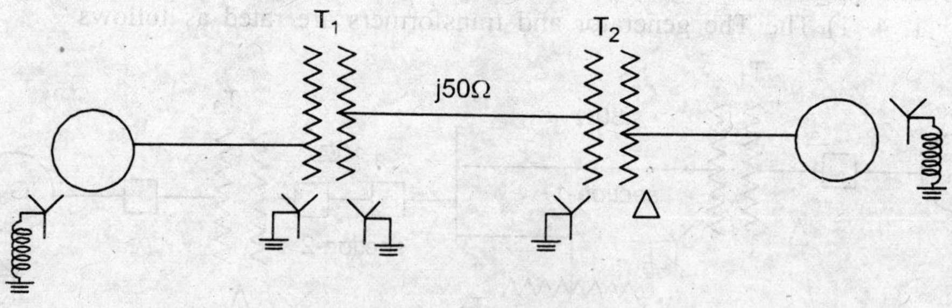

Fig. (E. 1. 5. 1)

Generator: 40 *MVA*, 25 *KV*, $X'' = 20\%$

Synchronous motor: 50 *MVA*, 11 *KV*, $X'' = 30\%$

Y-Y transformer: 40 *MVA*, 33/220 *KV*, $X = 15\%$

Y-Δ transformer: 30 *MVA*, 11/220 *KV* (Δ/*Y*), $X = 15\%$

Ans.

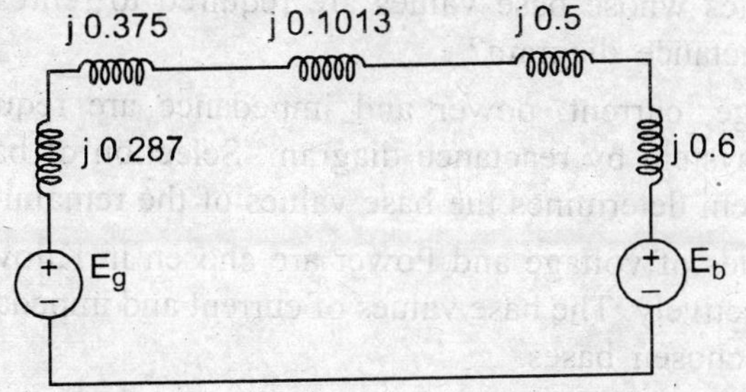

6. A 3 phase generator with 1000 KVA, 33 KV has it armature resistance and synchronous reactance as 20Ω/ phase and 70Ω/ phase Calculate P.u impedance of the generator.

Ans: $Z_{Pu} = 0.018 + j\ 0.064\ P.u$

7. A 3 Phase, Δ-*Y* transformer with rating 100 KVA, 11 KV/400V has its primary and secondary leakage reactance as 20Ω/ phase and .05Ω/ Phase respectively. Calculate the P.u reactance of transformer.

Ans: $X_{pu} = .0411\ P.u$

8. A 50 KW, 3 phase, *Y* connected load is fed by a 200 KVA transformer with voltage rating 11KV/400V through feeder. The length of the feeder is 0.5 Km and the impedance of the feeder is 0.1 + *j* 0.2Ω/km. If the load *P.f* is 0.8, calculate the *P.u* impedance of the load and feeder.

Ans: $Z_{pu,\ feeder} = 0.0625 + j\ 0.125\ P.u$

$Z_{pu,\ Load} = 2.56 + j\ 1.92\ P.u.$

Short Questions and Answers

Q. 1. What is single line diagram?

Ans. A single line diagram is a diagrammatic representation of power system in which the components are represented by their symbols and the inter connection between them are shown by a single straight line. The rating and the impedances of the components are also marked on the single line diagram.

Q. 2. What are the components of power system?

Ans. Generators, Power transformers, transmission Line, subastation transformers, Distribution transformers and loads.

Q. 3. Define Per unit value:

Ans. The Per unit value of any quantity is defined as the ratio of the actual value of the quantity to the base value expressed a decimal. The base value is an arbitrary chosen value of the quantity.

$$\text{Per unit value} = \frac{\text{Actual value}}{\text{Base value}}$$

Q. 4. What are the quantities whose base values are required to represent the power system by reactance diagram?

Ans. The values of voltage, current, power and impedance are required to represent the power system by reactance diagram. Selection of base values for any two of them determines the base values of the remaining two.

Usually the base values of voltage and Power are chosen in Kilovolt and KVA (or) MVA respectively. The base values of current and impedance are calculated using the chosen bases.

Q. 5. What is the need for base values?

Ans. The components or various sections of Power system may operate at different voltage and Power Levels. It will be convenient for analysis of Power system if the voltage, Power, current and impedance rating of components of Power system are expressed with reference to a commonvalue called base vale. Hence for analysis purpose a base value is chosen for voltage, Power, current and impedance. Then all the voltage, Power, current as a impedance ratings of the components are expressed as a percent (or) Per unit of the base value.

Q. 6. Write the equation for converting the P.u impudence expressed in one base to another?

Ans.
$$Z_{pu,\ new} = Z_{Pu,old} \times \left[\frac{KV_{b,old}}{KV_{b,new}}\right]^2 \left[\frac{MVA_{b,new}}{MVA_{b,old}}\right]$$

Q. 7. What are the advantages of Per-unit computations?

Ans.
1. Manufacturers usually specify the impedance of a device (or) machine in per unit on the base of the name plate rating.
2. The P.u values a of widely different rating machines lie within a narrow range, even though the onmic values has a very large range.
3. The Per unit impedance of circuit element connected by transformer expressed on a Proper base will be same if it is referred to either side of a transformer.
4. The per unit impedance of a 3-phase transformer is independent of the type of winding connection (Y or Δ)

Q. 8. Draw the equivalent circuit of a 3-Phase generator.

Ans.

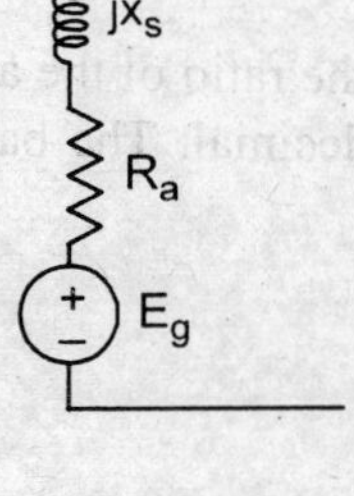

$X_s \Rightarrow$ Synonymous reactance/phase

$R_a \Rightarrow$ Armature resistance/phase

$E_g \Rightarrow$ Induced ems/Phase

$I_g \Rightarrow$ Current/Phase

Q. 9. How the Loads are represented in reactance or impedance diagram.

Ans. The resistive and reactive loads can be represented by any one of the following representation.

(*i*) Constant Power representation

Load Power, $S = P + jQ$

(*ii*) Constant current representation

$$\text{Load current, } I = \frac{\sqrt{P^2 + Q^2}}{|V|} \underline{|\delta - \theta}$$

(*iii*) Constant impedance representation

$$\text{Load impedance, } Z = \frac{|V|^2}{P - jQ}$$

Q. 10. How the induction motor is represented in reactance diagram?

Ans. For estimation of steady state fault current, the induction motor is neglected. But for estimation of fault current immediately after the fault (*i.e.* to estimate subtransient Fault currents), the induction motor can be represented by a source in Series with reactance. The value of the source is the induced emf per phase and the value of reactance is total reactance of induction motor per phase referred to stator.

Q. 11. What is impedance and reactance diagram?

Ans. (a) The impedance diagram is the equivalent circuit of power system in which the various components a of Power system are represented by their approximate or simplified equivalent impedance. The impedance diagram is used for load flow studies.

(*b*) The reactance diagram is the simplified equivalent circuit of Power system in which the various components are represented by their reactances. The reactance diagram can be obtained form impedance diagram if all the resistive components are neglected. The reactance diagram is used for fault calculation.

Q. 12. What are the approximations made in impedance diagram?

Ans. 1. The neural reactances are neglected.

2. The shunt brances in equivalent circuit of inducting motor and transformer are neglected.

Q. 13. What are the approximations made reactance diagram?

Ans. 1. The neutral reactances are neglected

2. The shunt branches in equivalent circuit of transformer is neglected

3. The resistances are neglected.
4. All static loads and induction motors are neglected.
5. The Capacitance of transmission lines are neglected.

Q. 14. Give equations for transforming base KV on LV side to HV side of transformer and Vice-versa.

Ans. *i.* Base *KV* on HT side = Base *KV* on LT side × $\frac{\text{HT Hegerating}}{\text{LT voltage rating}}$

ii. Base *KV* on LT, Side = Base *KV* on HT side × $\frac{\text{LT voltage rating}}{\text{HT voltage rating}}$

1.8 SYMMETRICAL COMPONENTS OF UNBALANCED PHASORS

1.8.1 Symmetrical Components

The symmetrical components are very effective and important tool in Power system analysis which can be used to analyse the unsymmetrical Polyphase network. This technique is introduced by Dr. C. Foretaste. He Proved An unbalanced system of n related vectors can be resolved into n system of balanced vectors called symmetrical components of original vectors The vectors of each set of components are equal in length and the phase angles between adjacent vectors of the set are equal. In a three phase system, the three unbalanced vectors [either V_a, V_b and V_c, I_a, I_b and I_c] can be resolved into three balanced system of vectors The vectors of the balanced system are called symmetrical components of the original system. The symmetrical components of three phase system are,

1. Positive sequence components
2. Negative sequence components
3. Zero sequence components

The Positive sequence consists of 3 vectors which are equal in magnitude, displaced from each other by 120° in phase, and having the same phase sequence as the original vectors.

The negative sequence components consists of three vectors equal in magnitude, displaced from each other by 120° in phase, and having the phase sequence opposite to that of the original vectors.

The zero sequence components consists of three vectors equal in magnitude and with zero phase displacement form each other.

Let V_a, V_b, and V_c be the set of unbalanced voltage vectors with phase sequence abc. Each voltage vector can be resolved into Positive, negative and zero sequence components.

(*i*) V_{a1}, V_{b1} and V_{c1} ⇒ Positive sequence components of V_a, V_b and V_c respectively with phase sequence abc.

(*ii*) V_{a2}, V_{b2}, V_{c2} ⇒ Negative sequence components of V_a, V_b and V_c respectively. with phase sequence abc.

(iii) V_{a0}, V_{b0}, and V_{c0} ⇒ Zero sequence components of V_a, V_b and V_c respectively with phase sequence abc.

The vector diagram. of positive, negative and zero sequence components are shown in Fig. (1.14)

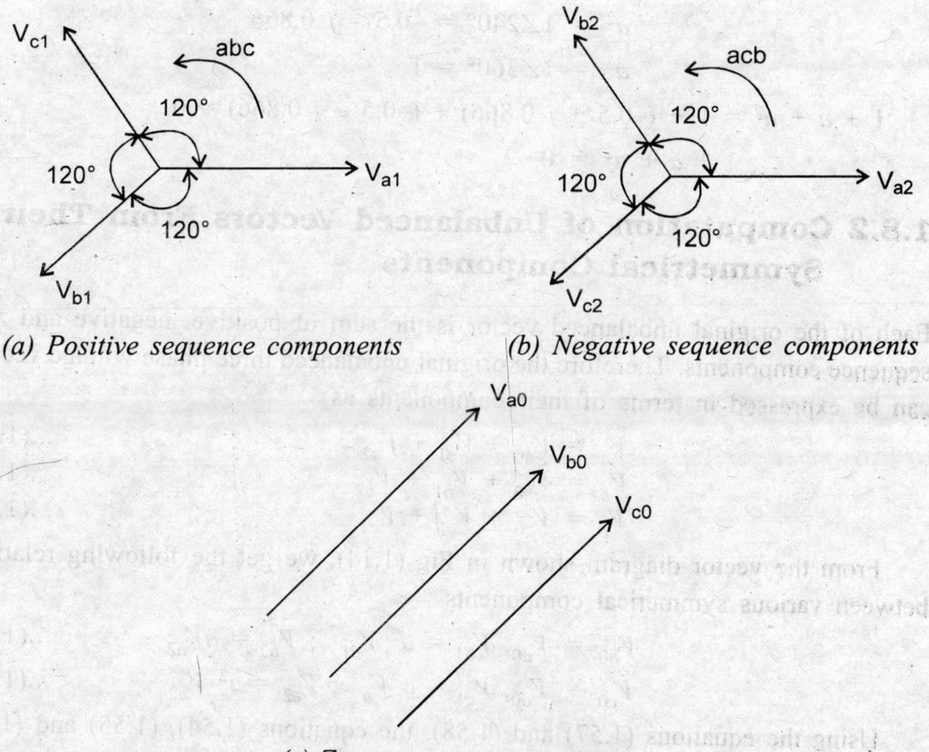

(a) Positive sequence components *(b) Negative sequence components*

(c) Zero sequence components

Figure 1.14 *Vector diagram of symmetrical components of unbalanced 3-phase voltage vectors*

From the vector diagram of symmetrical components the following points can be noted.

- On rotating the vector V_{a1} by 120° in anticlockwise direction we get V_{c1}
- On notating the vector V_{a1} by 240° in anticlockwise direction we get V_{b1}
- On rotating the vector V_{a2} by 120° in anticlockwise direction we get V_{b2}
- On rotating the vector V_{a2} by 240° in anticlockwise direction we get V_{c2}

Therefore, on rotating the sysmmetrical component of one vector by 120° (or) multiples of 120° we get the symmetrical components of other vectors. hence, we can define an operator which causes a rotation of 120° in the anticlockwise direction, such an operator is denoted by the letter "a".

The operator "a" is defined as,

$$a = 1\angle 120° = 1e^{+j\,2\pi/3} = \cos\frac{2\pi}{3} + j\sin\frac{2\pi}{3}$$

$$= -0.5 + j\,0.866$$

Since,

$$a = 1\angle 120° = -0.5 + j\,0.866$$

$$a^2 = 1\angle 240° = -0.5 - j\,0.866$$

$$a^3 = 1\angle 360° = 1$$

$$1 + a + a^2 = 1 + (-0.5 + j\,0.866) + (-0.5 - j\,0.866) = 0$$

$$\therefore \quad 1 + a + a^2 = 0$$

1.8.2 Computation of Unbalanced Vectors From Their Symmetrical Components

Each of the original unbalanced vector is the sum of positive, negative and zero sequence components. Therefore the original unbalanced three phase voltage vectors can be expressed in terms of their components as

$$V_a = V_{a0} + V_{a1} + V_{a2} \quad ...(1.54)$$

$$V_b = V_{b0} + V_{b1} + V_{b2} \quad ...(1.55)$$

$$V_c = V_{c0} + V_{c1} + V_{c2} \quad ...(1.56)$$

From the vector diagram shown in Fig (1.14), we get the following relations between various symmetical components.

$$V_{b0} = V_{a0},\ V_{b1} = a^2\,V_{a1}\ ;\ V_{b2} = aV_{a2} \quad ...(1.57)$$

$$V_{c0} = V_{a0},\ V_{c1} = a\,V_a\ ;\ V_{c2} = a^2\,V_{a2} \quad ...(1.58)$$

Using the equations (1.57) and (1.58), the equations (1.54), (1.55) and (1.56) can be written as,

$$V_a = V_{a0} + V_{a1} + V_{a2} \quad ...(1.59)$$

$$V_b = V_{a0} + a^2V_{a1} + aV_{a2} \quad ...(1.60)$$

$$V_c = V_{a0} + aV_{a1} + a^2V_{a2} \quad ...(1.61)$$

The equations (1.59), (1.60) and (1.61) can be arranged in the matrix form as,

$$\begin{bmatrix} V_a \\ V_b \\ V_c \end{bmatrix} = \begin{bmatrix} 1 & 1 & 1 \\ 1 & a^2 & a \\ 1 & a & a^2 \end{bmatrix} \begin{bmatrix} V_{a0} \\ V_{a1} \\ V_{a2} \end{bmatrix} \quad ...(1.62)$$

The equation (1.62) can be used to compute the unbalanced voltage vectors from the knowledge of symmetical components.

1.8.3 Computation of Symmetrical Components of Unbalanced Vectors

The equation (1.62) can be written in the vector notation as,

$$V = AV_{sy} \quad \text{...(1.63)}$$

Where, $V = \begin{bmatrix} V_a \\ V_b \\ V_c \end{bmatrix}$; $A = \begin{bmatrix} 1 & 1 & 1 \\ 1 & a^2 & a \\ 1 & a & a^2 \end{bmatrix}$ and

$$V_{sy} = \begin{bmatrix} V_{a0} \\ V_{a1} \\ V_{a2} \end{bmatrix}$$

On premultiplying the equation (1.63) by A^{-1} we get.

$$A^{-1}V = V_{sy}$$

$$\therefore \quad V_{sy} = A^{-1}V \quad \text{...(1.64)}$$

$$A^{-1} = \frac{\text{Adjoint of A}}{\text{Determinant of A}}$$

Let $\quad \Delta$ = Determinant of A

$$\therefore \quad \Delta = \begin{bmatrix} 1 & 1 & 1 \\ 1 & a^2 & a \\ 1 & a & a^2 \end{bmatrix} = 1 \times (a^4 - a^2) - 1 \times (a^2 - a) + 1(a - a^2)$$

$$= (a - a^2) + (a - a^2) + (a - a^2) = 3(a - a^2)$$

$$[\therefore \; a^4 = a^3 . \, a = a \text{ where } a^3 = 1]$$

Let $\quad \Delta ij$ = Cofactor of A_{ij}

$$\therefore \quad \Delta_{11} = a^4 - a^2 = a - a^2$$

$$\Delta_{12} = -(a^2 - a) = a - a^2$$

$$\Delta_{13} = a - a^2$$

$$\Delta_{21} = -(a^2 - a) = a - a^2$$

$$\Delta_{22} = a^2 - 1$$

$$\Delta_{23} = -(a - 1) = 1 - a$$

$$\Delta_{31} = a - a^2$$

$$\Delta_{32} = -(a - 1) = 1 - a$$

$$\Delta_{33} = a^2 - 1$$

$$\therefore \quad A^{-1} = \frac{1}{\Delta}\text{Adjoint of A}$$

$$= \frac{1}{\Delta}\begin{bmatrix} \Delta_{11} & \Delta_{12} & \Delta_{13} \\ \Delta_{21} & \Delta_{22} & \Delta_{23} \\ \Delta_{31} & \Delta_{32} & \Delta_{33} \end{bmatrix}$$

$$= \frac{1}{\Delta}\begin{bmatrix} \Delta_{11} & \Delta_{21} & \Delta_{31} \\ \Delta_{12} & \Delta_{22} & \Delta_{32} \\ \Delta_{13} & \Delta_{23} & \Delta_{33} \end{bmatrix}$$

$$A^{-1} = \frac{1}{3(a-a^2)}\begin{bmatrix} a-a^2 & a-a^2 & a-a^2 \\ a-a^2 & a^2-1 & 1-a \\ a-a^2 & 1-a & a^2-1 \end{bmatrix}$$

$$A^{-1} = \frac{(a-a^2)}{3(a-a^2)}\begin{bmatrix} 1 & 1 & 1 \\ 1 & \frac{a^2-1}{a-a^2} & \frac{1-a}{a-a^2} \\ 1 & \frac{1-a}{a-a^2} & \frac{a^2-1}{a-a^2} \end{bmatrix} \quad \text{...(1.65)}$$

$$\frac{a^2-1}{a-a^2} = \frac{a^3-a^3}{a-a^2} = \frac{a(a-a^2)}{(a-a^2)} = a \quad \text{...(1.66)}$$

$$\frac{1-a}{a-a^2} = \frac{a^2(1-a)}{a^2(a-a^2)} = \frac{a^2(1-a)}{(a^3-a^4)} = \frac{a^2(1-a)}{1-a} = a^2 \quad \text{...(1.67)}$$

➡ $a^3 = 1$; $a^4 = a^3 \, . \, a = a$

Using the equations the equations (1.66) and (1.67), the equation (1.68) can be written as

$$A^{-1} = \frac{1}{3}\begin{bmatrix} 1 & 1 & 1 \\ 1 & a & a^2 \\ 1 & a^2 & a \end{bmatrix} \quad \text{...(1.68)}$$

On substituting for A^{-1} from equation (1.68) in equation (1.64)

$$V_{sy} = \frac{1}{3}\begin{bmatrix} 1 & 1 & 1 \\ 1 & a & a^2 \\ 1 & a^2 & a \end{bmatrix} V$$

$$\begin{bmatrix} V_{ao} \\ V_{a1} \\ V_{a2} \end{bmatrix} = \frac{1}{3}\begin{bmatrix} 1 & 1 & 1 \\ 1 & a & a^2 \\ 1 & a^2 & a \end{bmatrix}\begin{bmatrix} V_a \\ V_b \\ V_c \end{bmatrix} \quad ...(1.69)$$

The equation (1.69) can be expressed as three independent linear equations as shown below.

$$V_{a0} = \frac{1}{3}[V_a + V_b + V_c] \quad ...(1.70)$$

$$V_{a0} = \frac{1}{3}\left[V_a + aV_b + a^2V_c\right] \quad ...(1.71)$$

$$V_{a0} = \frac{1}{3}\left[V_a + a^2V_b + aV_c\right] \quad ...(1.72)$$

The equations (1.69), (1.70) , (1.71) and (1.72) can be used to compute the symmetrical components of the unbalanced voltages.

1.8.4 Symmetrical Components of Unbalanced Current Vectors

The symmetrical components of unbalanced current vectors can be obtained by an analysis similar to that of voltage vectors. The current equations are same as voltage equation *i.e.*) voltage equations and current equations of symmetrical components of unbalanced vectors are similar if we replace V by I.

Let, $I_a, I_b, I_c \rightarrow$ Unbalanced current vectors with phase sequence *abc*.

$I_{a1}, I_{b1}, I_{c1} \rightarrow$ Positive sequence components of I_a, I_b, and I_c respectively with phase sequence abc.

$I_{a2}, I_{b2}, I_{c2} \rightarrow$ Negative sequence components of I_a, I_b, and I_c respectively with phase sequence *abc*.

$I_{a0}, I_{b0}, I_{c0} \rightarrow$ Zero sequence components of I_a, I_b, and I_c respectively. with phase sequence abc.

The vector diagram of Positive, negative and Zero sequence components are shown in Fig. 1.15.

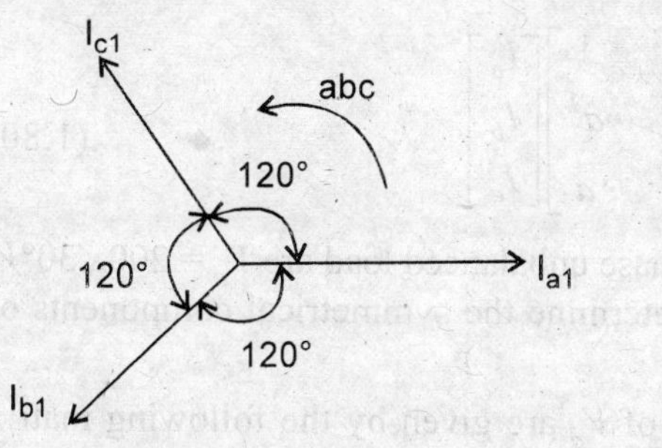

(a) *Positive sequence components*

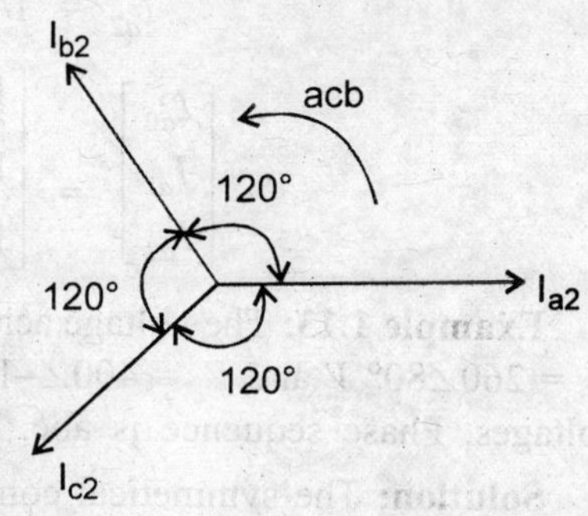

(b) *Negative sequence components*

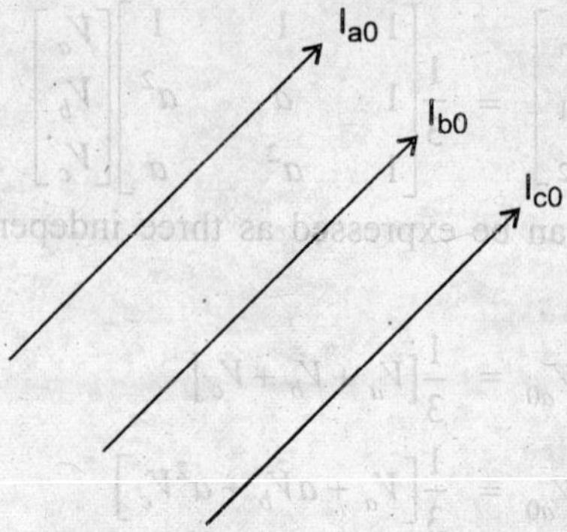

(c) Zero sequence components

Figure 1.15 *Vector diagram of symmetrical components of unbalanced 3-phase current vectors.*

The following equations are used to compute the unbalanced current vectors from the knowledge of their symmetrical components refer equations (1.59), (1.60), (1.61) and (1.62).

$$I_a = I_{a0} + I_{a1} + I_{a2} \quad ...(1.73)$$

$$I_b = I_{a0} + a^2 I_{a1} + aI_{a2} \quad ...(1.74)$$

$$I_c = I_{a0} + aI_{a1} + a^2 I_{a2} \quad ...(1.75)$$

$$\begin{bmatrix} I_a \\ I_b \\ I_c \end{bmatrix} = \begin{bmatrix} 1 & 1 & 1 \\ 1 & a^2 & a \\ 1 & a & a^2 \end{bmatrix} \begin{bmatrix} I_{a0} \\ I_{a1} \\ I_{a2} \end{bmatrix} \quad ...(1.76)$$

The symmetrical components of unbalanced current vectors can be written as, refer equations (1.69), (1.70), (1.71) and (1.72)

$$I_{a0} = 1/3[I_a + I_b + I_c] \quad ...(1.77)$$

$$I_{a1} = 1/3[I_a + aI_b + a^2 I_c] \quad ...(1.78)$$

$$I_{a2} = 1/3[I_a + a^2 I_b + aI_c] \quad ...(1.79)$$

$$\begin{bmatrix} I_{a0} \\ I_{a1} \\ I_{a2} \end{bmatrix} = \begin{bmatrix} 1 & 1 & 1 \\ 1 & a & a^2 \\ 1 & a^2 & a \end{bmatrix} \begin{bmatrix} I_a \\ I_b \\ I_c \end{bmatrix} \quad ...(1.80)$$

Example 1.13: The voltage across a 3 phase unbalanced load are $V_a = 200\angle 30° V$, $V_b = 260\angle 80°\ V$ and $V_c = 400\angle -120° V$. Determine the symmetrical components of voltages. Phase sequence is abc.

Solution: The symmetical components of V_a are given by the following matrix equations.

$$\begin{bmatrix} V_{a0} \\ V_{a1} \\ V_{a2} \end{bmatrix} = \frac{1}{3}\begin{bmatrix} 1 & 1 & 1 \\ 1 & a & a^2 \\ 1 & a^2 & a \end{bmatrix}\begin{bmatrix} V_a \\ V_b \\ V_c \end{bmatrix}$$

➡ Refer equation (1.67)

$$\therefore \quad V_{a0} = 1/3[V_a + V_b + V_c]$$

$$V_{a1} = 1/3[V_a + aV_b + a^2V_c]$$

$$V_{a2} = 1/3[V_a + a^2V_b + aV_c]$$

Given that, $V_a = 200\ \angle 30° V = 202.23 + j8.53$

$$V_b = 260\ \angle 80°\ V = 272.02 + j17.102$$

$$V_c = 400\ \angle{-120°} V = 417.61 - j16.69$$

$$\therefore \quad aV_b = 1\angle 120° \times 260\ \angle 80° = 260\ \angle 200° = 328.02 + j37.56$$

$$a^2V_b = 1\angle 240° \times 260\ \angle 80° = 260\ \angle 360° = 444 + j54.16$$

$$aV_c = 1\angle 120° \times 400\ \angle{-120°} = 400\ \angle 0° = 400 + j0$$

$$a^2V_c = 1\angle 240° \times 400\ \angle{-120°} = 400\ \angle 120° = 417.61 + j16.69$$

$$V_{a0} = 1/3[V_a + V_b + V_c] = 1/3[202.23 + j\ 8.53 + 272.02\ j17.102 + 417.61 - j16.69]$$

$$= \frac{1}{3}[891.86 + j8.86]$$

$$= \frac{1}{3}[891.9\underline{|.569°}]$$

$$\boxed{V_{a0} = 297.3\underline{|.569°}}$$

$$V_{a1} = 1/3[V_a + aV_b + a^2V_c]$$

$$= 1/3[202.23 + j8.53 + 328.02 + j37.56 + 417.61 + j16.69]$$

$$= 1/3[947.86 + j62.78]$$

$$= 1/3[949.936\ \angle 3.789°]$$

$$\boxed{V_{a1} = 316.645\underline{|3.789°}}$$

$$V_{a2} = \frac{1}{3}[V_a + a^2V_b + aV_c]$$

$$= \frac{1}{3}[202.23 + j8.53 + 444 + j54.16 + 400 + j0]$$

$$= \frac{1}{3}[\ 1046.23 + j\ 62.69]$$

$$= \frac{1}{3}[1048.106\ \angle 3.429°$$

$$V_{a2} = 349.36\ \angle 3.429°$$

We know that,

Zero sequence components, $V_{a0} = V_{b0} = V_{c0}$

+Ve sequence components, $V_{b1} = a^2V_{a1}$; $V_{c1} = aV_{a1}$

–Ve sequence components, $V_{b2} = aV_{a2}$; $V_{c2} = a^2V_{a2}$

∴ **Zero sequence components are,**

$$V_{a0} = 297.3 \angle .569°$$
$$V_{b0} = 297.3 \angle .569°$$
$$V_{c0} = 297.3 \angle .569°$$

+Ve sequence components are,

$$V_{al} = 316.645 \angle 3.789°$$
$$V_{b1} = a^2V_{al} = (1\angle 240°)\ (316.645\ \angle 3.789°) = 316.645\angle 243.78°$$
$$V_{c1} = aVa_1 = (1\angle 120°\ (316.645\ \angle 3.789°) = 316.645\ \angle 123.789°$$

–Ve sequence components are,

$$V_{a2} = 349.36\ \angle 3.429°$$
$$V_{b2} = aV_{a2} = (1\angle 120°)\ (349.36\angle 3.4290°) = 349.36\ \angle 123.429°$$
$$V_{c2} = a^2V_{a2} = (1\angle 240°)\ (349.36\ \angle 3.429°) = 349.36\ \angle 243.429°$$

Example 1.14: The symmetrical components of pahse-a voltage in *a* 3 phase unbalanced system are $V_{a0} = 20\angle 140°$, $V_{a1} = 40\angle 10°$ and $V_{a2} = 10\angle 80°$ *V*. Determine the phase voltage V_a, V_b and V_c.

Solution: The phase voltages of V_a, V_b and V_c are given by the following matrix equations.

$$\begin{bmatrix} V_a \\ V_b \\ V_c \end{bmatrix} = \begin{bmatrix} 1 & 1 & 1 \\ 1 & a^2 & a \\ 1 & a & a^2 \end{bmatrix} \begin{bmatrix} V_{a0} \\ V_{a1} \\ V_{a2} \end{bmatrix}$$

$$\therefore \quad V_a = V_{a0} + V_{a1} + V_{a2}$$
$$V_b = V_{b0} + a^2V_{a1} + aV_{a2}$$
$$V_c = V_{c0} + aV_{a1} + a^2V_{a2}$$

Given that,
$$V_{a0} = 20\angle 140° = -15.32 + j12.85$$
$$V_{al} = 40\angle 10° = 39.39 + j\ 6.94$$
$$V_{a2} = 10\angle 80° = 1.736 + j\ 9.84$$

$$\therefore \quad aV_{a1} = (1\angle 120°)\ (40\angle 10°) = 40\ \angle 130° = -25.71 + j30.64$$
$$a^2V_{a1} = (1\angle 240°)\ (40\angle 10°) = 40\angle 250° = -13.68 - j37.5$$
$$aV_{a2} = (1\angle 120°)\ (10\angle 80°) = 10\angle 200° = -9.39 - j3.42$$
$$a^2V_{a2} = (10\angle 240°)\ (10\angle 80°) = 10\angle 320° = 7.66 - j6.42$$
$$V_a = V_{a0} + V_{a1} + V_{a2} = -15.32 + j12.85 + 39.39 + j6.94 + 1.736 + j9.84$$

$$\boxed{V_a = 25.806 + j29.63}$$

$$V_b = V_{a0} + a^2V_{a1} + aV_{a2} = -15.32 + j12.85 - 13.68 - j37.5 - 9.39 - j3.42$$

$$\boxed{V_b = -38.39 - j28.07}$$

$$V_c = V_{a0} + aV_{a1} + a^2Va_2 = -15.32 + j12.85 - 25.71 + j30.64 + 7.66 - j6.42$$

$$\boxed{V_c = -33.37 + j37.07}$$

Ans:

$$V_a = 25.806 + j29.63$$
$$V_b = -38.39 - j28.07$$
$$V_c = -33.37 + j37.07$$

Example 1.15: The symmetrical components of phase-a fault current in a three phase unbalanced system are $I_{a0} = 320\angle 60° A$, $I_{a1} = 560\angle -60° A$ and $I_{a2} = 210\angle 70° V$. Determine the phase currents $I_{a,}$ I_b and I_c.

Solution: The currents I_a, I_b and I_c are given by the following matrix equations.

$$\begin{bmatrix} I_a \\ I_b \\ I_c \end{bmatrix} = \begin{bmatrix} 1 & 1 & 1 \\ 1 & a^2 & a \\ 1 & a & a^2 \end{bmatrix} \begin{bmatrix} I_{a0} \\ I_{a1} \\ I_{a2} \end{bmatrix}$$

$$\therefore \quad I_a = I_{a0} + I_{a1} + I_{a2}$$
$$I_b = I_{a0} + a^2I_{a1} + aI_{a2}$$
$$I_c = I_{a0} + aI_{a1} + a^2I_{a2}$$

Given that

$$I_{a0} = 320\angle 60° = 160 + j277.12$$
$$I_{a1} = 560\angle -60° = 280 - j\,484.97$$
$$I_{a2} = 210\angle 70° = 71.82 + j197.335$$

$$\therefore \quad aIa_1 = (1\angle 120°)\,(560\angle -60°) = 560\angle 60° = 280 + j484.97$$
$$a^2I_{a1} = (1\angle 240°)\,(560\angle -60°) = 560\angle 180° = -560 + j0$$
$$aI_{a2} = (1\angle 120°\ (210\angle 70° = 210\angle 190° = -206.8 - j36.46$$
$$a^2I_{a2} = (1\angle 240°)\,(210\angle 70°) = 210\angle 310° = 134.98 - j160.86$$

$$I_a = I_{a_o} + I_{a_1} + I_{a_2} = 160 + j277.12 + 280 - j484.97 + 71.82 + j197.335$$

$$\boxed{I_a = 511.82 - j10.515}$$

$$I_b = I_{a_o} + a^2I_{a_1} + aI_{a_2} = 160 + j277.12 - 560 + j0 - 206.8 - j36.46$$

$$\boxed{I_b = -606.8 + j240.66}$$

$$I_c = I_{a0} + aI_{a1} + a^2Ia_2$$
$$= 160 + j277.12 + 280 + j484.97 + 134.98 - j160.8$$
$$I_c = 574.98 + j601.23.$$

Ans:

$$I_a = 511.82 - j10.515$$
$$I_b = -606.8 + j240.66$$
$$I_c = 574.98 + j601.23$$

Example 1.16: Determine the symmetrical components of the unbalanced three phase currents.

$$I_a = 30\angle 10^\circ \; A, \; I_b = 24\angle 200^\circ \; A \text{ and } I_c = 50\angle 100^\circ \; A$$

Solution: The symmetrical components of I_a given by the following matrix equations.

$$\begin{bmatrix} I_{a0} \\ I_{a_1} \\ I_{a_2} \end{bmatrix} = \frac{1}{3}\begin{bmatrix} 1 & 1 & 1 \\ 1 & a & a^2 \\ 1 & a^2 & a \end{bmatrix}\begin{bmatrix} I_a \\ I_b \\ I_c \end{bmatrix}$$

$$\therefore \quad I_{a0} = \frac{1}{3}(I_a + I_b + I_c)$$

$$I_{a_1} = \frac{1}{3}\left(I_a + aI_b + a^2 I_c\right)$$

$$I_{a_2} = \frac{1}{3}\left(I_a + a^2 I_b + aI_c\right)$$

Given that,

$$I_a = 30\angle 10^\circ = 29.5 + j5.2$$
$$I_b = 24\angle 200^\circ = -22.55 - j8.2$$
$$I_c = 50\angle 100^\circ = -8.68 + j49.24$$

$$\therefore \quad aI_b = (1\angle 120^\circ)(24\angle 200^\circ) = 24\angle 320^\circ = 18.38 - j15.42$$

$$a^2 I_b = (1\angle 240^\circ)(24\angle 200^\circ) = 24\angle 440^\circ = 4.16 + j23.63$$

$$aI_c = (1\angle 120^\circ)(50\angle 100^\circ) = 50\angle 220^\circ \; - 38.3 - j32.13$$

$$a^2 I_c = (1\angle 240^\circ)(50\angle 100^\circ) = 50\angle 340^\circ = 46.98 - j17.10$$

$$I_{a0} = \frac{1}{3}(I_a + I_b + I_c) = \frac{1}{3}\,(29.5 + j5.2 - 22.55 - j8.2 - 8.68 + j49.24)$$

$$= \frac{1}{3}(-1.68 + j46.24)$$

$$= \frac{1}{3}(46.27\angle 92.08^\circ)$$

$$\boxed{I_{a0} = 15.42\angle 92.08^\circ}$$

$$I_{a_1} = \frac{1}{3}\left(I_a + aI_b + a^2 I_c\right) = \frac{1}{3}(29.5 + j5.2 + 18.38 - j15.42 + 46.98 - j17.10)$$

$$= \frac{1}{3}(94.86 - j27.32)$$

$$= \frac{1}{3}(98.715\underline{|16.06°})$$

$$\boxed{I_{a_1} = 32.905\underline{|-16.06°}}$$

$$I_{a_2} = \frac{1}{3}\left(I_a + a^2 I_b + aI_c\right) = \frac{1}{3}(29.5 + j5.2 + 4.16 + j23.63 - 38.3 - j32.13)$$

$$= \frac{1}{3}(-4.64 - j3.3)$$

$$= \frac{1}{3}(5.69\underline{|-144.57°})$$

$$\boxed{I_{a_2} = 1.89\underline{|-144.57°}}$$

we know that, $I_{a_0} = I_{b_0} = I_{c_0}$

∴ The zero sequence components are,

$$I_{a_0} = 15.42\underline{|92.08°}$$

$$I_{b_0} = 15.42\underline{|92.08°}$$

$$I_{c_0} = 15.42\underline{|92.08°}$$

We know that, $I_{b_1} = a^2 I_{a_1}$; $I_{c_1} = aI_{a_1}$

∴ The positive sequence components are

$$\boxed{I_{a_1} = 32.905\underline{|-16.06°}}$$

$$I_{b_1} = a^2 I_{a_1} = (1\underline{|240°})(32.905\underline{|-16.06°})$$

$$\boxed{I_{b_1} = 32.905\underline{|223.94°}}$$

$$I_{c_1} = aI_{a_1} = (1\underline{|120°})(32.905\underline{|-16.06°})$$

$$\boxed{I_{c_1} = 32.90.5\underline{|103.94°}}$$

We know that, $I_{b2} = a\, I_{a2}$; $I_{c2} = a^2 I_{a2}$

$\therefore$ The negative sequence components are

$$\boxed{I_{a2} = 1.89\underline{|-144.57°}}$$

$$I_{b2} = aI_{a2} = (1\underline{|120°})(1.89\underline{|-144.57°})$$

$$\boxed{I_{b2} = 1.89\underline{|-24.57°}}$$

$$I_{c2} = a^2 I_{a2} = (1\underline{|240°})(1.89\underline{|-144.57°})$$

$$\boxed{I_{c_2} = 1.89\underline{|95.43°}}$$

Ans: (*i*) $I_{a_0} = I_{b_0} = I_{c_0} = 15.42\ \underline{|92.08°}$

(*ii*) $I_{a1} = 32.905\ \underline{|-16.06°}$

$I_{b1} = 32.905\ \underline{|223.94°}$

$I_{c1} = 32.905\ \underline{|103.94°}$

(*iii*) $I_{a2} = 1.89\ \underline{|-144.57°}$

$I_{b2} = 1.89\ \underline{|-24.57°}$

$I_{c2} = 1.89\ \underline{|95.43°}$

EXERCISE

9. The voltages across a three phase unbalanced load are $V_a = 300\ \underline{|20°}\ V$, $V_b = 360\underline{|360°}\ V$ and $V_c = 500\ \underline{|-140°}\ V$. Determine the symmetrical components of Voltage. Phase sequence is *abc*.

Ans. (*i*) $V_{a0} = V_{b0} = V_{c0} = 57.89\ \underline{|126°}\ V$

(*ii*) $V_{a1} = 143.70\ \underline{|106°}\ V$

$V_{b1} = 143.76\ \underline{|346°}\ V$

$V_{c2} = 143.76\ \underline{|226°}\ V$

(*iii*) $V_{a2} = 364.05\ \underline{|-13°}\ V$

$V_{a2} = 364.05\ \underline{|107°}\ V$

$V_{a2} = 364.05\ \underline{|227°}\ V$

10. The symmetrical components of phase-a Voltage in a 3-phase unbalanced system are $V_{a0} = 10\angle 180° \ V$, $V_{a1} = 50 \ \angle 0° \ V$ and $V_{a2} = 20\angle 90° \ V$.

Determine the phase Voltages V_a V_b and V_c.

Ans. $V_a = 44.72 \ \angle 27° \ V$

$V_b = 74.69 \ \angle -134° \ V$

$V_c = 37.70 \ \angle 118° \ V$

11. The symmetrical components of phase-*a* fault current in a 3-phase unbalanced system are $I_{a0} = 350\angle 90° \ A$, $I_{a1} = 600 \ \angle -90° \ A$ and $I_{a2} = 250\angle 90°$. Determine the phase currents I_a, I_b and I_c.

Ans. $I_a = 0$

$I_b = 904.16 \ \angle 145° \ A$

$I_c = 904.16 \ \angle 35° \ A$

12. Determine the symmetrical components of the unbalanced three phase Current

$I_a = 10\angle 0° \ A$, $I_b = 12\angle 230°$ and $I_c = 10\angle 130° \ A$.

Ans. (*i*) $I_{a0} = I_{b0} = I_{c0} = 1.047 \ \angle -160° \ A$

(*ii*) $I_{a1} = 10.56\angle 0° \ A$

$I_{b1} = 10.56\angle 240° \ A$

$I_{c1} = 10.56\angle 120° \ A$

(*iii*) $I_{a2} = 1.04\angle 37° \ A$

$I_{b2} = 1.04\angle 157° \ A$

$I_{c2} = 1.04\angle 277° \ A$

1.9 SEQUENCE IMPEDANCES AND SEQUENCE NETWORKS

What is sequence impedances?

The sequence impedances are impedances offered by the circuit elements (or power system components) to positive, negative and zero sequence currents. In any element of a circuit, the voltage drop caused by current of a certain sequence depends on the impedance of the element to that sequence current.

The impedance of a circuit element when positive sequence currents alone are flowing is called the positive sequence impedance. Similarly, when only negative sequence currents are present, the impedance is called zero sequence impedance. The impedance of any element of a balanced circuit to current of one sequence may be different from impedance to currant of another sequence.

The single phase equivalent circuit of power system (impedance (or) reactance diagram) formed using the impedances of any one sequence only is called the sequence network for that particular sequence. Therefore the impedance or reactance diagram formed using positive sequence impedance is called positive sequence network. Similarly the impedance or reactance diagram formed using negative sequence impedance is called negative sequence network. The impedance or reactance diagram formed using sequence impedance is called zero sequence network.

The sequence impedances and networks are useful in the analysis of unsymmetrical faults in the power system. In unsymmetrical fault analysis of a power system, the positive, negative and zero sequence networks of the system are determined and then they are interconnected to represent the various unbalanced fault conditions.

1.9.1 Sequence Impedances and Networks of Generator

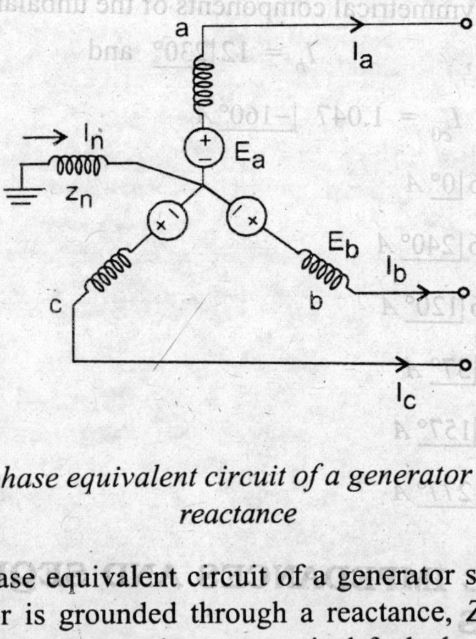

Figure 1.16 *Three phase equivalent circuit of a generator grounded through a reactance*

Consider the 3- phase equivalent circuit of a generator shown in Fig. 1.16. The neutral of the generator is grounded through a reactance, Z_n. When the generator is delivering a balanced load or under symmetrical fault the neutral current is zero. But when the generator is delivering an un balanced load or during unsymmetrical faults the neutral current flows through *Zn*.

The generator is designed to supply balanced 3 phase Voltages. Therefore the generated emps are of positive sequence only.

Let, E_a, E_b and $E_c \rightarrow$ Generated emt per phase in phase *a*, *b* and *c* respectively. (+ve sequence emf)

$Z_1 \rightarrow$ +ve sequence impedance/phase of generator.

Z_2 → –ve sequence impedance/phase of generator.

Z_{go} → zero sequence impedance/phase of generator.

Z_n → Neutral reactance.

Z_0 → Total zero sequence impedance/phase of zero sequence network of generator.

The positive sequence network consists of an emf in series with positive sequence impedance of the generator. The negative and zero sequence network will not have any sources but include their respective sequence impedance. The positive, negative and zero sequence current paths are shown in Fig. 1.17.

The positive, negative and zero sequence networks of the generator are shown in Fig. 1.18.

➡ The positive and negative sequence currents are balanced currents and so they will not pass through neutral reactance.

The reactance in positive sequence network is subtransient, transient or synchronous reactance depending on whether subtransient, transient (or) steady state conditions are being studied. Under no load condition the emf E_a is the induced emf per phase. Under load or fault conditions E_a is replaced by E_g' for transient state and E_a is replaced by E_g'' for subtransient state. On examining the zero sequence current paths (refer Fig. (1.17c) we can say that the current through neutral reactance is $3I_{a0}$. The zero sequence voltage drop from point-*a* to ground is $-3I_{a0} Z_n - I_{a0} Z_{g0}$.

The zero sequence network is a single phase network and assumed to carry only the zero sequence current of one phase. Hence the zero sequence current of one phase, must have an impedance of $3Z_n + Z_{g0}$.

∴ The total zero sequence impedance per phase of a generator grounded through reactance $\Bigg\}$, $Z_0 = 3Z_n + Z_{g0}$ ...(1.81)

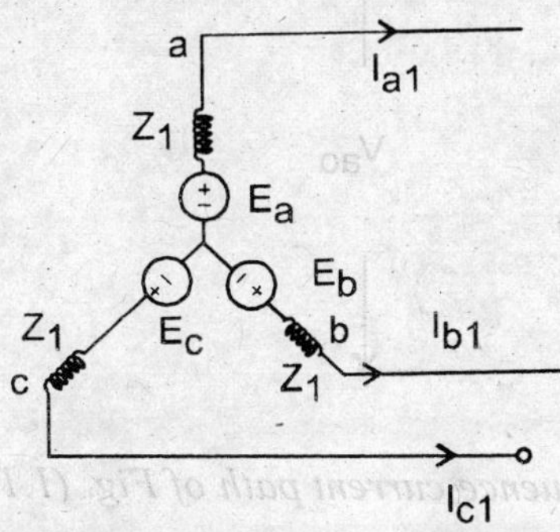

Fig (1.17(a)) *+Ve sequence current paths*

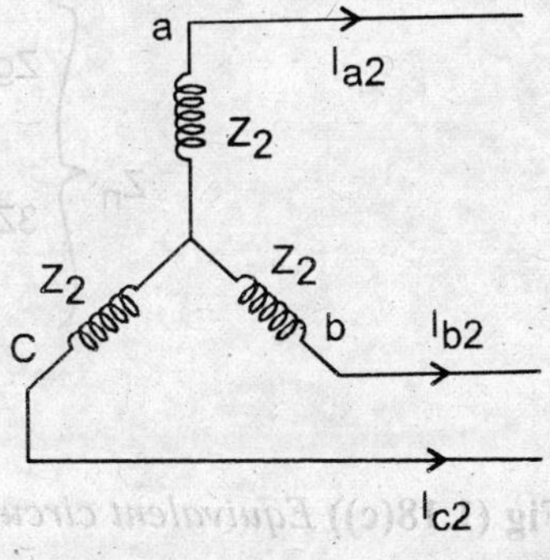

Fig (1.17(b)) *–Ve sequence current paths*

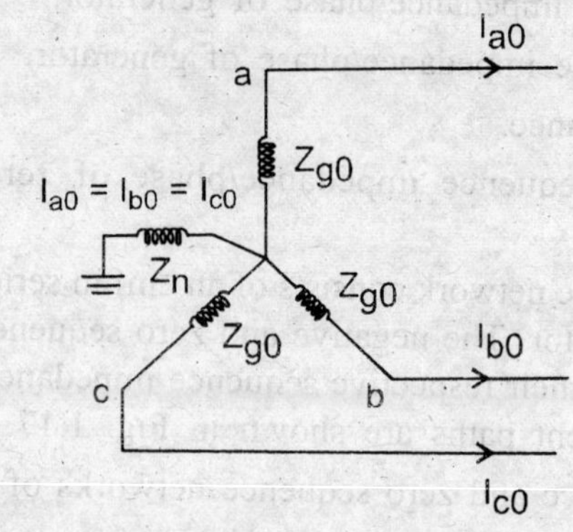

Fig (1.17 (c)) *Zero sequence current paths*

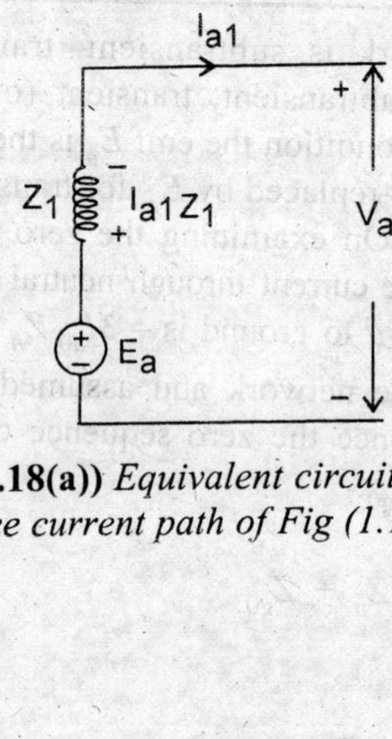

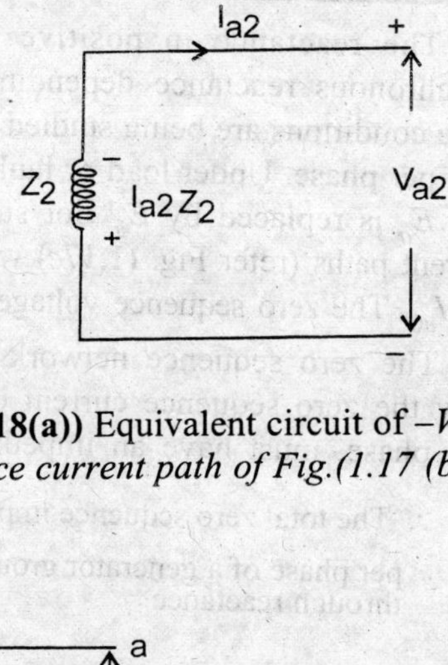

Fig (1.18(a)) *Equivalent circuit of + Ve sequence current path of Fig (1.17(a))*

Fig (1.18(a)) Equivalent circuit of *–V e sequence current path of Fig.(1.17 (b))*

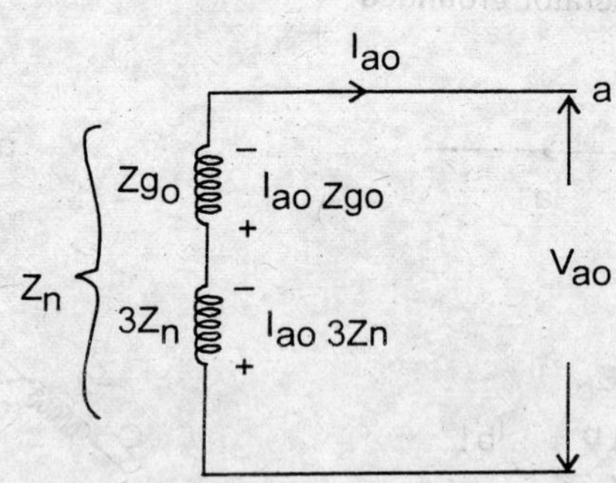

Fig (1.18(c)) *Equivalent circuit of Zero sequence current path of Fig. (1.17(c))*

with reference to Fig. (1.18a, b and c) the equations for the phase-*a* component voltages are,

$$V_{a1} = E_a - I_{a1} Z_1 \quad ...(1.82)$$
$$V_{a2} = - I_{a2} Z_2 \quad ...(1.83)$$
$$V_{a0} = - I_{a0} Z_0 \quad ...(1.84)$$

The zero sequence network of generator when the neutral is solidly grounded (i.e. directly grounded) and when the neutral is ungrounded are shown in Fig. 1.19 and Fig. 1.20. respectively. In these cases there is no change in positive and negative sequence network.

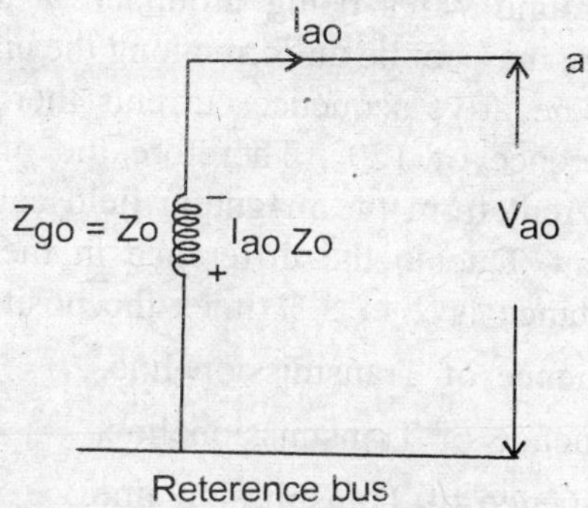

Figure 1.29 *Zero-sequence network of a generator when the neutral is solidly grounded*

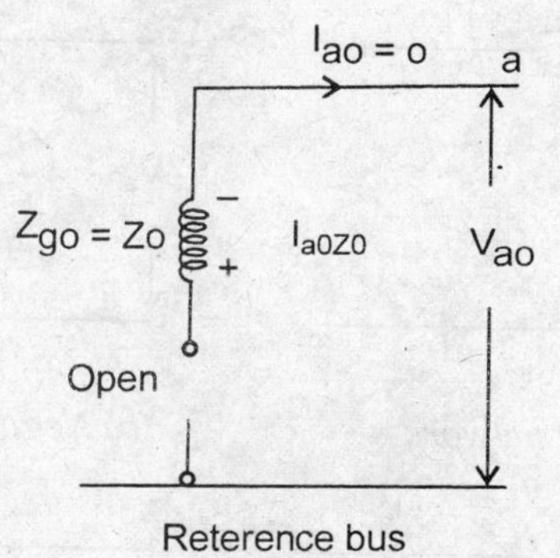

Figure 1.20 *Zero-sequence network of a generator when the neutral is solidly is ungrounded*

➡ The sequence networks of synchronous motor is same as that of generator when the directions of currents in the sequence network of generator are reversed.

1.9.2 Sequence Impedances and Networks of Transmission Lines

The impedance per phase of transmission line for balanced currents is independent of phase sequence. This is due to the symmetry of transposed transmission lines. Therefore, the impedances offered by the transposed transmission lines for positive and negative sequence currents are identical.

The zero sequence current is identical (both in magnitude and phase) in each phase conductor and returns through the ground, through over head ground wires or through both. The ground wires being grounded at several towers, the return currents in the ground wire may not be uniform along the entire length of transmission line. But for positive or negative sequence currents there is no return current and they have a phase difference of 120°. Therefore the magnetic field due to zero sequence current is different from the magnetic field caused by either positive or negative sequence current. Due to the difference in the magnetic field, the zero sequence inductive reactance is 2 to 3.5 times the positive sequence reactance.

Let, $Z_1 \rightarrow$ +ve sequence of Transmission line.

$Z_2 \rightarrow$ –ve sequence of Transmission line.

$Z_0 \rightarrow$ zero sequence of Transmission line.

The positive, negative and zero sequence impedances of transmission lines are represented as a series impedance in their respective sequence networks as shown in Fig. 1.21.

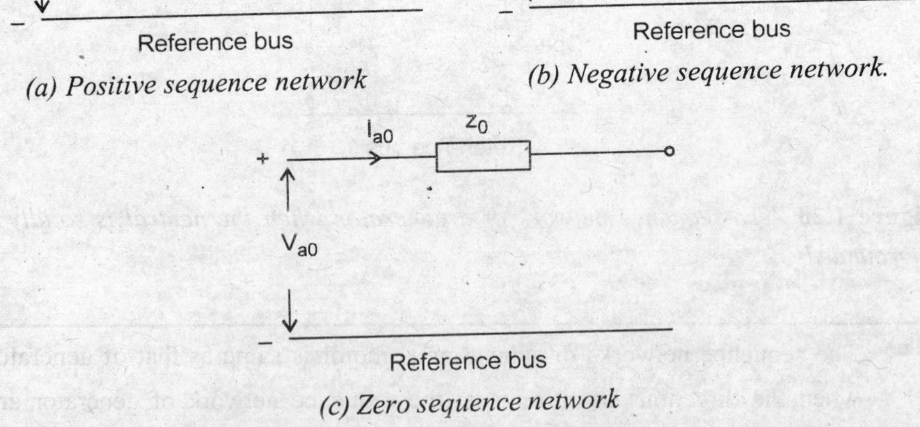

Figure 1.21 *+ve, –ve and zero sequence network of transmission line*

1.9.3 Sequence Impedances and Networks of Transformer

When the applied voltage is balanced, the positive and negative sequence of linear symmetrical, static devices are identical. therefore in a transformer the +ve and –ve sequence impedances are identical.

Eventhough the zero sequence impedance may slightly differ from +ve and –ve sequence impedances, it is normal practice to assume the zero sequence impedance as equal to positive or negative sequence impedance.

> (*i*) For all types of transformers the series impedances of all sequences are assumed equal.
>
> (*ii*) when the neutral of star connection is grounded through reactance Z_n then $3Z_n$ should be added to zero sequence impedance of transformer to get the total zero sequence impedance.

Let, $Z_1 \rightarrow$ +ve sequence impedance of transformer

$Z_2 \rightarrow$ –ve sequence impedance of transformer

$Z_0 \rightarrow$ zero sequence impedance of transformer

The positive and negative sequence impedances of transformer are represented as a series impedance in their respective sequence networks as shown in Fig. 1.22.

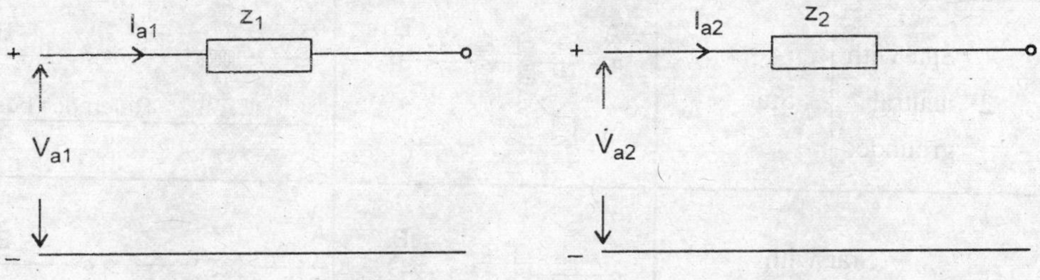

(a) +Ve sequence network *(b) –Ve sequence network*

Figure 1.22 *+Ve and –Ve sequence networks of transformer*

The zero sequence network of the transformer depends on the type of connections (Y or Δ) of the primary and secondary windings and also on the grounding of neutral in Y connection.

The following general points to be noted for zero sequence currents in transformers.

- The zero sequence current will flow in the primary winding only when there is a path for zero sequence current in secondary winding.

(➡ The above said point can be justified as, when magnetizing current is negected, the primary winding will carry current only if there is a current flow on the secondar winding.)

- There is no zero sequence current in star connected winding unless the neutral point of Y connected winding is grounded.
- The zero sequence current will circuilate in the delta connected winding but not in the lines which is connected to the delta connected winding.

By considering the above points, Let we draw the zero sequence network of 3-phase transformer for different configurations refer figure 1.23.

➡
- The arrows on the winding indicate the direction of zero sequence current
- If no indication then there is no zero sequence currents.

Configuration	Representation	Zero sequence network
1 Star : Star	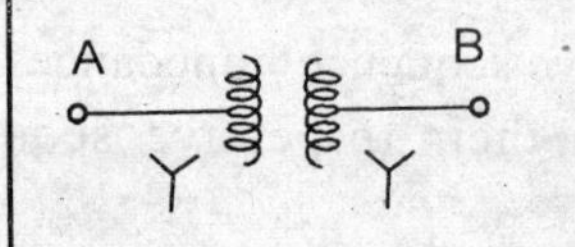	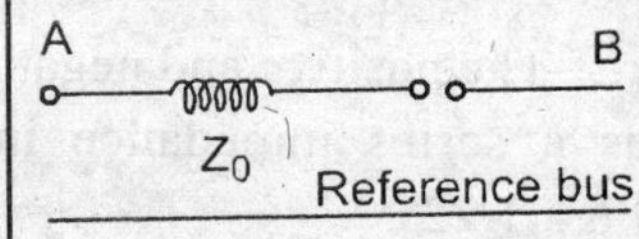
2 Star with neutral grounded : Star	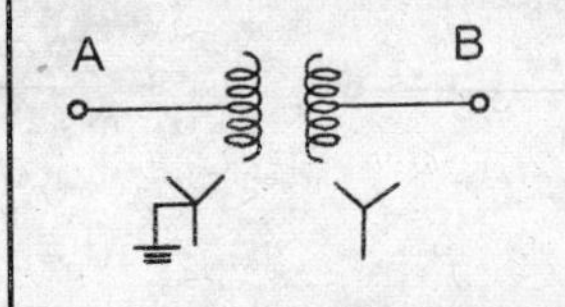	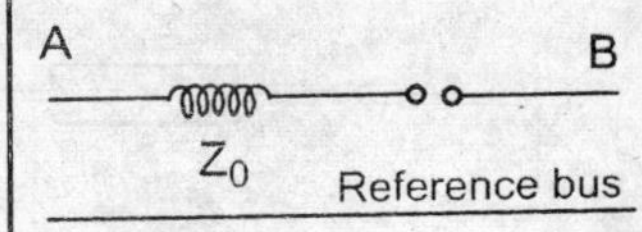
3 Star : Star with neutral grounded	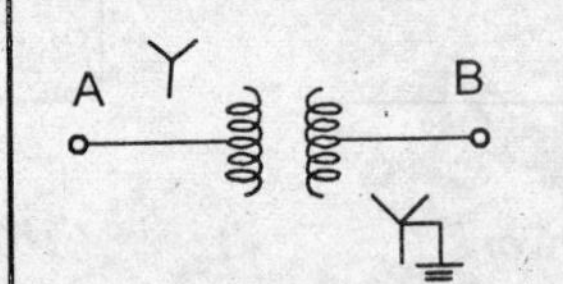	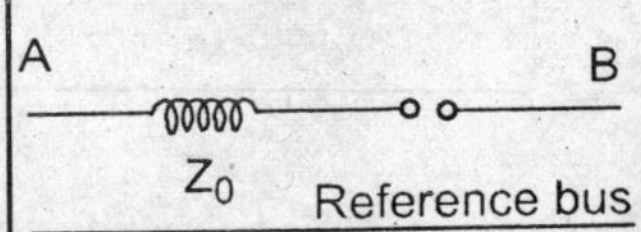
4 Star with neutral grounded : Star with neutral grounded	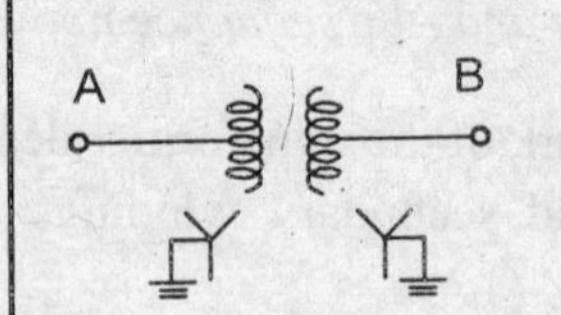	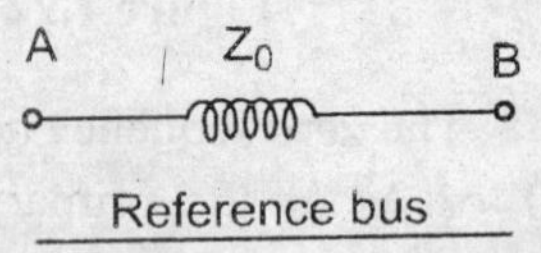
5 Star : Delta	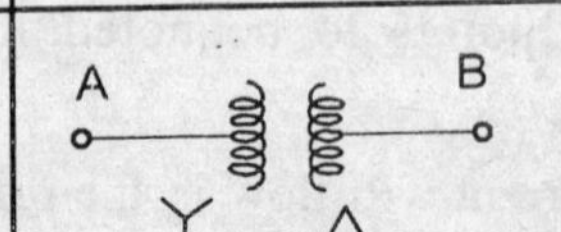	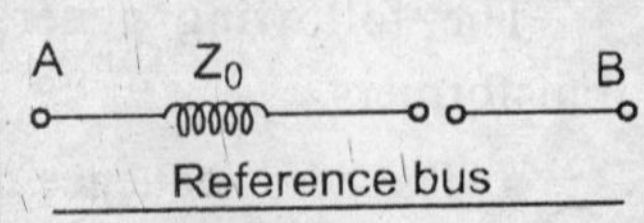

Configuration	Connection	Zero sequence network
6 Delta : Star	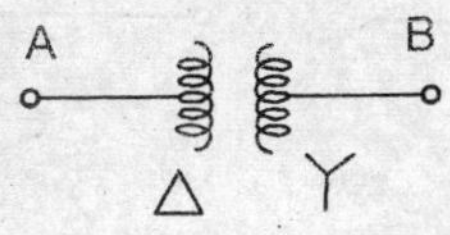	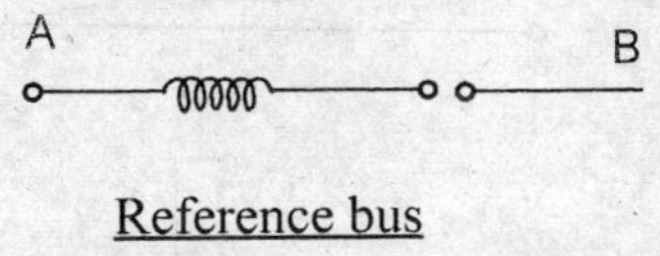
7 Star with neutral grounded : Delta	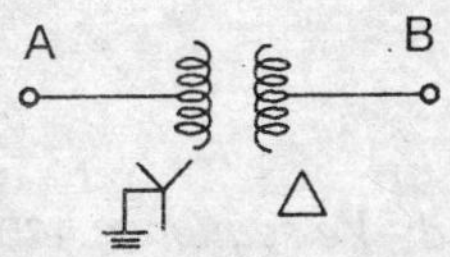	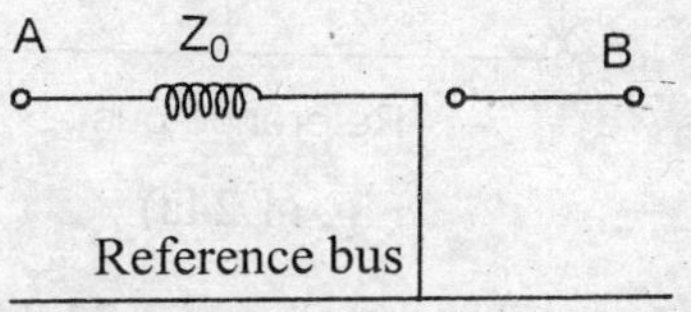
8 Delta : Star with neutral grounded	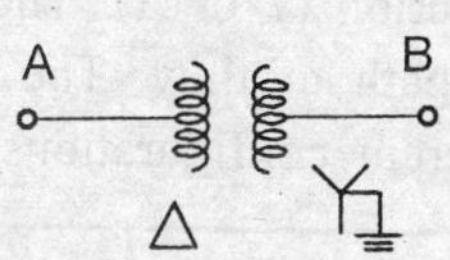	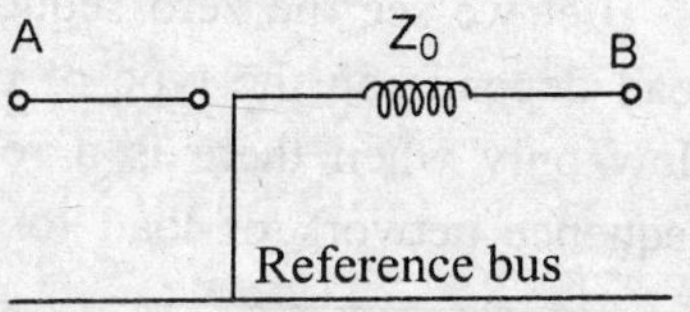
9 Delta : Delta	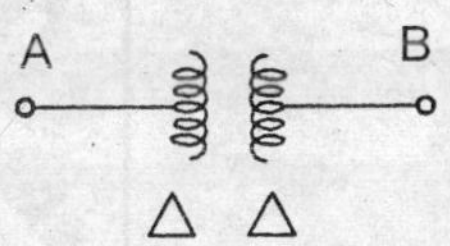	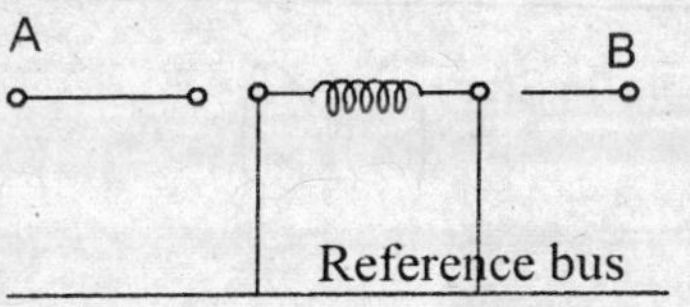

Fig(1.23): *Zero sequence network of 3 phase transtormer for different configurations.*

1.9.4 Sequence Impedance and networks of Loads

As we know that, there are two kinds of loads (*i*) balanced load (*ii*) unbalanced load. In case of balanced loads the positive, negative and zero sequence impedances are equal. When the neutral point of star connected load is grounded through a reactance Z_n then $3Z_n$ is added to the zero sequence impedance of load to get the total zero sequence impedance of load.

Let,

Z_{L0} → Zero sequence impedance of Load

Z_{L1} → + Ve sequence impedance of Load

Z_{L2} → – Ve sequence impedance of Load

Now let we see the positive and negative sequence networks.

> ➡ The +*Ve* and –*Ve* sequence impedances of load are represented as a shunt impedance in their respective sequence network. [refer figure 1.24]

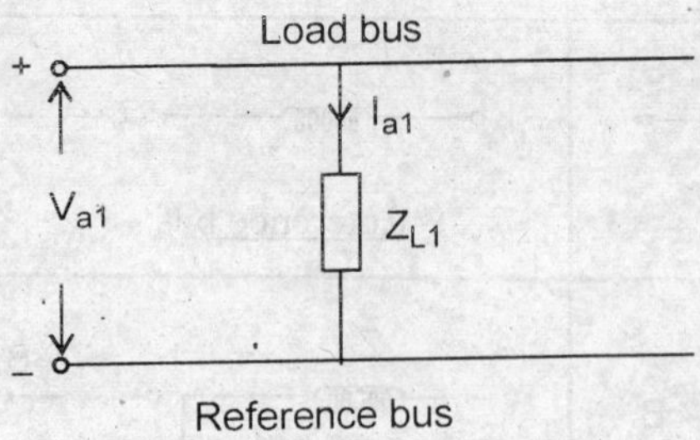

Fig. (1.24a)

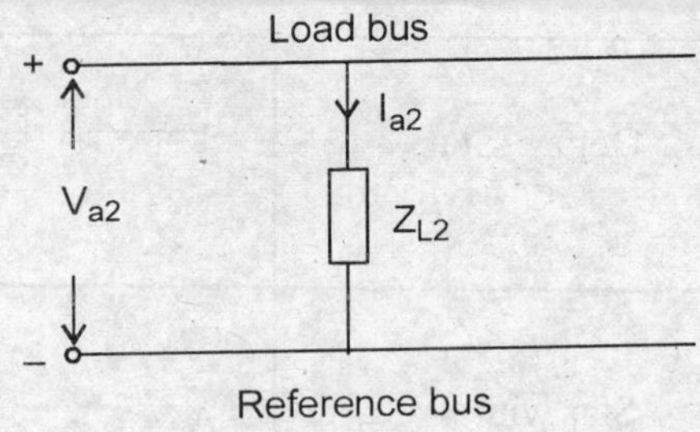

Fig. (1.24b)

Figure 1.24 *+Ve and –Ve sequence network of load*

Let we see the zero sequence network of load. The zero sequence of 3-phase load depends on the type of connection (Y or Δ). The zero sequence current will flow only when there is a return path to flow. The figure 1.25 shows the zero sequence network of load for different configurations.

Configuration	Zero sequence
1 Star 	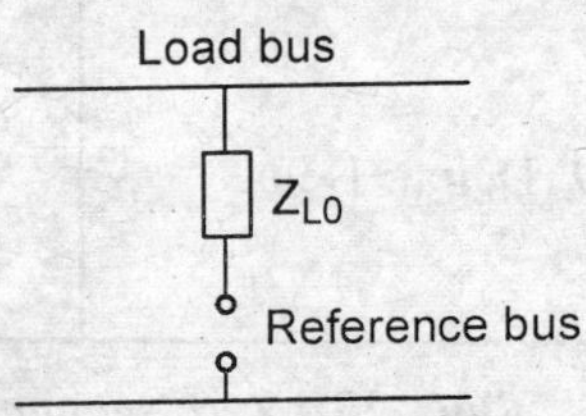
2 Star with neutral grounded 	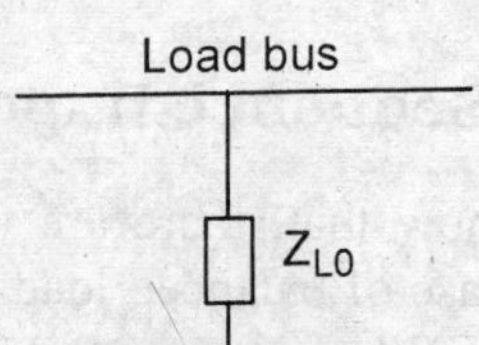
3 Star with neutral grounded with reactance 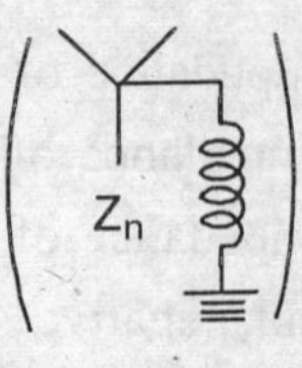	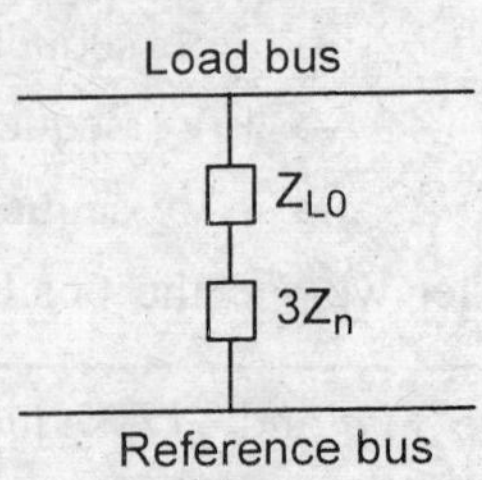

4 Delta △	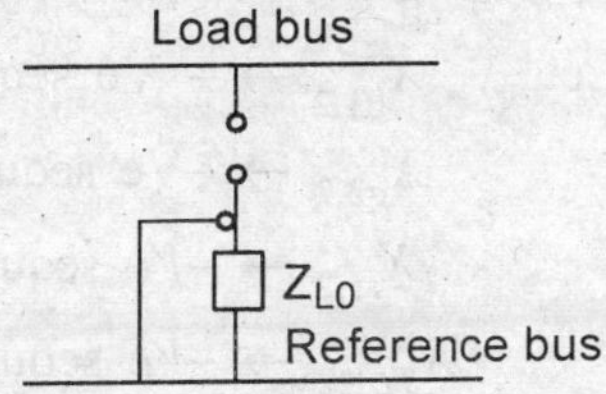

Fig. (1.25): Zero sequence network of load for different configurations.

Example 1.17: Draw positive, negative and zero sequence reactance diagram of the power system shown in Fig. (1.16.1)

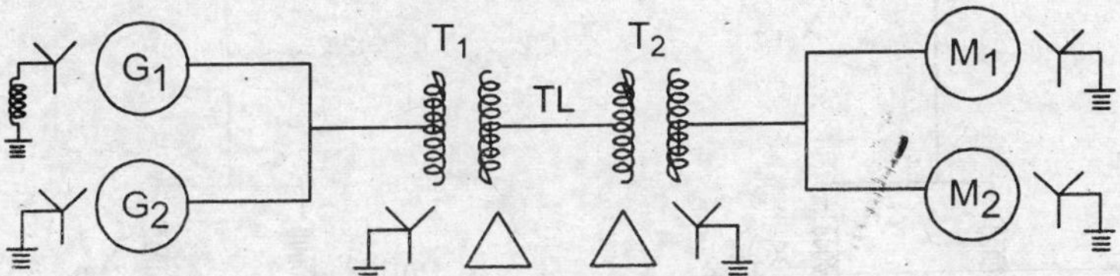

Figure (1.17.1)

Solution: (*i*) Positive sequence network

Let, $X_{g1,1} \rightarrow +Ve$ sequence reactance of G_1

$X_{g2,1} \rightarrow$ +Ve sequence reactance of G_2

$X_{t1,1} \rightarrow$ +Ve sequence reactance of T_1

$X_{TL,1} \rightarrow$ +Ve sequence reactance of TL

$X_{t2,1} \rightarrow$ +Ve sequence reactance of T_2

$X_{m1,1} \rightarrow$ +Ve sequence reactance of M_1

$X_{m2,1} \rightarrow$ +Ve sequence reactance of M_2

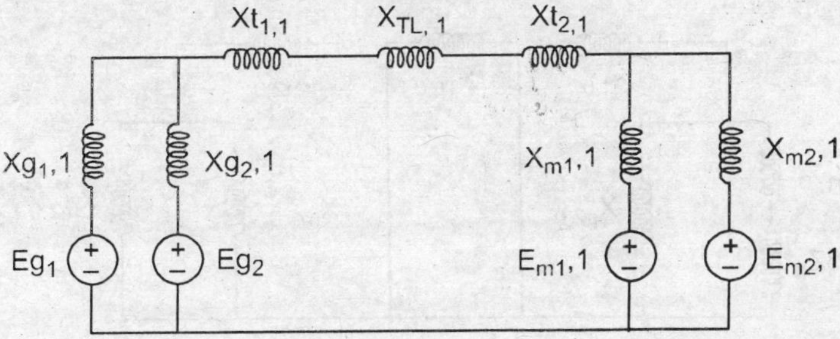

Figure 1.17.2

(*ii*) Negative sequence network

Let, $X_{g1,2}$ → – Ve sequence reactance of G_1

$X_{g2,2}$ → –Ve sequence reactance of G_2

$X_{t1,2}$ → –*Ve* sequence reactance of T_1

$X_{TL1,2}$ → –*Ve* sequence reactance of *TL*

$X_{t2,2}$ → –*Ve* sequence reactance of T_2

$X_{m1,2}$ → –*Ve* sequence reactance of M_1

$X_{m2,2}$ → –*Ve* sequence reactance of M_2

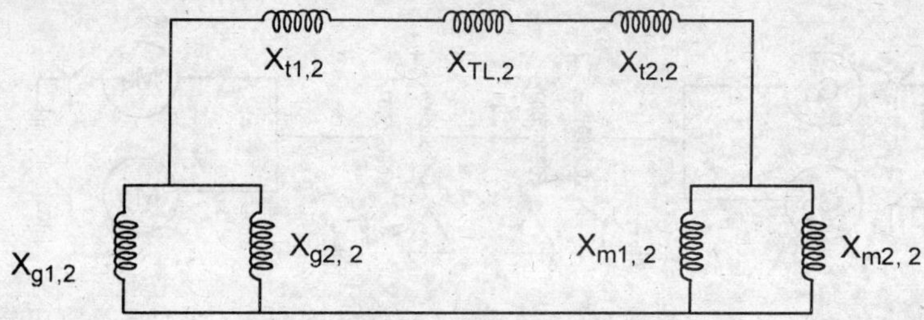

Figure (1.17.3)

(*iii*) Zero sequence network

Let, $X_{g1,0}$ → Zero sequence reactance of G_1

$X_{g2,0}$ → Zero sequence reactance of G_2

$X_{t1,0}$ → Zero sequence reactance of T_1

$X_{t2,0}$ → Zero sequence reactance of T_2

$X_{TL,0}$ → Zero sequence reactance of *TL*

$X_{m1,0}$ → Zero sequence reactance of M_1

$X_{m2,0}$ → Zero sequence reactance of M_2

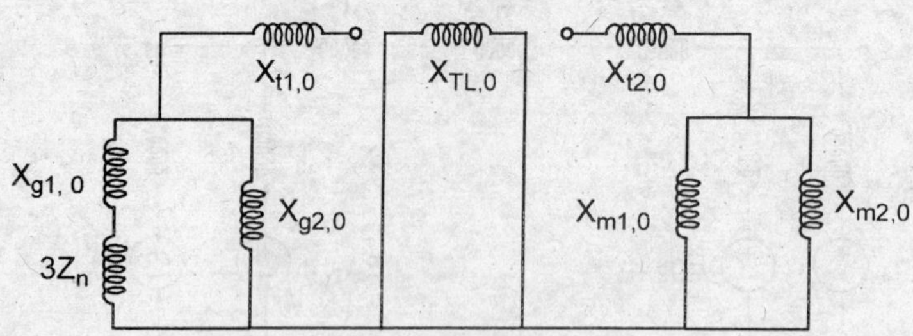

Fig. (1. 17. 4)

Example 1.18: Draw positive, negative and zero sequence reactance diagram of the power system shown in Fig. 1.17.1.

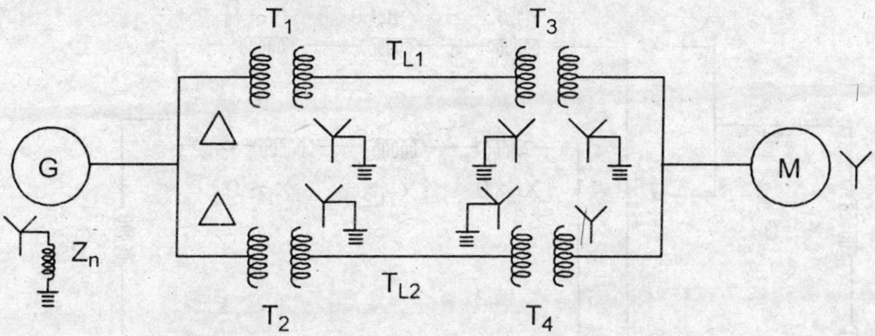

Fig. (1. 18. 1)

Solution: (*i*) Positive sequence network:

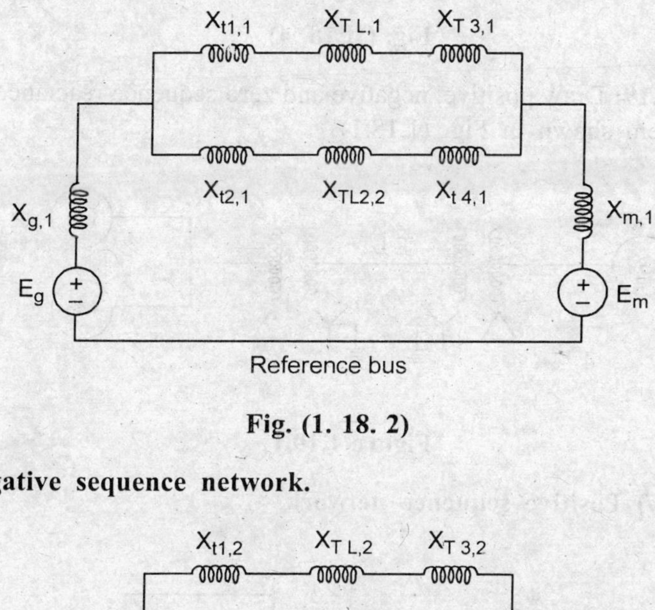

Fig. (1. 18. 2)

(*ii*) Negative sequence network.

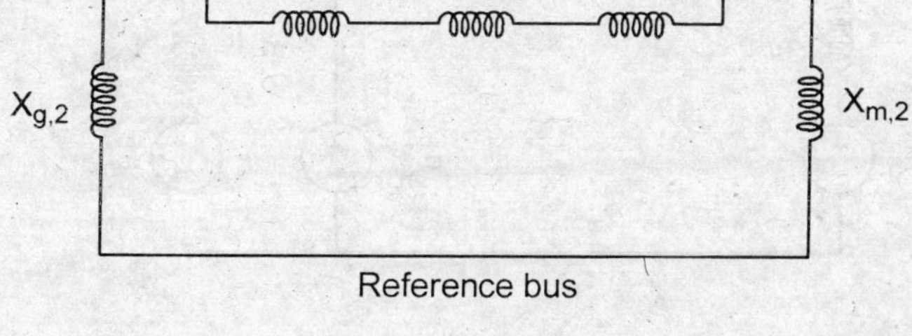

Fig. (1. 18. 3)

***(iii)* Zero sequence network:**

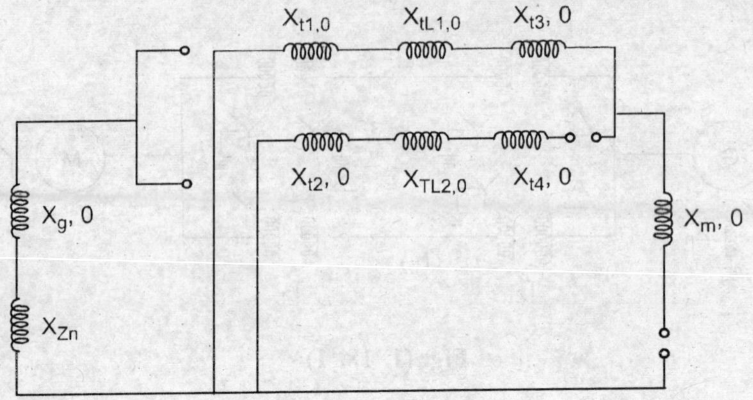

Fig. (1. 18. 4)

Example 1.19: Draw positive, negative and zero sequence reactance diagram of the power system shown in Fig. (1.18.1)

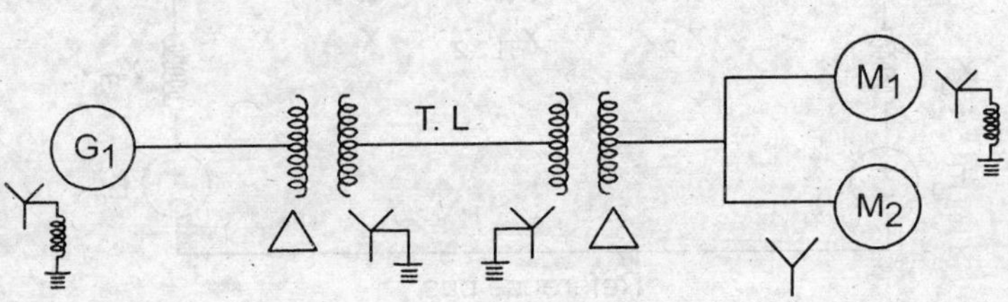

Figure 1.19.1

Solution: *(i)* Positive sequence network

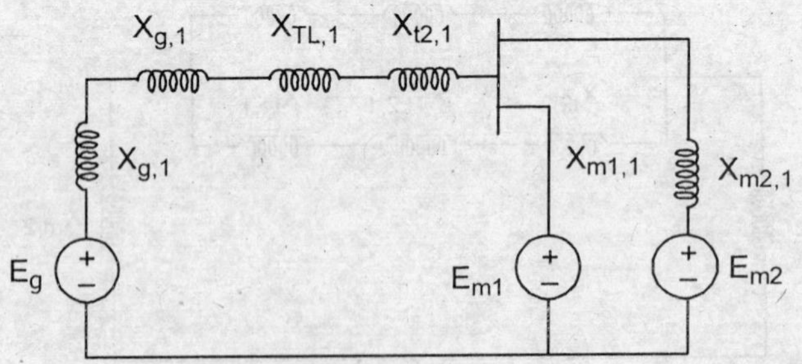

Fig. (1. 19. 2)

(*ii*) Negative sequence network

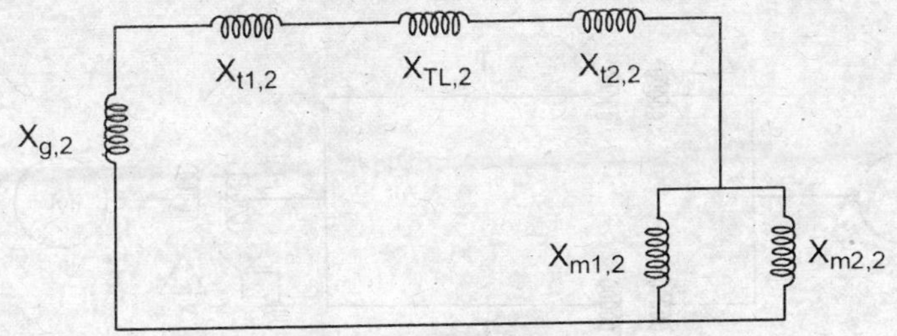

Fig. (1. 18. 3)

(*iii*) Zero sequence network

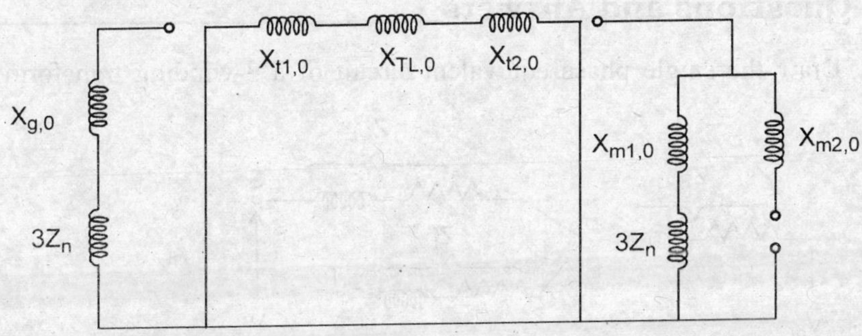

EXERCISE

13. Draw positive, negative and zero sequence reactance network for the power system shown in Fig. (*E.* 1.13.1)

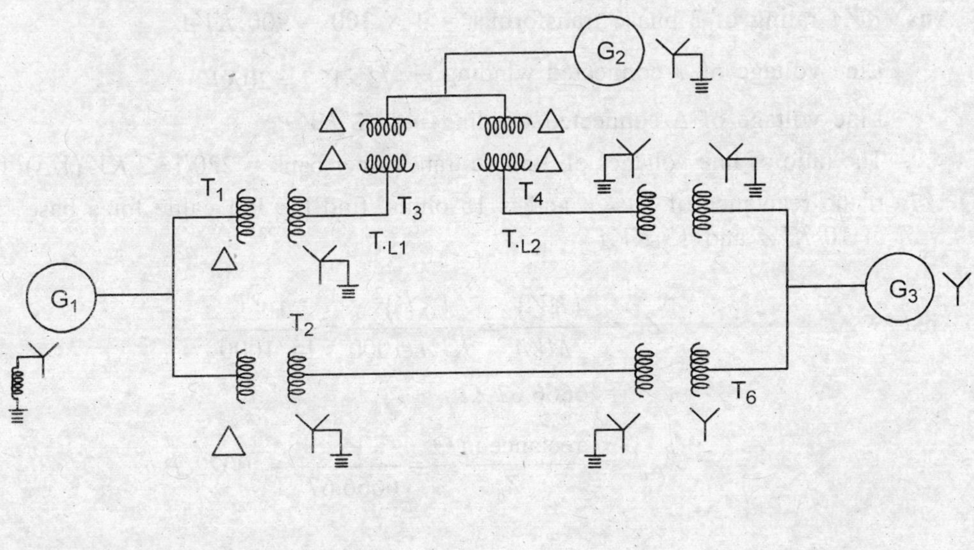

14. Draw positive, negative and zero sequence reactance network for the power system shown in Fig. (*E*.1.14.1)

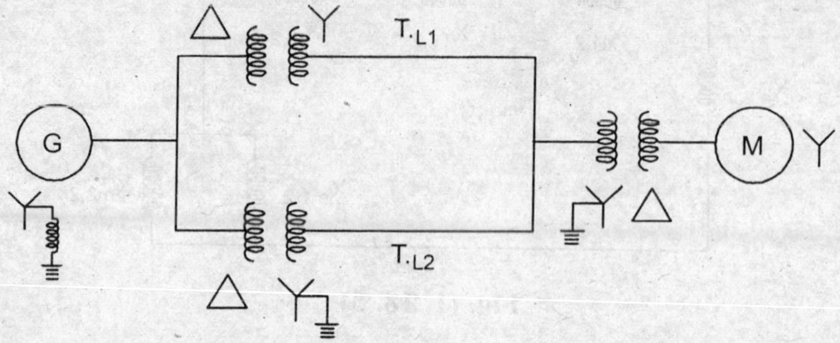

Short Questions and Answers

Q. 15. Draw the single phase equivalent circuit of a 3-winding transformer.

Ans.

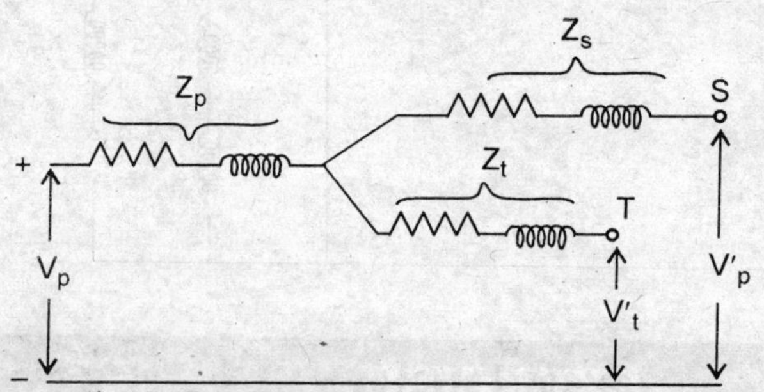

Q. 16. A *Y*-Δ, 3 phase transformer bank is formed using three numbers of single phase transformers each rated at 300 *KVA,* 127/13.2 *KV*. What is the *KVA* and *KV* rating of the 3-phase bank?

Ans. *KVA* rating of 3-phase transformer = 3 × 300 = 900 *KVA*

Line voltage of *Y* connected winding = $\sqrt{3} \times 127 = 20KV$

Line voltage of Δ connected winding = 13.2 *KV*

∴ The ratio of line voltages of 3-phase transformer bank = 220/13.2 *KV* (*Y*/Δ)

Q. 17. If the reactance of 10 *KV* and is 15 ohms, find the P.u value for a base of 10 *KVA* and 15 *KVA*

Ans.

$$Z_b = \frac{(KV)^2}{MVA} = \frac{(KV)^2}{KVA/1000} = \frac{10^2}{15/1000}$$

$$= 6666.67\ \Omega$$

$$X_{P.u} = \frac{\text{rectance in } \Omega}{Z_b} = \frac{15}{6666.67} = .0022\ P.u$$

Q. 18. A generator rated at 30 *MVA*, 11*KV* has a reactance of 20%. Calculate its P.u reactance for a base of 50 *MVA* and 10 *KV*.

Ans. New *P.u* reactance of generator

$$= X_{P.u,old} \times \left[\frac{KV_{b,old}}{KV_{b,new}}\right]^2 \times \frac{MVA_{b,new}}{MVA_{,old}}$$

$$X_{P.u,g} = 0.2 \times \left[\frac{11}{10}\right]^2 \times \frac{50}{30} = .403 P.u$$

Q. 19. *A Y* connected generator rated at 300 MVA, 33KV , has a reactance of 1.24 P.u find the ohmic value of reactance.

Ans.

$$Z_b = \frac{(KV)^2}{MVA} = \frac{(33)^2}{300} = 3.63\Omega / \text{phase}$$

$$X_{p.u} = \frac{\text{Actual reactance}}{\text{Base reactance.}}$$

$$\therefore \text{ Actual reactance} = X_{p.u} \times \text{Base reactance}$$

$$= 1.24 \times 3.63 = 4.5012 \text{ W/phase}$$

Q. 20. The base *KV* and base *MVA* of a three phase transmission line is 33 *KV* and 10 *MVA*. Calculate base current and base impedance.

Ans.

$$I_b = \frac{KVA_b}{\sqrt{3}KV_b} = \frac{MVA_b \times 1000}{\sqrt{3}KV_b} = \frac{10 \times 1000}{\sqrt{3} \times 33}$$

$$I_b = 174.95 \text{ A.}$$

$$Z_b = \frac{(KV_b)^2}{MVA_b} = \frac{(33)^2}{10} = 108.9 \ \Omega/\text{phase}$$

Q. 21. What are symmetrical components?

An unbalanced system of *N* related vectors can be resolved into *N* systems of balanced vectors. The N-sets of balanced vectors are called symmetrical components. Each set consist of *N* vectors which are equal in length having equal phase angles between adjacent vectors.

Q. 22. Write the symmetrical components of three phase system.

1. Positive sequence components
2. Negative sequence components
3. Zero sequence components

➡ The above 3 components applicable to both current and voltage vectors.

Q. 23. What are positive sequence components?

The positive sequence components of a 3-phase unbalanced vectors consists of three vectors of equal magnitude displaced from each other by 120° in Phase and having the same phase sequence as the original vectors.

Q. 24. What are negative sequence components ?

The negative sequence components of a 3-phase unbalanced vector consists of 3-vectors of equal magnitude and displaced from each other by 120° in phase and having the phase sequence opposite to that of the original vectors.

Q. 25. What are zero sequence components?

The zero sequence components of a 3-phase unbalanced vectors consists of three vectors of equal magnitude and with zero phase diaplacement from each other.

Q. 26. Prove that $1 + a + a^2 = 0$

Ans.

$$a = 1\angle 120° = -0.5 + j\,0.866$$

$$a^2 = 1\angle 240° = -0.5 - j\,0.866$$

$$\therefore \quad 1 + a + a^2 = 1 - 0.5 + j0.866 - 0.5 - j\,0.866 = 0$$

Q. 27. Express the unbalanced voltage and currents ((V_a, V_b and V_c) and (I_a, I_b and I_c)) in terms of symmetrical components. V_{a1}, V_{a2} , and V_{a0}) and (I_{a1}, I_{a2}, and I_{a0})).

Ans.

$$V_a = V_{a0} + V_{a1} + V_{a2}$$

$$V_b = V_{a0} + a^2V_{a1} + aV_{a2}$$

$$V_c = V_{a0} + aV_{a1} + a^2V_{a2}$$

or

$$\begin{bmatrix} V_a \\ V_b \\ V_c \end{bmatrix} = \begin{bmatrix} 1 & 1 & 1 \\ 1 & a^2 & a \\ 1 & a & a^2 \end{bmatrix} \begin{bmatrix} V_{a0} \\ V_{a1} \\ V_{a2} \end{bmatrix}$$

Similarly,

$$\begin{bmatrix} I_a \\ I_b \\ I_c \end{bmatrix} = \begin{bmatrix} 1 & 1 & 1 \\ 1 & a & a^2 \\ 1 & a^2 & a \end{bmatrix} \begin{bmatrix} I_{a0} \\ I_{a1} \\ I_{a2} \end{bmatrix}$$

Q. 28. Express symmetrical components in terms of unbalanced vectors.

$$\begin{bmatrix} V_{a0} \\ V_{a1} \\ V_{a2} \end{bmatrix} = \frac{1}{3} \begin{bmatrix} 1 & 1 & 1 \\ 1 & a & a^2 \\ 1 & a^2 & a \end{bmatrix} \begin{bmatrix} V_a \\ V_b \\ V_c \end{bmatrix}$$

and $$\begin{bmatrix} I_{a0} \\ I_{a1} \\ I_{a2} \end{bmatrix} = \frac{1}{3}\begin{bmatrix} 1 & 1 & 1 \\ 1 & a^2 & a \\ 1 & a & a^2 \end{bmatrix}\begin{bmatrix} I_a \\ I_b \\ I_c \end{bmatrix}$$

1.10 SYMMETRICAL FAULT ANALYSIS

1.10.1 Introduction

A fault in a circuit is any failure which interferes with the normal flow of current. The faults occur in power system due to insulation failure of equipments, flash over of lines initiated by a lightning stroke, due to permanent damage to conductors and towers (or) due to accidental faulty operations.

1.10.2 Types of Faults

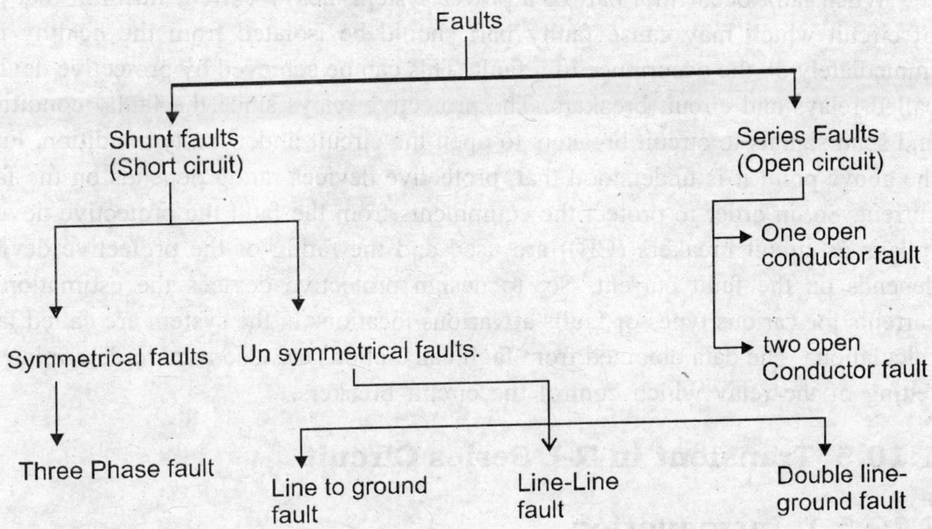

1.10.3 Effect of Faults

The shunt type of faults involves short circuit between conductor and ground (or) short circuit between two (or) more conductors. The shunt faults are characterized by increase in current and fall in voltage and frequency.

The series fault may occur with one (or) two broken conductors which creates open circuits.

It also happens in circuits controlled by fuses (or) breakers which do not open all the three phases, *i.e.*, one (or) two phases of the circuit may open and the other phases may be closed. The series faults are characterized by increase in voltage and frequency and fall in current in the faulted phase

The symmetrical fault conditions are analyzed on per phase basis using the venins theorem (or) by bus impedance matrix. The unsymmetrical faults are analyzed using symmetrical components. The relative frequency of occurrence of various types of faults in the power systems are,

3 – phase faults – 5%

Double line – ground fault – 10%

Line-line fault – 15%

Single – line to ground fault – 70%

1.10.4 Fault calculations

Why we need to do fault calculation?

When fault occur in a part of a power system, heavy current flows in that part of circuit which may cause faulty part should be isolated from the healthy part immediately on the occurrence of a fault. This can be achieved by protective devices called relays and circuit breakers. The protective relays since the faulty conditions and send signals to circuit breakers to open the circuit under faulty condition. From the above point it is understood that, protective devices rating depends on the fault current. So, in order to protect the equipments from the fault the protective devices (relays & circuit breakers (CB)) are used and the rating of the protective devices depends on the fault current. So, to design protective devices the estimation of currents for various types of faults at various locations in the system are called fault calculations. The data obtained from fault calculations are also used to determine the setting of the relay which control the circuit breakers.

1.10.5 Transient in R-L Series Circuit

1.10.5.1 Introduction

What is transient?

Sudden change in current due to switching operation of the circuit.

Most of the components of the power system have inductive property which give rise to transients when there is a sudden change in current. Faults in power system are accom panied by sudden change in current which give rise to transient conditions in power system.

Transients in *R-L* series circuit

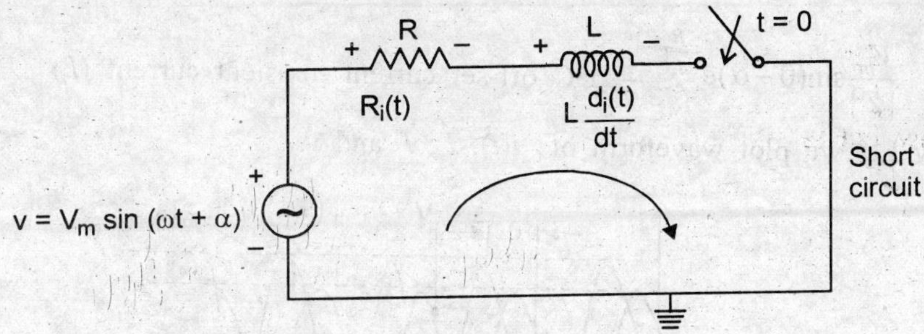

Figure 1.26 *R-L series circuit*

An unloaded transmission line can be represented by an *R-L* circuit exited by a sinusoidal source.

(**Note:** The capacitance of the transmission line is neglected)

In Fig. 1.26

Let, $i(t) \rightarrow$ current in transmission line under short circuit condition.

$Z \rightarrow$ Impedance of the transmission line.

Where,

$$Z = R + j X_L = R + j\omega L$$

$$= \sqrt{R^2 + \omega^2 L^2} \angle \tan^{-1} \frac{\omega L}{R} = |Z| L\theta$$

$$\therefore \quad |Z| = \sqrt{R^2 + \omega^2 L^2}$$

$$\theta = \tan^{-1}\left(\frac{\omega L}{R}\right)$$

The fault condition can be simulated by closing the switch at $t = 0$. When the switch is closed at $t = 0$, fault current (short circuit current) flows in the circuit. The differential equation governing the circuit can be obtained using *KVL*.

For Fig. 1.26 after applying ***KVL,***

$$Ri(t) + L\frac{di(t)}{dt} = V_m \sin(\omega t + \alpha) \qquad ...(1.85)$$

The equation (1.85) can be solved using any technique which is used in circuit theory to find $i(t)$.

On solving equation (1.85), the short circuit current $i(t)$ is given by the equation (1.86)

$$i(t) = \frac{V_m}{|Z|}\sin(\omega t + \alpha - \theta) + \frac{V_m}{|Z|}\sin(\theta - \alpha)e^{-\frac{R}{L}t} \qquad ...(1.86)$$

When, $\frac{V_m}{|Z|}\sin(\omega t + \alpha - \theta) \rightarrow$ symmetrical short circuit current (or) steady state currents (I_s)

$$\frac{V_m}{|Z|}\sin(\theta-\alpha)e^{-\frac{R}{L}t} \rightarrow \text{DC off-set current transient current } (I_t)$$

Now Let we plot waveform of $i(t)$, I_s, I_t and v.

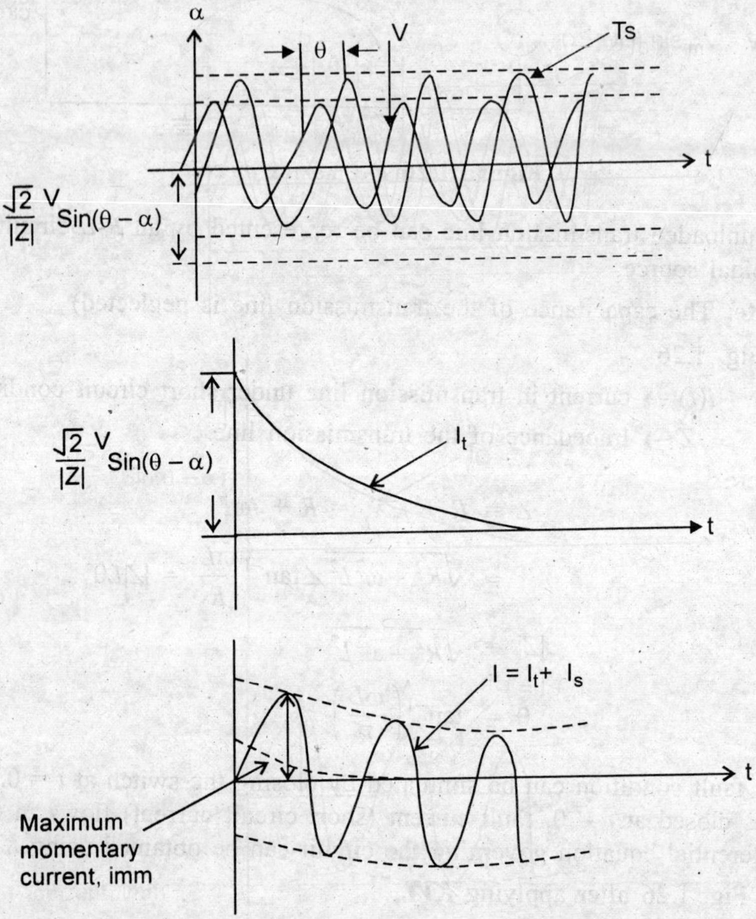

Figure 1.27 *Waveform of currents in a transmission line under short circuit condition*

In the short circuit current $i(t)$, the value corresponding to the first peak is called the maximum momentary short circuit (i_{mm}). In equation (1.86) the first peak value is obtained when $\sin(\omega t + \alpha - \theta) = 1$

$$\therefore \qquad i_{mm} = \frac{V_m}{|Z|}+\frac{V_m}{|Z|}\sin(\theta-\alpha)e^{-\frac{R}{L}t} \qquad ...(1.87)$$

If the delay of transient current in the interval between $t = 0$ and time at which first peak occurs is neglected then equation (1.87) can be written as

$$i_{mm} = \frac{V_m}{|Z|} + \frac{V_m}{|Z|}\sin(\theta - \alpha) \qquad ...(1.88)$$

In transmission lines the resistance is very low when compared to reactance and so $\theta \approx 90°$

$(i.e.,\ \tan^{-1}\left(\frac{\omega L}{R}\right) \approx \tan^{-1}\infty = 90°)$

Hence by considering q value as 90°, the equation (1.88) can be written as,

$$\therefore \quad i_{mm} = \frac{V_m}{|Z|} + \frac{V_m}{|Z|}\sin(90 - \alpha)$$

$$i_{mm} = \frac{V_m}{|Z|} + \frac{V_m}{|Z|}\cos\alpha \qquad ...(1.89)$$

From equation (1.89), we can say that, i_{mm} has maximum possible value when a = 0. This implies that the effect of short circuit will be severe if the fault occurs when the voltage wave is going through zero.

When $\alpha = 0$, the equation (1.89) can be written as

$$i_{mm(\text{maximum possible})} = \frac{V_m}{|Z|} + \frac{V_m}{|Z|} = 2\frac{V_m}{|Z|} \qquad ...(1.90)$$

From equation (1.86) the maximum value of symmetrical short circuit current is $\frac{V_m}{|Z|}$ end from equation (1.90) the maximum possible value of maximum momentary short circuit is double the value of maximum symmetrical short circuit current. It such a condition exists in a transmission line then this effect is called doubling effect.

1.10.6 Calculation of 3-Phase Short Circuit Current and Reactance of Synchronous Machine

Consider a 3-phase synchronous machine running on no-loud. If a three phase fault occurs at the terminals of the asynchronous machine (alternator) then a heavy short circuit current flows in the armature circuit (stator). The oscillogram of the short circuit current after removing the DC-off set current is shown in Fig. (1.28). At the time of a short circuit in a 3-phase alternator, the voltage wave of the 3-phases will have different phases. Therefore the DC-offset current will be different in each phase of a 3-phase machine and so it is accounted separately on an empirical basis. The ac component of the short circuit current, whose wave form is shown in Fig. 1.28.

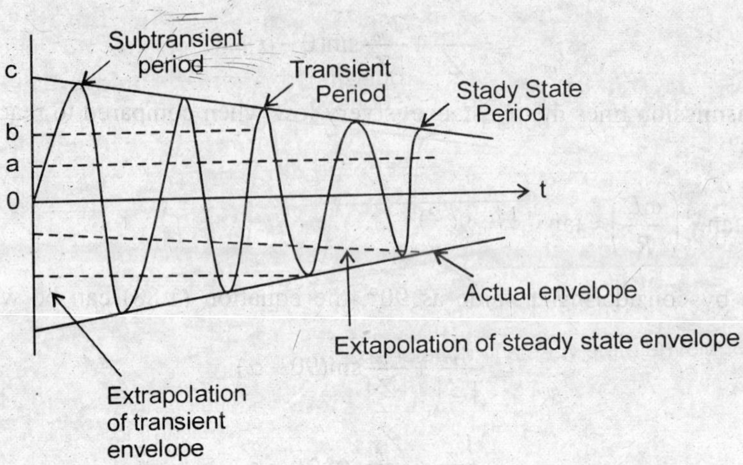

Figure 1.28 *Symmetrical short circuit amateur current in an alternator*

The symmetrical short circuit current shown in Fig. 1.28 can be divided into **3** regions

1. Subtransiout region
2. Transient region
3. Steady state region

Under steady state short circuit conditions, the armature reaction of a synchronous generator produces a demagnetizing flux. This effect is represented as a reactance called armature reactance X_a. The sum of Leakage reactance, X_l and X_a is called synchronous reactance, X_s. In case of salient pole machines the synchronous reactance is called direct axis reactance and denoted by X_d. On neglecting the armature resistance the steady state short circuit model of an alternator will be as shown in Fig. (1.29). At the instant of short circuit the DC-offset current appears in all the three phases of stator. This dc-offset current can induce. Currents in rotor field winding and damper winding by transformer action. The increase in field current and damper winding current will set up flux in a direction to argument the main flux. This effect can be represented by two reactances in parallel with X_a as shown in Fig. (1.30). Here X_f represents the flux created by induced current in the field winding and X_{dw} represents the flux created by induced currents in the damper winding. The combined effect of all the three reactances is to reduced the total reactance of the machine and so the short circuit current is very high in this state which is called as sub-transient state. The total reactance under this condition is called subtransient reactance and denoted by X_d''

$$\therefore \quad X''_d = X_l \frac{1}{\frac{1}{xa}+\frac{1}{xf}+\frac{1}{xdw}}$$

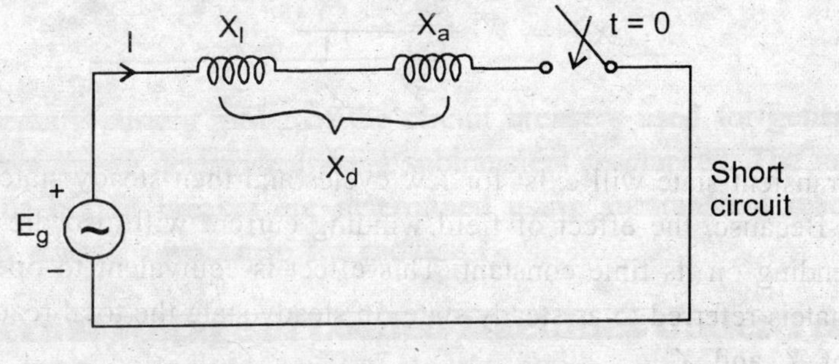

Figure 1.29 *Steady state short circuit model of an alternator*

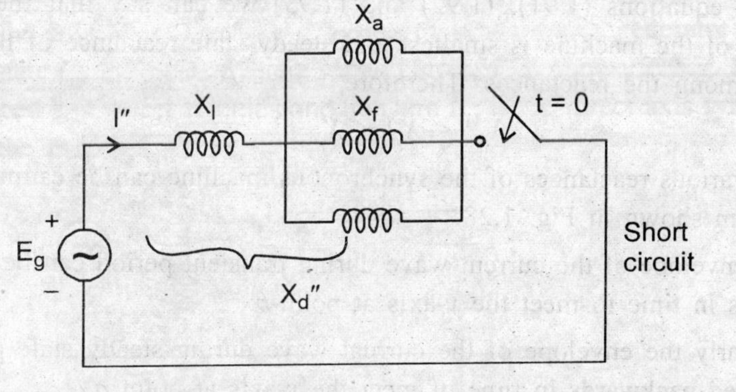

Figure 1.30 *Circuit model of subtransient state of an alternator*

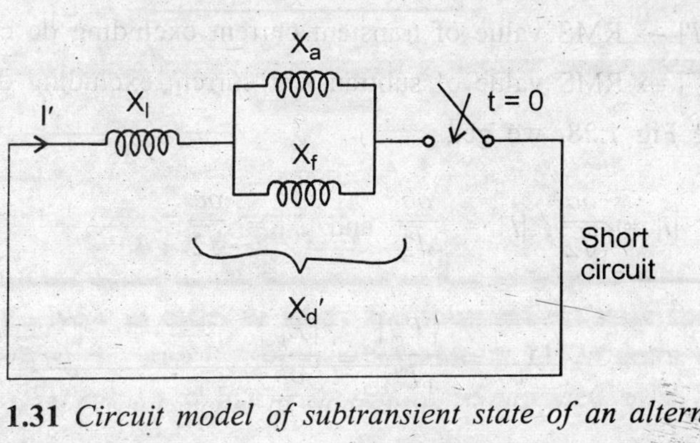

Figure 1.31 *Circuit model of subtransient state of an alternator*

The induced currents in the damper winding disappears after few cycles from the instant of fault. Because the time constant of the damper winding is smaller than the field winding. This effect is equivalent to open circuited X_{dw} and this state is called transient state. The transient state of the alternator is shown in Fig. 1.31. The total reactance in transient state is called transient reactance and it is denoted by X_d'.

$$\therefore \qquad X_d' = X_l + \frac{1}{\frac{1}{X_a} + \frac{1}{X_f}} \qquad ...(1.92)$$

The transient state will exist for few cycles and then steady state conditions ae achieved. Because, the effect of fiefd winding current will also die-out in a short time depending on its time constant. This effect is equivalent to open circuited X_f and this stateis referred to as steady state. In steady state the total reactance is given by sum of X_l and X_a.

$$\therefore \qquad X_d = X_l + X_a \qquad ...(1.93)$$

From equations (1.91), (1.92) and (1.93) we can say that the subtransient reactance of the machine is smallest and steady state reactance of the machine is highest among the reactances. Therefore,

$$X_d'' < X_d' < X_d$$

The various reactances of the synchronous machine can be estimated from the oscillogram shown in Fig. 1.28.

The envelope of the current wave during transient period can be extra polated backwards in time to meet the y-axis at point-b.

Similarly the envelope of the current wave during steady state period can be extrapolated backwards in time of meet the y-axis at point a.

Let, $|I| \rightarrow$ RMS value of steady state current.

$|I'| \rightarrow$ RMS value of transient current excluding dc component.

$|I''| \rightarrow$ RMS value of subtransient current excluding dc component

From the Fig 1.28, we get,

$$|I| = \frac{oa}{\sqrt{2}}; \; |I'| = \frac{ob}{\sqrt{2}} \text{ and } |I''| = \frac{o\dot{c}}{\sqrt{2}}$$

$$\therefore \qquad X_d'' = \frac{|E_g|}{|I''|} = \frac{|E_g|}{\frac{ob}{\sqrt{2}}} \qquad ...(1.94)$$

$$X_d' = \frac{|E_g|}{|I'|} = \frac{|E_g|}{\frac{ob}{\sqrt{2}}} \quad ...(1.95)$$

$$X_d = \frac{|E_g|}{|I|} = \frac{|E_g|}{\frac{oc}{\sqrt{2}}} \quad ...(1.96)$$

The momentary current rating of the circuit breakers used for generators and synchronous motors are determined using subtransient reactances. The interrupting Capacity of the circuit breaker are determined using subtransient reactance for generators and transient reactance for motors.

1.10.7 Internal Voltage of Loaded Machines Under Transient Conditions.

Consider a generator connected to a bus with voltage V_t. Let I_L be the current delivered by the generator. The circuit model of synchronous generator operating under steady state conditions supplying a load current I_L is shown in Fig. 1.32. Here E_g is the induced emf under loaded condition and X_d is the direct axis synchronous reactance of the machine.

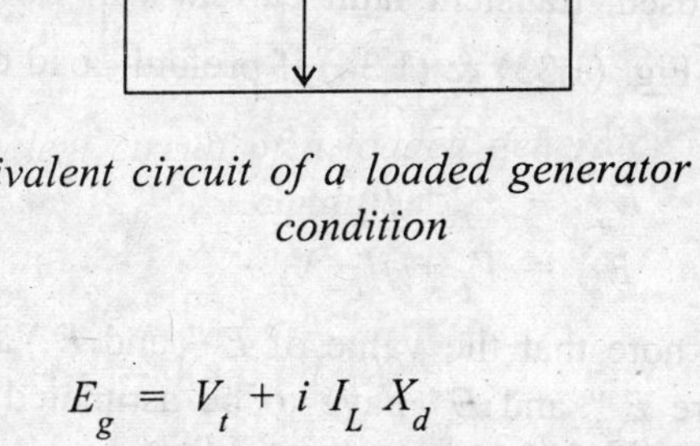

Figure 1.32 *Equivalent circuit of a loaded generator under steady state condition*

From Fig. 1.32,

$$E_g = V_t + i\, I_L\, X_d \quad ...(1.97)$$

Let a short circuit occur at the terminals of the generator while delivering the load current I_L. Now in order to study the subtransient state the E_g and X_d should be replaced by E_g'' and X_d'' as shown in Fig. 1.33. In order to study the transient state the E_g and X_d of Fig. 1.32 should be replaced by E_g' and X_d' as shown in Fig. 1.33

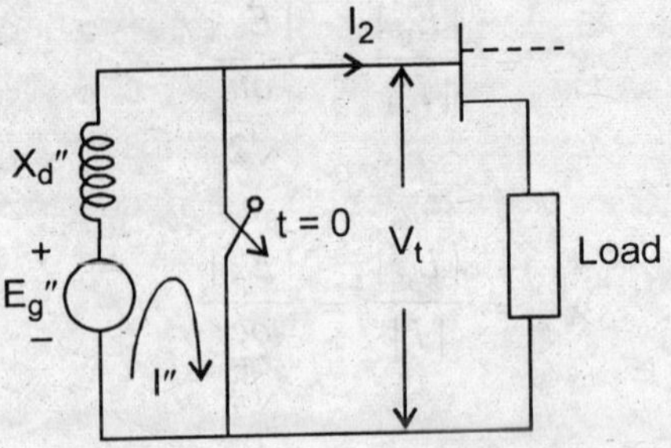

Figure 1.33 *Circuit model for computing subtransient current*

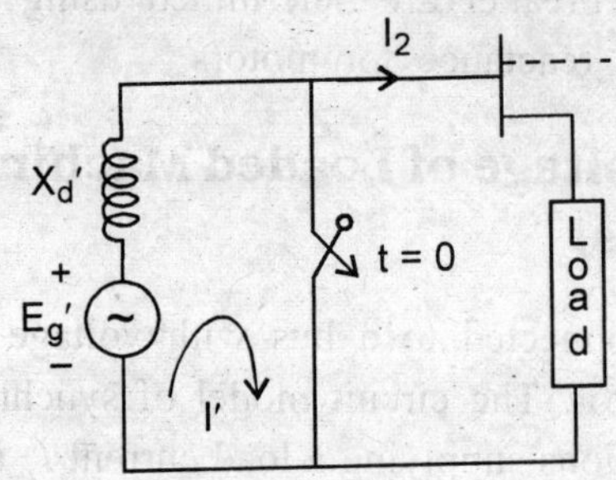

Figure 1.34 *Circuit model for computing subtransient current*

Here, E_g'' = Subtransient internal voltage
E_g' = Transient internal voltage

In Fig. 1.33 and 1.34 the prefault current is I_L when the switch is not closed. The fault condition can be represented by closing the switch. In Fig. 1.33 when the switch is closed subtransient fault current will flow in the fault path. In Fig. 1.34 when the switch is closed, transient fault current will flow in the fault path.

With reference to Fig. (1.33) & (1.34) if prefault load current is I_L then we can write,

$$E_g'' = V_t + jI_L X_d'' \quad \text{...(1.98)}$$

$$E_g' = V_t + jI_L X_d' \quad \text{...(1.99)}$$

It is important to note that the value of E_g'' and E_g' are functions of prefault load current. Therefore E_g'' and E_g' have to be astimated for each value of load current.

The synchronous motors have internal emfs and reactances similar to that of a generator except that the current direction is reversed. Hence for short circuit studies the circuit models similar to that of Fig. 1.32 to 1.34 can be used for synchronous motor with reversed direction of current. Therefore for synchronous motor,

$$E_g'' = V_t - j\, I_L\, X_d'' \quad \text{...(1.100)}$$

$$E_g' = V_t - j\, I_L\, X_d' \quad \text{...(1.101)}$$

Short Questions and Answers

Q. 29. What is meant by a fault?

Ans. A fault in a circuit is any failure which interferes with the normal flow of current. The faults are associated with abnormal change in current, voltage and frequency of the power system. The faults may cause damage to the equipments if is allowed to persist for a long time. Hence every part of a system has been protected by means of relays and circuit breakers to sense the faults and to isolate the faulty part from the healthy part in the event of fault.

Q. 30. Why faults occur in power system?

Ans. The faults occur in a power system due to insulation failure of equipments, flashover of lines initiated by a lightning stroke, due to permanent damage to conductors and towers (or) due to accidential faulty operations.

Q. 31. How the faults are classified?

Ans. In one method of classification, the faults are classified as shunt and series faults. The shunt faults are due to short circuits in conductors and the series faults are due to open conductors.

In another method of classification, the faults are classified into symmetrical and unsymmetrical faults. In symmetrical faults the fault currents are equal in all the phases and can be analysed on per phase basis. In unsymmetrical faults the fault currents are unbalanced and so they can be analysed only using symmetrical components.

Q. 32. List the various types of shunt & series faults.

Shunt faults

1. Line to ground fault
2. Double line to ground fault
3. Line-line fault
4. Three phase fault

Series faults

1. Two open conductor fault
2. One open conduction fault

Q. 33. List the symmetrical and un symmetrical faults

Symmetrical faults

1. Three phase faults

Unsymmetrical faults

1. Line-line fault
2. Line-ground fault
3. Double line to ground fault

Q. 34. What is symmetrical and unsymmetrical fault?

Ans.
- If the fault current is equal in all the phases then it is known as symmetrical fault
- If the fault current is not equal in all the phases then it is known as unsymmetrical fault.

Q. 35. Name any two methods of reducing short circuit current.

Ans. 1. By providing neutral reactance

2. By providing Large value of shunt reactance between buses.

Q. 36. Write the main difference in representation of power system for load flow and short circuit studies.

Ans.

Load flow	Short circuit
• Both resistance and reactances are considered	• Resistances are neglected
• Admittance matrix is used	• Impedance matrix is used

Q. 37. What is meant by fault calculation?

Ans. The fault condition of a power system can be divided into subtransient, transient and steady state periods. The current in the various parts of the system and in the fault and different in these periods. The estimation of these currents for various types of faults at various locations in the system are commonly referred to as fault calculation.

Q.38. What is the need for short circuit studies (or) fault analysis?

Ans. The short circuit studies are essential in order to design or develop the protective schemes for various parts of two system. The protective schemes consists of current and voltage sensing devices, protective relays and circuit breakers. The selection of these devices mainly depends on various currents that may flow in the fault conditions.

Q. 39. What is the reason for transients during short circuits?

Ans. The faults or short circuits are associated with sudden change in currents. Most of the components of the power system have inductive property which opposes any sudden change in currents and so the faults are associated with transients.

Q. 40. What is meant by doubling effect?

Ans. In symmetrical fault occurs when the voltage wave is going through zero then the maximum momentary short circuit current will be double the value of maximum symmetrical short circuit current. This effect is called doubling effect.

Q. 41. Define DC off-set current.

Ans. The unidirectional transient component of short circuit current is called DC off-set current.

Q. 42. What is synchronous reactance?

Ans. The synchronous reactance is the ratio of induced emf and the steady state rms current. It is the sum of leakage current reactance and the reactance representing armature reaction.

Q. 43. Define subtransient reactance.

Ans. The subtransient reactance is the ratio of induced emf on no-load and the subtransient symmetrical rms current.

Q. 44. Define transient reactance.

Ans. The transient reactance is the ratio of induced emf on no load and the transient symmetrical rms current.

Q. 45. What is the significance of subtransient reactance in short circuit studies?

Ans. The subtransient reactance can be used to estimate the initial value of fault current immediately on the occurrence of the fault. The maximum momentary short circuit current rating of the circuit breaker used for protection or fault clearing should be less than this initial fault current.

Q. 46. What is the significance of transient reactance in short circuit studies?

Ans. The transient reactance is used to estimate the transient state fault currents. Most of the circuit breakers open their contacts only during this period. Therefore for a circuit breaker used for fault clearing, its interrupting short circuit current rating should be less then the transient fault current.

Q. 47. Distinguish between the transient and subtransient reactances of a synchronous machine.

Ans. The transient reactance is the ratio of induced emf and transient current. But the subtransient reactance is the ratio of induced emf and subtransient current.

Q. 48. When a synchronous machine is suddenly short circuit, what is the maximum possible instantaneous current in terms of symmetrical short circuit current?

Ans. The initial short circuit current will have its maximum possible value if the fault occurs when the voltage wave is going through zero. Then maximum possible value will be double that of maximum symmetrical short circuit current.

Q. 49. How symmetrical faults are analyzed?

Ans. The symmetrical faults are analysed using per unit reactance diagram of the power system. Once the reactance diagram of the power system is formed then the fault is simulated by short circuit. The currents and voltages at various parts of the system can be eliminated by any one of the following method–

1. using kirchoff's law
2. using thevenir's theorem
3. By bus impedance matrix

Objective Type Questions

1. A balanced 3 phase system is analysed by which method?

(a) per phase basis (b) for all 3 phase basis at one time
(c) both a & b (d) none of the above

2. How the base values of power, voltage, current and impedance can be selceted?

(a) independently (b) dependently
(c) both a & b (d) None of the above

3. The impedance of a device is specified in the name plate is

(a) per-unit basis (b) in actual value
(c) both a and b (d) None of the above

4. In 3 winding transformer, the three windings may have

(a) different KVA rating
(b) same KVA rating
(c) both a and b
(d) None of the above

5. In reactance diagram the resistance are

(a) neglected (b) considered
(c) opened (d) shorted

6. In forming reactance diagram the base KV depends on

(a) Transformer ratio (b) motor
(c) Generator (d) Transmission line

7. Three phase fault is

(a) Symmetrical fault (b) Unsymmetrical fault
(c) short circuit fault (d) None of the above

8. In power system 70 to 80% faults are

(a) single line to ground fault
(b) Double line to ground fault
(c) Neutral fault
(d) Line-line fault

9. The series faults are associated with

(a) low voltage (b) high current
(c) both a & b (d) None of the above

10. The transients during short circuit is due to

(a) Inductor (b) capacitor

(c) both a & b (d) None of the above

ANSWERS

1. a	2. b	3. a	4. c	5. a
6. a	7. a	8. a	9. a	10. a

Fill in the Blanks

1. In a ________ the components of power system are represented by symbols and the infer connections between them are shown by straight lines.
2. The impedance diagrams are used for ________ and reactance diagrams are used for________.
3. ________ components consists of 3 vectors equal in magnitude & phase.
4. ________ and ________ sequence network does not have any sources.
5. The meeting point of various components in a power system is called ________.
6. ________ in a circuit is any failure which interferes with the normal current.
7. ________ faults are due to short circuit
8. ________ faults are due to open conductors
9. ________ faults are more severe faults
10. ________ faults is most commonly occuring faults

ANSWERS

1. Single line diagram
2. Load flow studies, fault calculation
3. zero sequence
4. Negative and zero
5. Bus
6. Fault
7. Shunt
8. Series
9. 3-phase
10. Single line to ground fault

Chapter-2

UNSYMMETRICAL FAULT ANALYSIS

Analysis of single line to ground fault, Line to line fault and double line to ground fault on an unloaded generators and power system network with and without fault impedance.

Formation of Z_{bus} using singular Transformation and algorithm, computer method for short circuit calculations.

2.1 INTRODUCTION

The unsymmetrical faults are the faults in which the fault current in the three phases are unequal. The different types of unsymmetrical faults are,

1. Single line to ground fault
2. Line to line fault
3. Double line to ground fault
4. One (or) two open conductor faults.

Since any unsymmetrical fault causes unbalanced current to flow in the system, the unsymmetrical faults are analyzed using unsymmetrical components. The concepts of Symmetrical components and sequence network which is in unit-1 will be used for the analysis of unsymmetrical faults.

2.2 ANALYSIS OF SINGLE LINE TO GROUND FAULT ON AN UNCOADED GENERATE

The circuit diagram for a single line to ground fault on an unloaded Y connected generator with its neutral grounded through a reactance which is shown in Fig. (2.1) Here it is assumed that phase-a is shorted to ground. Now the fault current, $I_f = I_a$. Since the generator is unloaded the current in other phases are zero.

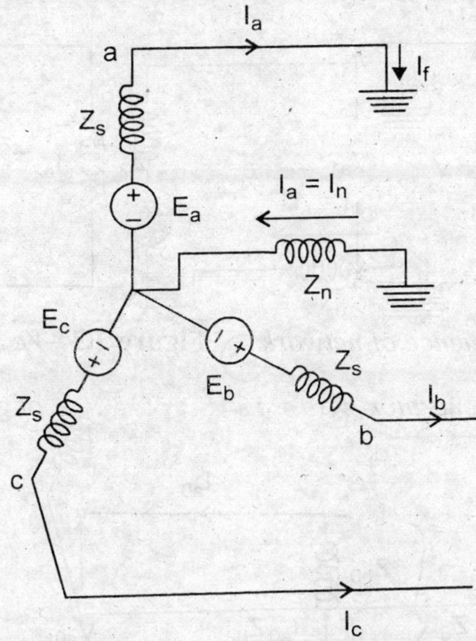

Fig.(2.1): Circuit diagram for single line to ground fault on phase-a of an unloaded generator.

The condition at the fault is expressed by the following equations.

$$I_b = 0;\ I_c = 0;\ V_a = 0 \qquad ...(2.2)$$

The symmetrical components of the currents are given by,

$$\begin{bmatrix} I_{a0} \\ I_{a1} \\ I_{a2} \end{bmatrix} = \frac{1}{3}\begin{bmatrix} 1 & 1 & 1 \\ 1 & a & a^2 \\ 1 & a^2 & a \end{bmatrix}\begin{bmatrix} I_a \\ I_b \\ I_c \end{bmatrix} \qquad ...(2.2)$$

On substituting the conditions $I_b = 0$ in the symmetrical components of currents in equation (2.2) we get,

$$\begin{bmatrix} I_{a0} \\ I_{a1} \\ I_{a2} \end{bmatrix} = \frac{1}{3}\begin{bmatrix} 1 & 1 & 1 \\ 1 & a & a^2 \\ 1 & a^2 & a \end{bmatrix}\begin{bmatrix} I_a \\ 0 \\ 0 \end{bmatrix} \qquad ...(2.3)$$

on the equation (2.3) we get, expanding

$$I_{a0} = I_{a1} = I_{a2} = I_{a/3}$$

From the sequence network of the generator we get the following matrix. (refer note)

$$\begin{bmatrix} V_{a0} \\ V_{a1} \\ V_{a2} \end{bmatrix} = \begin{bmatrix} 0 \\ E_a \\ 0 \end{bmatrix} - \begin{bmatrix} Z_0 & 0 & 0 \\ 0 & Z_1 & 0 \\ 0 & 0 & Z_2 \end{bmatrix}\begin{bmatrix} I_{a0} \\ I_{a1} \\ I_{a2} \end{bmatrix} \qquad ...(2.4)$$

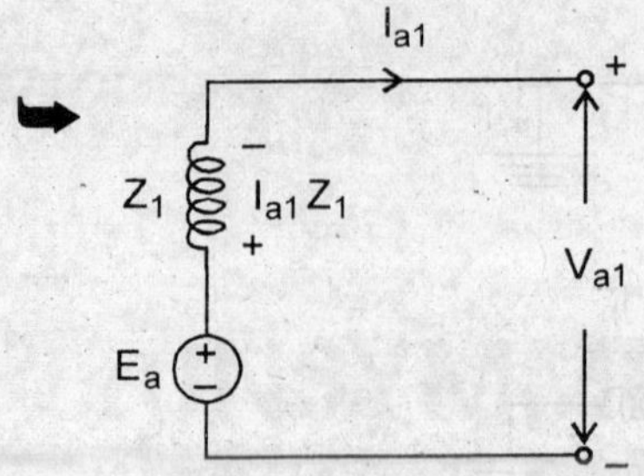

Figure 2.2 *+Ve sequence of network of generator*

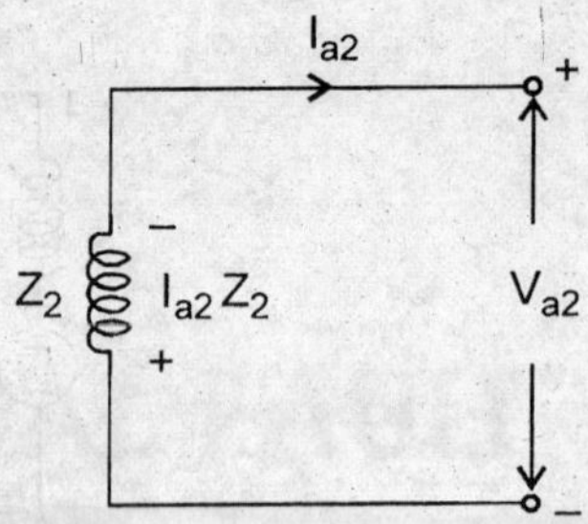

Figure 2.3 *–Ve sequence of network of generator*

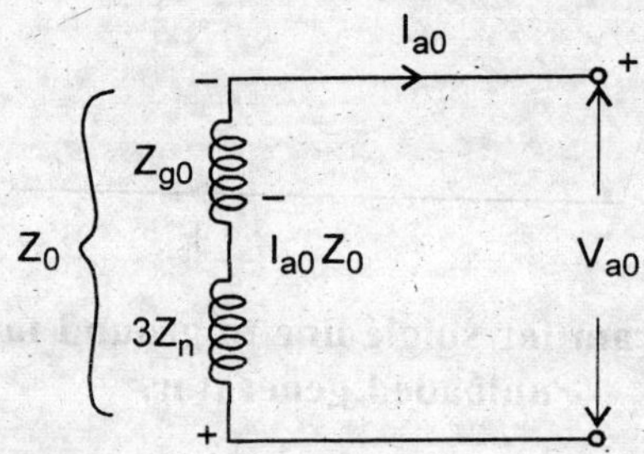

Figure 2.4 *Zero sequence of network of generator*

I_n Fig. (2.2),

$$V_{al} = E_a - I_{a1} Z_1 \qquad ...(2.5)$$

In Fig. (2.3)

$$V_{a2} = -I_{a2} Z_2 \qquad ...(2.6)$$

In Fig. (2.4)

$$V_{a0} = -I_{a0} Z_0 \qquad ...(2.7)$$

by combing the equation (2.5) to (2.7) in the form of matrix equation is,

$$\begin{bmatrix} V_{a0} \\ V_{al} \\ V_{a2} \end{bmatrix} = \begin{bmatrix} 0 \\ E_a \\ 0 \end{bmatrix} - \begin{bmatrix} Z_0 & 0 & 0 \\ 0 & Z_1 & 0 \\ 0 & 0 & Z_2 \end{bmatrix} \begin{bmatrix} I_{a0} \\ I_{al} \\ Ia2 \end{bmatrix} \qquad ...(2.8)$$

On substituting $I_{a0} = I_{a1}$ and $I_{a2} = I_{a1}$ in equation (2.4)

$$\begin{bmatrix} V_{a0} \\ V_{al} \\ V_{a2} \end{bmatrix} = \begin{bmatrix} 0 \\ E_a \\ 0 \end{bmatrix} - \begin{bmatrix} Z_0 & 0 & 0 \\ 0 & Z_1 & 0 \\ 0 & 0 & Z_2 \end{bmatrix} \begin{bmatrix} I_{al} \\ I_{al} \\ I_{al} \end{bmatrix} \qquad ...(2.9)$$

On expanding the equation (2.9),

$$V_{a0} = -Z_0 I_{a1} \qquad (2.10)$$

$$V_{a1} = E_a - Z_1 I_{a1} \qquad ...(2.11)$$

$$V_{a2} = -Z_2 I_{a1} \qquad ...(2.12)$$

On adding (2.10), (2.11) and (2.12) we get,

$$V_{a0} + V_{a1} + V_{a2} = -I_{a1} Z_0 + E_a - I_{a1} Z_1 - I_{a1} Z_2 \qquad ...(2.13)$$

Since phase a is shorted to ground,

$$V_a = V_{a0} + V_{a1} + V_{a2} = 0$$

∴ The equation (2.13) can be written as

$$-I_{a1} Z_0 + E_a - I_{a1} Z_1 - I_{a1} Z_2 = 0$$

$$I_{a1} = E_a / Z_1 + Z_2 + Z_0 \qquad ....(2.14)$$

Using equation (2.14) The equivalent circuit of generator during single line to g round fault is drawn as shown in Fig. (2.5). Here the positive, negative and Zero sequence network of the generators are connected in series.

If the neutral of the generator is not grounded, the Zero-sequence network is open circuited and Z_0 is infinite. Under this condition $I_{a,}$ is Zero and So I_{a2} and I_{a0} must be zero. Therefore no path exists for the flow of current in the fault unless the generator neutral is grounded.

Here the fault current is given by,

$$\boxed{I_f = I_a = 3I_{a1}} \qquad ...(2.15)$$

$$[\therefore \quad I_a = I_{a0} + I_{a1} + I_{a2}$$

$$\text{Since } I_{a0} = I_{a1} = I_{a2} = I_{a1}$$

$$I_a = I_{a1} + I_{a1} + I_{a1}$$

$$I_a = 3I_{a1}$$

$$\therefore \quad I_f = I_a = 3I_{a1}$$

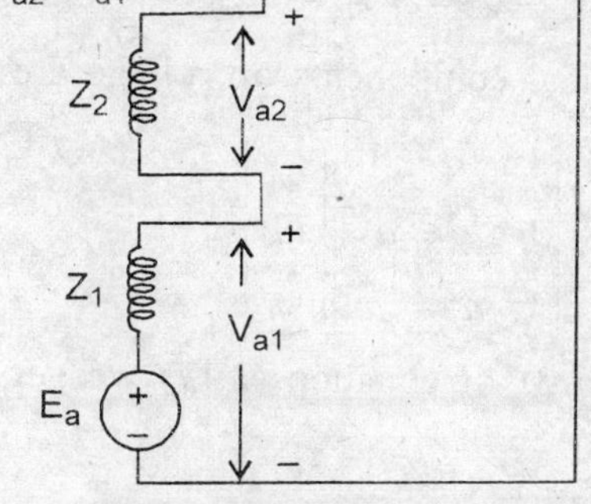

Figure 2.5 *Connection of sequence network of an unloaded generator for line-to-ground fault on phase-a.*

2.3 ANALYSIS OF LINE TO LINE FAULT ON AN UNLOADED GENERATOR

The circuit diagram for a line to line fault on an unloaded, Y connected generator is shown in Fig. (2.6). Here it is assumed that phase-b and phase are shorted. Now the fault current $I_f = I_b = -I_c$.

Since the generater is unloaded the current in phase-a is Zero.

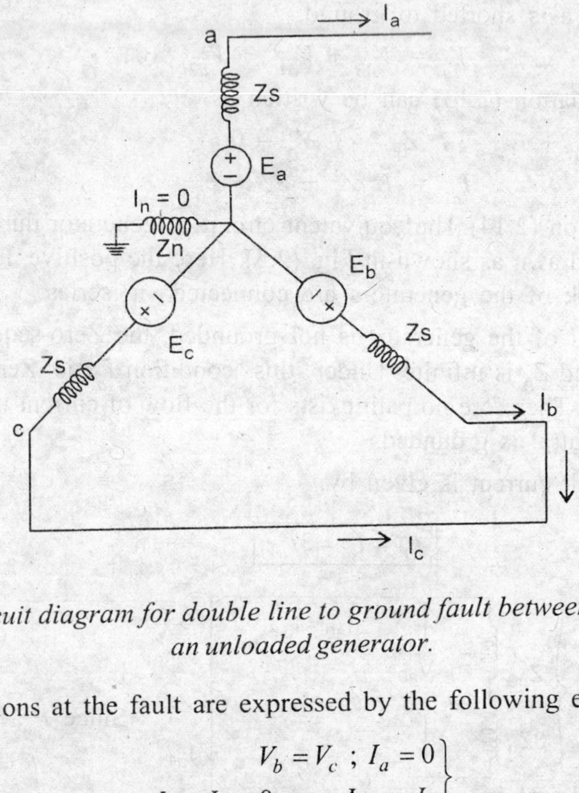

Figure 2.6 *Circuit diagram for double line to ground fault between phase b and c in an unloaded generator.*

The conditions at the fault are expressed by the following equations

$$\left.\begin{aligned} V_b = V_c \ ; \ I_a = 0 \\ I_b + I_c = 0 \ \Rightarrow \ I_b = -I_c \end{aligned}\right\} \qquad ...(2.16)$$

With $V_c = V_b$, the symmetrical components of voltage are given by

$$\begin{bmatrix} V_{a0} \\ V_{a1} \\ V_{a2} \end{bmatrix} = \frac{1}{3}\begin{bmatrix} 1 & 1 & 1 \\ 1 & a & a^2 \\ 1 & a^2 & a \end{bmatrix}\begin{bmatrix} V_a \\ V_b \\ V_b \end{bmatrix} \qquad ...(2.17)$$

On multiplying of row 2 an 3 of equation (2.17) we get,

$$V_{a1} = \frac{1}{3}(V_a + aV_b + a^2V_b) \qquad ...(2.18)$$

$$V_{a2} = \frac{1}{3}(V_a + a^2V_b + aV_b) \qquad ...(2.19)$$

From equations (2.18) and (2.19) we can write

$$V_{a1} = V_{a2} \qquad \text{...(2.20)}$$

The symmetrical components of current are given by

$$\begin{bmatrix} I_{a0} \\ I_{a1} \\ I_{a2} \end{bmatrix} = \frac{1}{3}\begin{bmatrix} 1 & 1 & 1 \\ 1 & a & a^2 \\ 1 & a^2 & a \end{bmatrix}\begin{bmatrix} I_a \\ I_b \\ I_c \end{bmatrix} \qquad \text{...(2.21)}$$

on substituting $I_b = -I_c$ and $I_a = 0$ in equation (2.21) we get,

$$\begin{bmatrix} I_{a0} \\ I_{a1} \\ I_{a2} \end{bmatrix} = \frac{1}{3}\begin{bmatrix} 1 & 1 & 1 \\ 1 & a & a^2 \\ 1 & a^2 & a \end{bmatrix}\begin{bmatrix} 0 \\ -I_c \\ I_c \end{bmatrix} \qquad \text{...(2.22)}$$

On expanding the equation (2.22) we get,

$$I_{a0} = 1/3[-I_c + I_c] = 0$$

$$I_{a1} = 1/3[-aI_c + a^2 I_c] \qquad \text{...(2.23)}$$

$$I_{a2} = 1/3[-a^2 I_c + aI_c] = 1/3[-aI_c + a^2 I_c] \qquad \text{...(2.24)}$$

From equation (2.23) & (2.24) we can write

$$I_{a2} = -I_{a1} \qquad \text{...(2.25)}$$

From the sequence network of the generator we get the following matrix equations.

(Refer note in section (2.2))

$$\begin{bmatrix} V_{a0} \\ V_{a1} \\ V_{a2} \end{bmatrix} = \begin{bmatrix} 0 \\ E_a \\ 0 \end{bmatrix} - \begin{bmatrix} Z_0 & 0 & 0 \\ 0 & Z_1 & 0 \\ 0 & 0 & Z_2 \end{bmatrix}\begin{bmatrix} I_{a0} \\ I_{a1} \\ I_{a2} \end{bmatrix} \qquad \text{...(2.26)}$$

On substituting $I_{a0} = 0$ and $I_{a2} = -I_{a1}$ in equation(2.26) we get,

$$\begin{bmatrix} V_{a0} \\ V_{a1} \\ V_{a2} \end{bmatrix} = \begin{bmatrix} 0 \\ E_a \\ 0 \end{bmatrix} - \begin{bmatrix} Z_0 & 0 & 0 \\ 0 & Z_1 & 0 \\ 0 & 0 & Z_2 \end{bmatrix}\begin{bmatrix} 0 \\ I_{a1} \\ -I_{a1} \end{bmatrix} \qquad \text{...(2.27)}$$

On expanding equation (2.27) we get,

$$V_{a0} = 0 \qquad \text{(2.28)}$$

$$V_{a1} = E_a - I_{a1} Z_1 \qquad \text{...(2.29)}$$

$$V_{a2} = Z_2 I_{a2} \qquad \text{...(2.30)}$$

From equation (2.20) we know that $V_{a1} = V_{a2}$, Hence on equation the equations (2.29) and (2.30) we get,

$$E_a - Z_1 I_{a1} = Z_2 I_{a2}$$

or

$$I_{a1}(Z_1 + Z_2) = E_a$$

$$\therefore \quad I_{a1} = \frac{E_a}{Z_1 + Z_2} \qquad \text{...(2.31)}$$

Using the equation (2.31), the equivalent circuit of generator during line to line fault is drawn as shown in Fig.. (2.7) since $V_{a1} = V_{a2}$, Linen the positive and negative sequence networks of the generator are in parallel. Since $V_{a0} = 0$ the zero sequence network is shorted and so if need not be considered. Since this type of fault does not involve ground, the neutral current $I_n = 0$. Hence the presence (or) absence of a grounded neutral at generator does not affect the fault current.

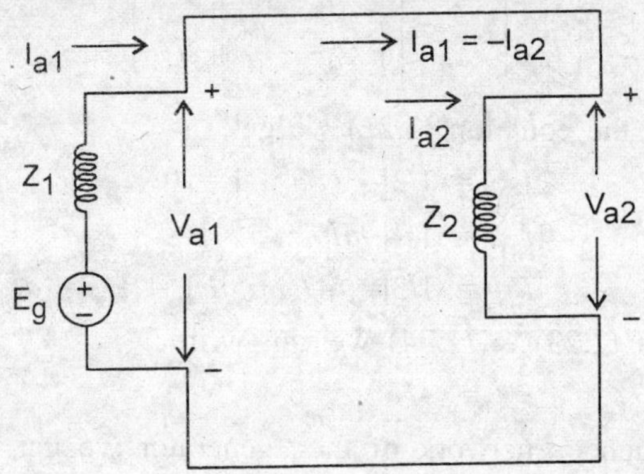

Figure 2.7 *Connection of tue sequence networks of an unloaded generator for a line to line fault between b and c.*

Hence the fault, If = $I_b = -I_c$

The unbalanced currents I_a, I_b and I_c are related to the symmetrical components of currents by the following matrix equation.

$$\begin{bmatrix} I_a \\ I_b \\ I_c \end{bmatrix} = \begin{bmatrix} 1 & 1 & 1 \\ 1 & a^2 & a \\ 1 & a & a^2 \end{bmatrix} \begin{bmatrix} I_{a0} \\ I_{a1} \\ I_{a2} \end{bmatrix} \qquad \text{...(2.32)}$$

From equation (2.32) we get.

$$I_b = I_{a0} + a^2 I_{a1} + a I_{a2}$$

On substituting the conditions $I_{a0} = 0$ and $I_{a2} = -I_{a1}$, which exists during line to line fault in equation (2.33) we get,

$$I_b = a^2 I_{a1} - a I_{a2} = I_{a1}(a^2 - a)$$

$$\therefore \text{ Fault current,} \quad I_f = I_b = I_{a1}(a^2 - a) \qquad \text{...(2.33)}$$

2.4 ANALYSIS OF DOUBLE LINE TO GROUND FAULT ON AN UNLOADED GENERATOR;

The circuit diagram for a double line to ground fault on an unloaded Y connected generator having a grounded neutral is shown in Fig.. (2.8). The fault phases are b and c. Now the fault current $I_f = I_b + I_c$ since the generator is is unloaded the current in phase-a is Zero.

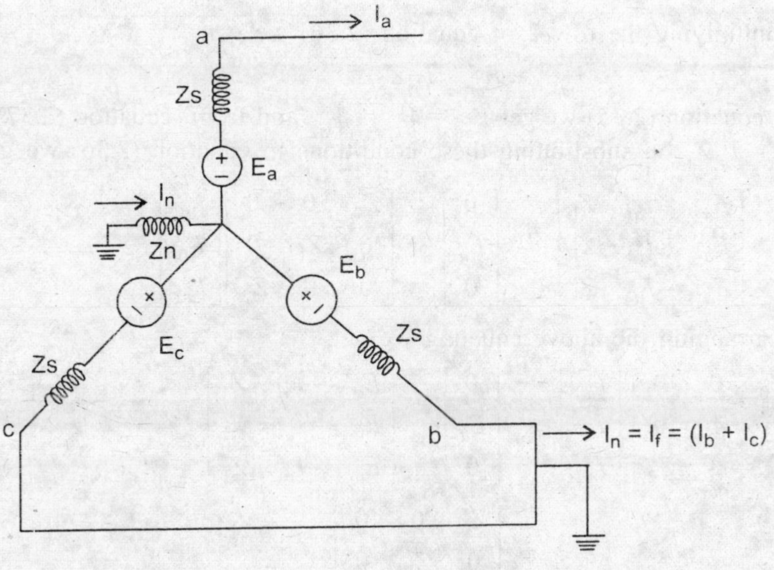

Figure 2.8 *Circuit diagram for double line to ground fault between phase b and c in an unloaded generator.*

The conditions at the fault are expressed by the following equations

$$V_b = 0$$
$$V_c = 0$$
$$I_a = 0$$

With $V_b = V_c = 0$, the symmetrical components of voltage are given by

$$\begin{bmatrix} V_{a0} \\ V_{a1} \\ V_{a2} \end{bmatrix} = \frac{1}{3}\begin{bmatrix} 1 & 1 & 1 \\ 1 & a & a^2 \\ 1 & a^2 & a \end{bmatrix}\begin{bmatrix} V_a \\ 0 \\ 0 \end{bmatrix} \qquad ...(2.34)$$

On expanding the equation (2.34) we get,

$$V_{a0} = V_{a1} = V_{a2} = \frac{V_a}{3} \qquad ...(2.35)$$

From the sequence networks of the generator we get the following matrix equation.

(refer note in section (2.2))

On rearranging the equation from note in section 2.2

$$\begin{bmatrix} V_{a0} \\ V_{a1} \\ V_{a2} \end{bmatrix} = \begin{bmatrix} 0 \\ E_a \\ 0 \end{bmatrix} - \begin{bmatrix} Z_0 & 0 & 0 \\ 0 & Z_1 & 0 \\ 0 & 0 & Z_2 \end{bmatrix} \begin{bmatrix} I_{a0} \\ I_{a1} \\ I_{a2} \end{bmatrix} \quad ...(2.36)$$

On multiplying the row 2 of equation (2.36) we get,

$$V_{a1} = E_a - I_{a1} Z_1 \quad ...(2.37)$$

From equation (2.35) we get $V_{a0} = V_{a1} = V_{a2}$ and from equation (2.37) we get $V_{a1} = E_a - I_{a1} Z_1$ on substituting these conditions in equation (2.36) we get,

$$\begin{bmatrix} E_a - I_{a1} Z_1 \\ E_a - I_{a1} Z_1 \\ E_a - I_{a1} Z_1 \end{bmatrix} = \begin{bmatrix} 0 \\ E_a \\ 0 \end{bmatrix} - \begin{bmatrix} Z_0 & 0 & 0 \\ 0 & Z_1 & 0 \\ 0 & 0 & Z_2 \end{bmatrix} \begin{bmatrix} I_{a0} \\ I_{a1} \\ I_{a2} \end{bmatrix} \quad ...(2.38)$$

On rearranging the above equating we get,

$$\begin{bmatrix} Z_0 & 0 & 0 \\ 0 & Z_1 & 0 \\ 0 & 0 & Z_2 \end{bmatrix} \begin{bmatrix} I_{a0} \\ I_{a1} \\ I_{a2} \end{bmatrix} = \begin{bmatrix} 0 \\ E_a \\ 0 \end{bmatrix} - \begin{bmatrix} E_a - I_{a1} Z_1 \\ E_a - I_{a1} Z_1 \\ E_a - I_{a1} Z_1 \end{bmatrix} \quad ...(2.39)$$

Let,

$$Z = \begin{bmatrix} Z_0 & 0 & 0 \\ 0 & Z_1 & 0 \\ 0 & 0 & Z_2 \end{bmatrix}$$

$$\therefore \quad Z^{-1} = \begin{bmatrix} 1/Z_0 & 0 & 0 \\ 0 & 1/Z_1 & 0 \\ 0 & 0 & 1/Z_2 \end{bmatrix}$$

On premultiplying the equation (2.39) by Z^{-1} we get,

$$\begin{bmatrix} I_{a0} \\ I_{a1} \\ I_{a2} \end{bmatrix} = \begin{bmatrix} 1/Z_0 & 0 & 0 \\ 0 & 1/Z_1 & 0 \\ 0 & 0 & 1/Z_2 \end{bmatrix} \begin{bmatrix} 0 \\ E_a \\ 0 \end{bmatrix} - \begin{bmatrix} 1/Z_0 & 0 & 0 \\ 0 & 1/Z_1 & 0 \\ 0 & 0 & 1/Z_2 \end{bmatrix} \begin{bmatrix} E_a - I_{a1} Z_1 \\ E_a - I_{a1} Z_1 \\ E_a - I_{a1} Z_1 \end{bmatrix}$$

$$\therefore \quad \begin{bmatrix} I_{a0} \\ I_{a1} \\ I_{a2} \end{bmatrix} = \begin{bmatrix} 0 \\ \dfrac{E_a}{Z_1} \\ 0 \end{bmatrix} - \begin{bmatrix} \dfrac{E_a - I_{a1} Z_1}{Z_0} \\ \dfrac{E_a - I_{a1} Z_1}{Z_1} \\ \dfrac{E_a - I_{a1} Z_1}{Z_2} \end{bmatrix} \quad ...(2.40)$$

On expanding the equation (2.40) we get,

$$I_{a0} = -\left[\frac{E_a - I_{a1}Z_1}{Z_0}\right] = -\frac{E_a}{Z_0} + \frac{I_{a1}Z_1}{Z_0} \quad (2.41)$$

$$I_{a1} = \frac{E_a}{Z_1} - \left[\frac{E_a - I_{a1}Z_1}{Z_1}\right] = \frac{E_a}{Z_1} - \frac{E_a}{Z_1} + \frac{I_{a1}Z_1}{Z_1} = I_{a1} \quad ...(2.42)$$

$$I_{a2} = \left[\frac{E_a - I_{a1}Z_1}{Z_2}\right] = -\frac{E}{Z_2} + \frac{I_{a1}Z_1}{Z_2} \quad ...(2.43)$$

Here $I_a = 0$, on expressing I_a by its symmetrical component we get,

$$I_a = I_{a0} + I_{a1} + I_{a2} = 0 \quad ...((2.44)$$

On substituting for I_{a0}, I_{a1}, and I_{a2} from equations (2.41), (2.42) and (2.43) in equation (2.44) we get

$$-\frac{E_a}{Z_0} + \frac{I_{a1}Z_1}{Z_0} + I_{a1} - \frac{E_a}{Z_2} + \frac{I_{a1}Z_1}{Z_2} = 0$$

$$I_{a1}\left[\frac{Z_1}{Z_0} + 1 + \frac{Z_1}{Z_2}\right] = \frac{E_a}{Z_0} + \frac{E_a}{Z_2}$$

$$I_{a1}\left(1 + Z_1\left(\frac{1}{Z_0} + \frac{1}{Z_2}\right)\right) = E_a\left(\frac{1}{Z_0} + \frac{1}{Z_2}\right)$$

$$I_{a1}\left(1 + Z_1\left(\frac{Z_0 + Z_2}{Z_0 Z_2}\right)\right) = E_a\left(\frac{Z_0 + Z_2}{Z_0 Z_2}\right) \quad ...(2.45)$$

On multiplying the equation (2.45) throughout by $\dfrac{Z_0 Z_2}{Z_0 + Z_2}$ we get,

$$I_{a1}\left(\frac{Z_0 Z_2}{Z_0 + Z_2} + Z_1\right) = E_a$$

$$\therefore \qquad I_{a1} = \frac{E_a}{Z_1} + \frac{Z_0 Z_2}{Z_0 + Z_2} \quad ...(2.46)$$

Using the equation (2.46) the equivalent circuit of generator under double line to ground fault can be drown as shown in Fig. (2.9)

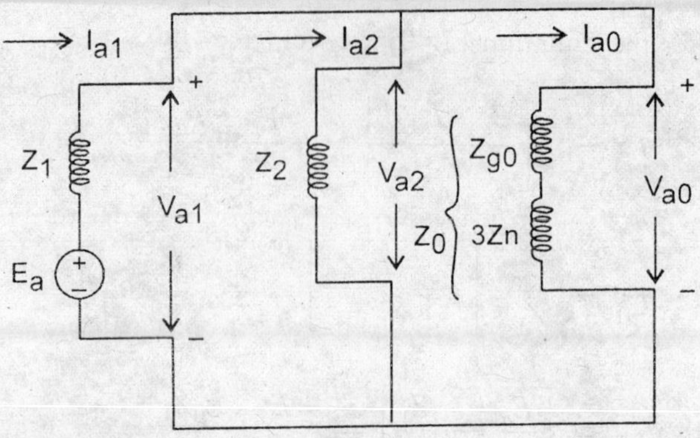

Figure 2.9 *Connection of the sequence networks of an unloaded generator for a double line to ground fault on phase b and c.*

Under this fault condition the sequence networks should be connected in parallel since the Positive, negative, and zero sequence voltages are equal during this fault. In the absence of a ground connection at the generator no current can flow into the ground at the fault. In this case Z_0 would be infinite and I_{a0} would be zero and so the fault will be similar to line to line fault.

Here the fault current, $I_a = I_b + I_c$ ...(2.47)

To determine the fault current, first the symmetrical components I_{a1}, I_{a0}, and I_{a2} are calculated using equations (2.46), (2.41) and (2.43).

Alternatively the I_a, can be calculated using the equation (2.46) and then I_{a2} and I_{a0} can be calculated using current division rule as,

$$I_{a2} = -I_{a1} \times \frac{Z_0}{Z_1 + Z_2} \text{ and } I_{a0} = -(I_{a1} - I_{a2})$$

The unbalanced currents I_a, I_b and I_c are related to the symmetrical components currents by the following equation.

$$\begin{bmatrix} I_a \\ I_b \\ I_c \end{bmatrix} = \begin{bmatrix} 1 & 1 & 1 \\ 1 & a^2 & a \\ 1 & a & a^2 \end{bmatrix} \begin{bmatrix} I_{a0} \\ I_{a1} \\ I_{a2} \end{bmatrix} \quad \text{...(2.48)}$$

From equation (2.48) we get,

$$I_b = I_{a0} + a^2 Ia_1 + aI_{a2}$$
$$I_c = I_{a0} + aI_{a1} + a^2 Ia_2$$

∴ Fault current,

$$I_f = I_b + I_c = I_{a0} + a^2 Ia_1 + aIa_2 + Ia_0 + aI_{a1} + a^2 I_{a2}$$
$$= 2I_{a0} + (a + a^2)I_{a1} + (a + a^2)I_{a2}$$

$$\boxed{I_f = 2I_{a0} + (a + a^2)(I_{a1} + I_{a2})} \quad \text{...(2.49)}$$

2.5 UNSYMMETRICAL FAULTS ON POWER SYSTEM:

The unsymmetrical faults on the power system are analyzed using the Thevenin's equivalent of positive, negative and zero sequence networks of the power system. The thevenin's equivalent is obtained with respect to fault point and reference.

In Power system the prefault voltage and currents are assumed to be balanced. Hence the prefault voltage at the fault point is balanced voltage and so it has only positive sequence component. The negative and zero sequence component of prefault voltage at the fault point is zero therefore the prefault voltage at the fault point is the thevenins' voltage of positive sequence network.

Let, $Z_1 \rightarrow$ Thevenin's impedance of positive sequence network.

$Z_2 \rightarrow$ Thevenin's impedance of negative sequence network.

$Z_0 \rightarrow$ Thevenin's impedance of zero sequence network.

$V_{pf} \rightarrow$ Prefault voltage at the fault Point.

Thevenin's equivalent of Positive, negative and zero sequence network of the power system with respect to fault point will be shown in Fig.. (2.10), (2.11) and (2.12) respectively.

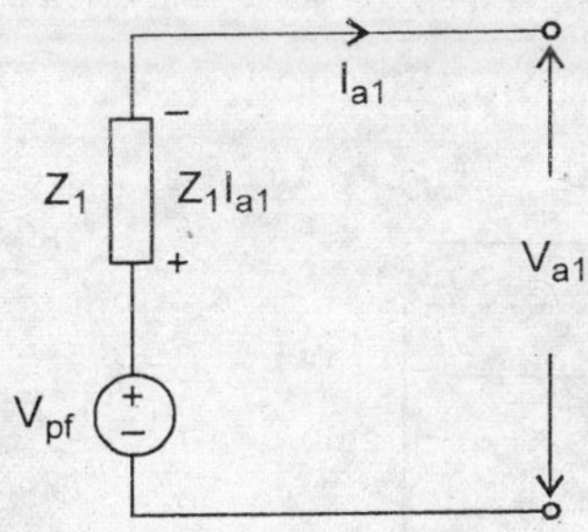

Figure 2.10 *Thevenin's equivalent of positive sequence network*

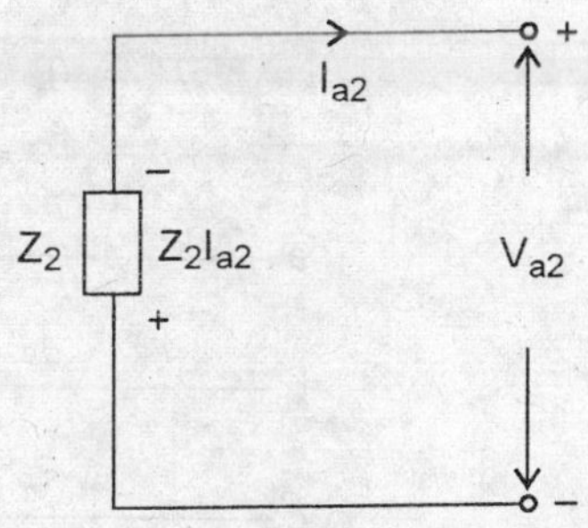

Figure 2.11 *Thevenin's equivalent of negative sequence network*

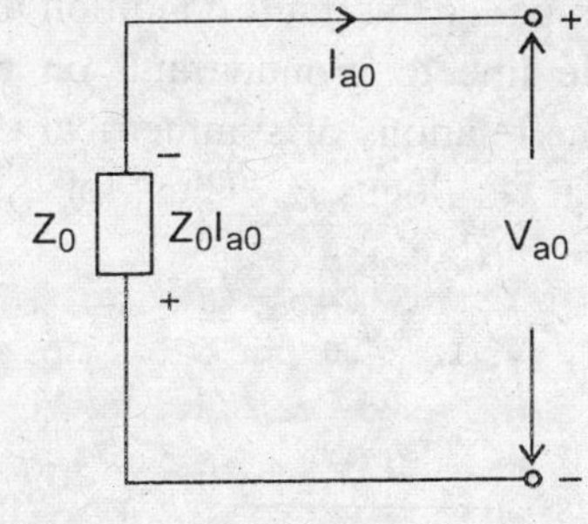

Figure 2.12 *Thevenin's equivalent of zero sequence network.*

From Fig. (2.10), (2.11), and (2.12) using kirchoff's voltage law we get,

$$V_{a0} = -Z_0 I_{a0} \quad ...(2.50)$$

$$V_{a1} = V_{pf} - Z_1 I_{a1} \quad ...(2.51)$$

$$V_{a2} = -Z_2 I_{a2} \quad ...(2.52)$$

On arranging the equations in the matrix form we get,

$$\begin{bmatrix} V_{a0} \\ V_{a1} \\ V_{a2} \end{bmatrix} = \begin{bmatrix} 0 \\ V_{pf} \\ 0 \end{bmatrix} - \begin{bmatrix} Z_0 & 0 & 0 \\ 0 & Z_1 & 0 \\ 0 & 0 & Z_2 \end{bmatrix} \begin{bmatrix} I_{a0} \\ I_{a1} \\ I_{a2} \end{bmatrix} \quad ...(2.53)$$

The equation (2.53) will be useful in the analysis at unsymmetrical faults an the power system.

(i) Single line to ground fault on a power system

The single line to ground fault, on a power system can be represented by connecting three stub on the three transmission lines as shown in Fig. (2.13), The following relations exist at the fault.

$$I_b = 0;$$

$$I_c = 0;$$

$$V_a = 0 \quad ...(2.54)$$

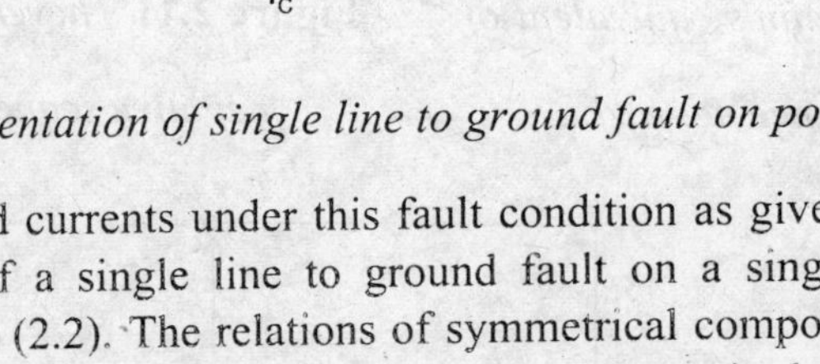

Figure 2.13 *Representation of single line to ground fault on power system with stub*

The voltage and currents under this fault condition as given be equation (2.54) are same as that of a single line to ground fault on a single generator. As we discussed in section (2.2). The relations of symmetrical components must also have the some solutions except that V_{pf} locus $E_{a.}$ Thus, for a single live to ground fault,

$$I_{a1} = I_{a2} = I_{a0} = \frac{I_a}{3} \quad ...(2.55)$$

and

$$I_{a1} = \frac{V_{pf}}{Z_1 + Z_2 + Z_0} \quad ...(2.56)$$

The equations (2.55) and (2.56) indicate that the three sequence networks should be connected in series through the fault point in order to simulate a single line to ground fault as shown in Fig.. (2.14). The calculation of symmetrical components of current and fault currents are similar to that of section (2.2)

i.e) $$If = I_a = 3I_{a1}$$

Let we see the connection of sequence network

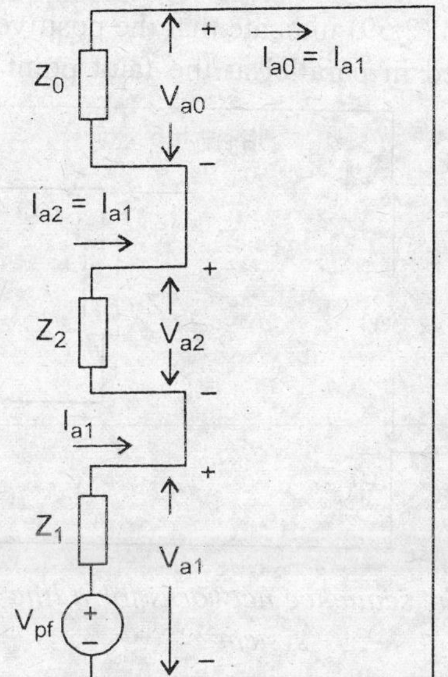

Figure 2.14 *Connection of sequence network for a single line to ground fault in a power system.*

(ii) Line to line fault on a power system

The line to line faul on a power system can be represented by connecting three stubs on the three tr nsmission lines as shown in Fig. (2.15)

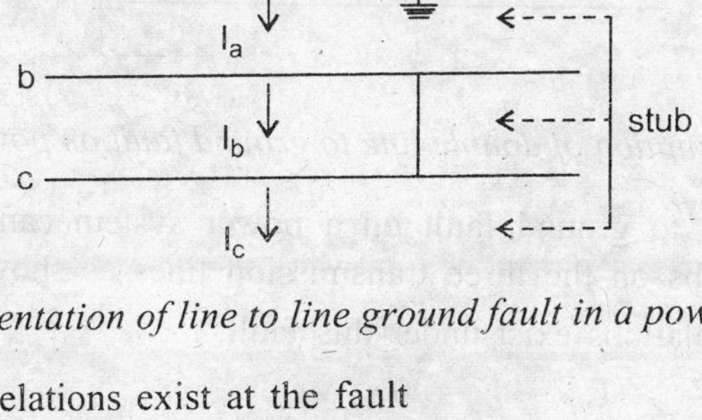

Figure 2.15 *Representation of line to line ground fault in a power system with strip*

The following relations exist at the fault

$$V_b = V_c \qquad I_a = 0$$

and $$I_b = -I_c \qquad \text{...(2.57)}$$

The voltage and currents under this fault condition as given by equation (2.57) are same as that of a line to line fault on an isolated generator as discussed in section 2.3. Hence for a line to line fault,

$$V_{a1} = V_{a2} \quad \text{...(2.58)}$$

$$I_{a1} = \frac{V_{pf}}{Z_1 + Z_2} \quad \text{...(2.59)}$$

The equations (2.58) and (2.59) indicate that the positive and negative sequence networks should be connected in parallel at the fault point.

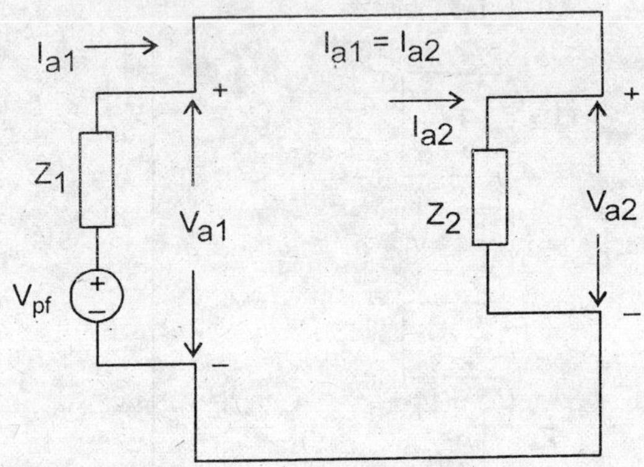

Figure 2.16 *Connection of the sequence networks for a line to line fault in a power system*

The calculations of symmetrical components of currents and fault currents are similar to that of section (2.3).

(iii) Double line to ground fault on a power system

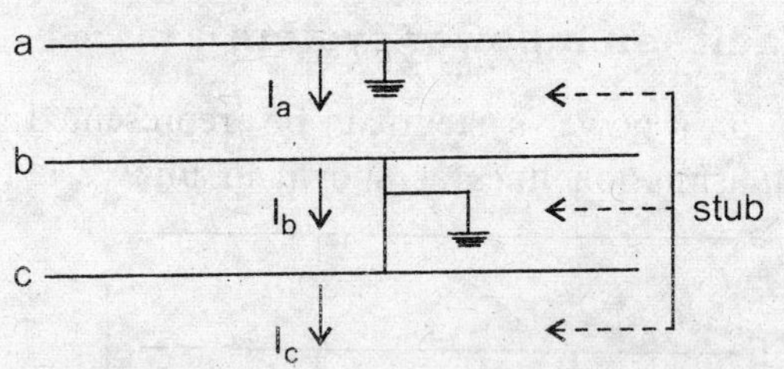

Figure 2.17 *Representation of double line to ground fault on power system with stub*

The double line to ground fault on a power system can be represented by connecting three stubs on the three transmission lines as shows in Fig. (2.17).

The following relations exist under this fault,

$$V_b = V_c = 0;$$

$$I_a = 0 \quad \text{...(2.60)}$$

The voltage and currents under this fault condition is given by the equation (2.60) is same as that of a double line to ground fault on an isolated generator discussed in section (2.4). Hence for a double line to ground fault.

$$V_{al} = V_{a2} = V_{a0} \qquad ...(2.61)$$

$$I_{a1} = \frac{V_{pf}}{Z_1 + \left[\frac{Z_0 Z_2}{Z_0 + Z_2}\right]} \qquad ...(2.62)$$

The equations (2.61) and (2.62) indicate that the three sequence networks should be connected in parallel at the fault point refer Fig. (2.18).

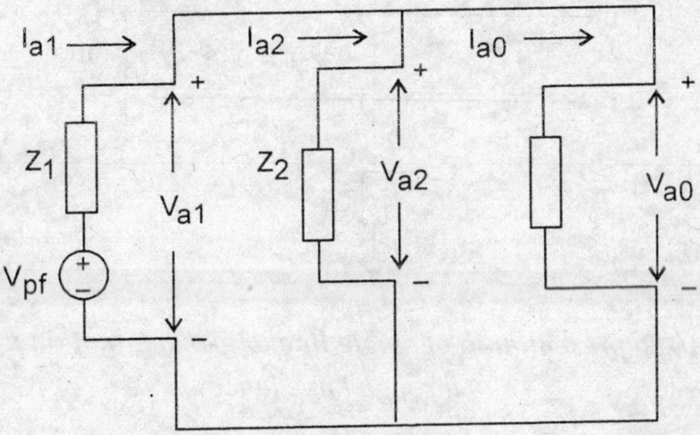

Figure 2.18 *Connection of the sequence network for a double line to ground fault in a power system*

The calculations of symmetrical components of currents and fault current *s* are similar to that of section (2.4).

- Now we will see unsymmetrical faults on power system with impedance.

2.6 FAULTS THROUGH IMPEDANCE:

The fault considered in section (2.5) are direct short circuits between lines and between one (or) two lines to ground. In these fault analysis the impedance between the fault point is considered as zero. There may be situations in which the fault path incudes an impedance between the faulted points.

In these situations the analysis is carried similar to that of section (2.5), except that the fault impedances are included at appropriate Points in the circuits obtained by connecting sequence networks.

(i) Single line to ground fault:

A single line to ground fault at point F in a Power system through a fault impedance Z_f can be represented by connecting three stubs as shown in Fig. (2.19). The fault is on phase a

At the fault point F, the following relation exist.

$$I_b = 0;$$
$$I_c = 0;$$
$$V_a = Z_f Ia \qquad ...(2.63)$$

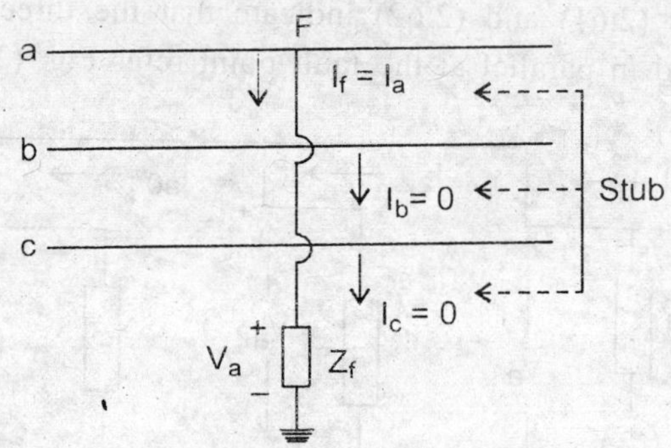

Figure 2.19 *Representation of single line to ground fault on power system through impedance*

The symmetrical components of the fault current are

$$\begin{bmatrix} I_{a0} \\ I_{a1} \\ I_{a2} \end{bmatrix} = \begin{bmatrix} 1 & 1 & 1 \\ 1 & a & a^2 \\ 1 & a^2 & a \end{bmatrix} \begin{bmatrix} I_a \\ 0 \\ 0 \end{bmatrix}$$

$$\therefore \quad I_{a1} = I_{a2} = I_{a0} = 1/3\ I_a \qquad ...(2.64)$$

or
$$I_a = 3I_{a1} \qquad ...(2.65)$$

Expressing the voltage V_a in terms of symmetrical components, we have

$$V_a = V_{a1} + V_{a2} + V_{a0} \qquad ...(2.66)$$

using equations (2.63), (2.65) and (2.66) we can write

$$V_{a1} + V_{a2} + V_{a0} = Z_f I_a = 3Z_f I_{a1} \qquad ...(2.67)$$

From equations (2.64) and (2.67) we can say that all sequence currents are equal and the sum of sequence voltages equals $3Z_f I_{a1}$. Therefore, these equations suggest a series connection of sequence networks through an impedance $3Z_f$ as shown in Fig. (2.20)

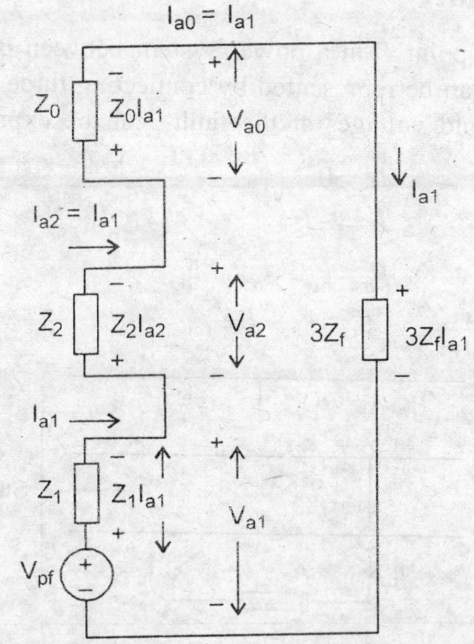

Figure 2.20 *Connection of sequence networks for a single line to ground fault through an impedance* Z_f

From Fig.. (2.20) we can write

$$I_{a1} = \frac{V_{pf}}{(Z_1 + Z_2 + Z_0) + 3Z_f} \quad ...(2.68)$$

Fault current, $$I_f = I_a = 3I_{a1} = \frac{3V_{pf}}{(Z_1 + Z_2 + Z_0) + 3Z_f} \quad ...(2.69)$$

The symmetrical components of phase-a voltage are calculated by writing kirchoff's voltage law equations in the sequence networks of Fig. (2.20) as shown below.

$$V_{a1} = V_{pf} - Z_1 I_{a1}$$
$$V_{a2} = -Z_2 I_{a1}$$
$$V_{a0} = -Z_0 I_{a1}$$

From the symmetrical components of phase-a voltage, the phase voltage V_a, V_b and V_c are calculated using the following matrix equation.

$$\begin{bmatrix} V_a \\ V_b \\ V_c \end{bmatrix} = \begin{bmatrix} 1 & 1 & 1 \\ 1 & a^2 & a \\ 1 & a & a^2 \end{bmatrix} \begin{bmatrix} V_{a0} \\ V_{a1} \\ V_{a2} \end{bmatrix}$$

(ii) Line to line fault

A line-to-line fault at point F in a power system between phases b and c through a fault impedance Z_f can be represented by connecting three stubs as shown in Fig. (2.21). The currents and voltages at the fault can be expressed as

$$\left.\begin{aligned} I_b &= 0 \\ I_b &= -I_c \end{aligned}\right\} \quad \text{...(2.70)}$$

and

$$\left.\begin{aligned} V_b - V_c &= I_b Z_f \\ V_c &= V_b - I_b Z_f \end{aligned}\right\} \quad \text{...(2.71)}$$

Also

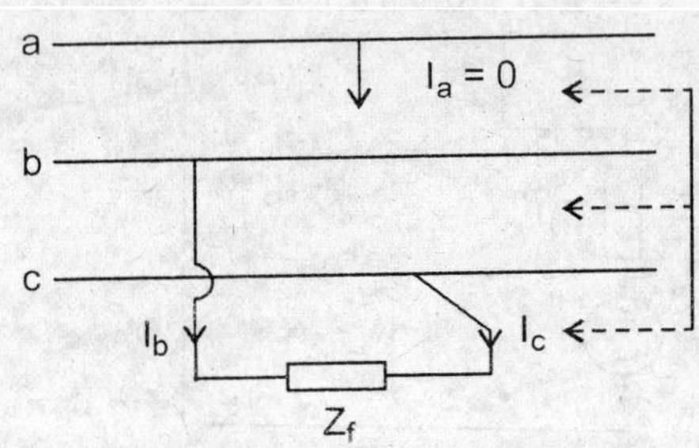

Figure 2.21 *Representation of line to line faulton power system through impedance*

The symmetrical components of the currents after substituting

$$I_a = 0$$

and

$$I_c = -I_b \text{ we get,}$$

$$\begin{bmatrix} I_{a0} \\ I_{a1} \\ I_{a2} \end{bmatrix} = \frac{1}{3}\begin{bmatrix} 1 & 1 & 1 \\ 1 & a & a^2 \\ 1 & a^2 & a \end{bmatrix}\begin{bmatrix} 0 \\ I_b \\ -I_b \end{bmatrix} \quad \text{...(2.72)}$$

On expanding the equation (2.72) we get,

$$I_{a0} = \frac{1}{3}(I_b - I_b) = 0$$

$$I_{a1} = \frac{1}{3}(aI_b - a^2 I_b)$$

$$I_{a2} = \frac{1}{3}(a^2 I_b - aI_b) = -I_{a1} \quad \text{...(2.73)}$$

The line currents can be obtained from the following matrix equation, after substituting $I_{a0} = 0$ and $I_{a2} = -I_{a1}$

$$\begin{bmatrix} I_a \\ I_b \\ I_c \end{bmatrix} = \frac{1}{3}\begin{bmatrix} 1 & 1 & 1 \\ 1 & a^2 & a \\ 1 & a & a^2 \end{bmatrix} = \begin{bmatrix} 1 & 1 & 1 \\ 1 & a^2 & a \\ 1 & a & a^2 \end{bmatrix}\begin{bmatrix} 0 \\ I_{a1} \\ -I_{a1} \end{bmatrix} \quad \text{...(2.74)}$$

From row-2 of matrix equation (2.74) we get,

$$I_b = a^2 I_{a1} - aI_{a1} = I_{a1}(a^2 - a)$$

$$= I_{a1}\left[-\frac{1}{2} - j\frac{\sqrt{3}}{2} + \frac{1}{2} - j\frac{\sqrt{3}}{2}\right] = -j\sqrt{3}I_{a1} \quad ...(2.75)$$

The symmetrical components of phase voltage are

$$\begin{bmatrix} V_{a0} \\ V_{a1} \\ V_{a2} \end{bmatrix} = \frac{1}{3}\begin{bmatrix} 1 & 1 & 1 \\ 1 & a & a^2 \\ 1 & a^2 & a \end{bmatrix}\begin{bmatrix} V_a \\ V_b \\ V_c \end{bmatrix} \quad ...(2.76)$$

On substituting $V_c = V_b - I_b Z_f$ in equation (2.76) we get,

$$\begin{bmatrix} V_{a0} \\ V_{a1} \\ V_{a2} \end{bmatrix} = \frac{1}{3}\begin{bmatrix} 1 & 1 & 1 \\ 1 & a & a^2 \\ 1 & a^2 & a \end{bmatrix}\begin{bmatrix} V_a \\ V_b \\ V_b - I_b Z_f \end{bmatrix} \quad ...(2.77)$$

From row-2 of matrix equation (2.77) we get,

$$V_{a1} = \frac{1}{3}[V_a + aV_b + a^2 V_b - a^2 I_b Z_f]$$

$$\therefore \quad 3V_{a1} = V_a + V_b(a + a^2) - a^2 Z_f I_b \quad ...(2.78)$$

From row-3 of matrix equation (2.77) we get,

$$N_{a2} = \frac{1}{3}[V_a + a^2 V_b + aV_b - aI_b Z_f]$$

$$\therefore \quad 3V_{a2} = V_a + V_b(a^2 + a) - aZ_f I_b \quad ...(2.79)$$

On subtracting equation (2.79) from equation (2.78) we get,

$$3V_{a1} - 3V_{a2} = V_a + V_b(a + a^2) - a^2 Z_f I_b - V_a - V_b(a^2 + a) + aZ_f I_b$$

$$= (-a^2 + a)Z_f I_b$$

$$= \left[\frac{1}{2} + j\frac{\sqrt{3}}{2} - \frac{1}{\sqrt{2}} + j\frac{\sqrt{3}}{2}\right]Z_f I_b = j\sqrt{3}Z_f I_b$$

$$\therefore \quad V_{a1} - V_{a2} = \frac{1}{3}j\sqrt{3}Z_f I_b = j\frac{1}{\sqrt{3}}Z_f I_b \quad ...(2.80)$$

The equations (2.73) and (2.80) suggest parallel connection of positive and negative sequence networks through a series impedance Z_f. Therefore the equivalent circuit of power system under line to line fault can be drawn as shown in Fig. (2.22)

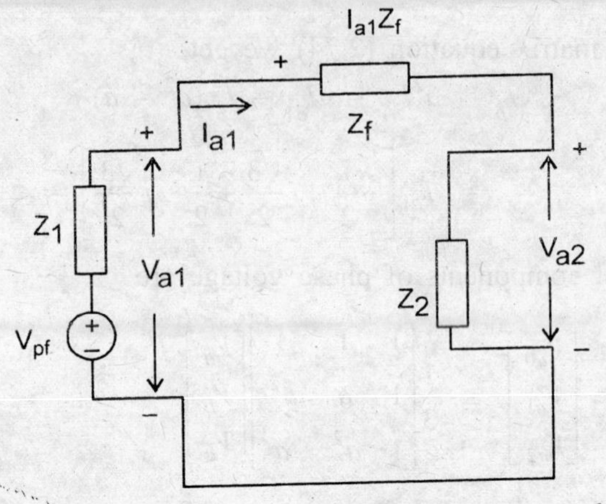

Figure 2.22 *Connection of sequence networks for a line to line to line fault through an impedance Z_f*

From Fig. (2.22) we get, $I_{a1} = \dfrac{V_{pf}}{Z_1 + Z_2 + Z_f}$

The fault current $I_f = I_b = -j\sqrt{3}I_{a1}$ ...(2.81)

(iii) Double line to ground fault

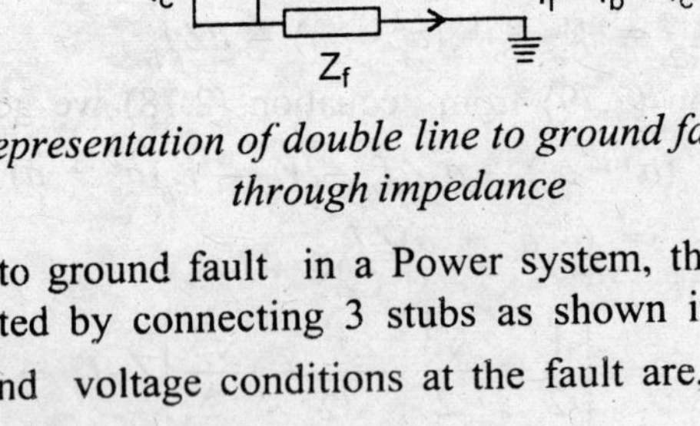

Figure 2.23 *Representation of double line to ground fault on power system through impedance*

A double line to ground fault in a Power system, through a fault impedance Z_f can be represented by connecting 3 stubs as shown in Fig.. (2.23).

The current and voltage conditions at the fault are,

$$I_a = 0 \qquad ...(2.83)$$

$$V_a = V_c = Z_f(I_b + I_c) \qquad ...(2.83)$$

The line currents are given by

$$\begin{bmatrix} I_a \\ I_b \\ I_c \end{bmatrix} = \begin{bmatrix} 1 & 1 & 1 \\ 1 & a^2 & a \\ 1 & a & a^2 \end{bmatrix} \begin{bmatrix} I_{a0} \\ I_{a1} \\ I_{a2} \end{bmatrix} \qquad ...(2.84)$$

From row-1 of equation (2.84) we get,

$$I_a = I_{a0} + I_{a1} + I_{a2}$$

From equation (2.83) we get $I_a = 0$

$$\therefore \quad I_{a0} + I_{a1} + I_{a2} = 0$$

or

$$I_{a0} = -(I_{a1} + I_{a2}) \qquad ...(2.85)$$

from row 2 and 3 of equation (2.84) we get,

$$I_b = I_{a0} + a^2 I_{a1} + aI_{a2}$$

$$I_c = I_{a0} + aI_{a1} + a^2 I_{a2}$$

$$I_b + I_c = I_{a0} + a^2 I_{a1} + aI_{a2} + I_{a0} + aI_{a1} + a^2 I_{a2}$$

$$= 2I_{a0} + (a^2 + a)I_{a1} + (a^2 + a)I_{a2} \qquad ...(2.86)$$

We know that,

$$1 + a + a^2 = 0$$

$$\therefore \quad a + a^2 = -1 \qquad ...(2.87)$$

From equation (2.86) and (2.87) we get,

$$I_b + I_c = 2I_{a0} - I_{a1} - I_{a2}$$

$$= 2I_{a0} - (I_{a1} + I_{a2}) \qquad ...(2.88)$$

On substituting for $I_{a1} + I_{a2}$ from equation (2.85) in equation (2.88) we get,

$$I_b + I_c = 2I_{a0} - (-I_{a0}) = 3I_{a0} \qquad ...(2.89)$$

The symmetrical components of voltages after substituting

$V_c = V_b$ are given by,

$$\begin{bmatrix} V_{a0} \\ V_{a1} \\ V_{a2} \end{bmatrix} = \frac{1}{3}\begin{bmatrix} 1 & 1 & 1 \\ 1 & a & a^2 \\ 1 & a^2 & a \end{bmatrix}\begin{bmatrix} V_a \\ V_b \\ V_c \end{bmatrix} \qquad ...(2.90)$$

From row-1 of equation (2.90) we get,

$$V_{a0} = \frac{1}{3}[V_a + V_b + V_b] = \frac{1}{3}[V_a + 2V_b]$$

From row-2 and 3 of equation (2.90) we get,

$$V_{a1} = V_{a2} = \frac{1}{3}[V_a + aV_b + a^2 V_b] \qquad (\therefore\ a + a^2 = -1)$$

$$= \frac{1}{3}(V_a + (a + a^2)V_b) = \frac{1}{3}(V_a - V_b) \qquad ...(2.9)$$

From equation (2.90) and (2.91) we car write,

$$V_{a0} - V_{a1} = \frac{1}{3}(V_a + 2V_b) - \frac{1}{3}(V_a - V_b)$$

$$= \frac{1}{3}(V_a + 2V_b - V_a + V_b) = V_b$$

$$\therefore \quad V_{a0} - V_{a1} = V_b \qquad \text{...(2.92)}$$

On substituting for V_b from equation (2.83) in equation (2.92) we get,

$$V_{a0} - V_{a1} = Z_f (I_b + I_c) \qquad \text{...(2.93)}$$

On substituting $(I_b + I_c)$ from equation (2.89) in equation (2.93) we get,

$$V_{a0} - V_{a1} = Z_f 3I_{a0}$$

$$\therefore \quad V_{a0} = V_{a1} + 3Z_f I_{a0} \qquad \text{...}(\because V_{a1} = V_{a2})$$

Also, $$V_{a0} = V_{a2} + 3Z_f I_{a0} \qquad \text{...(2.95)}$$

Equations (2.85) and (2.94) suggest the connection of sequence networks as shown in Fig. (2.24) for a double line to ground fault.

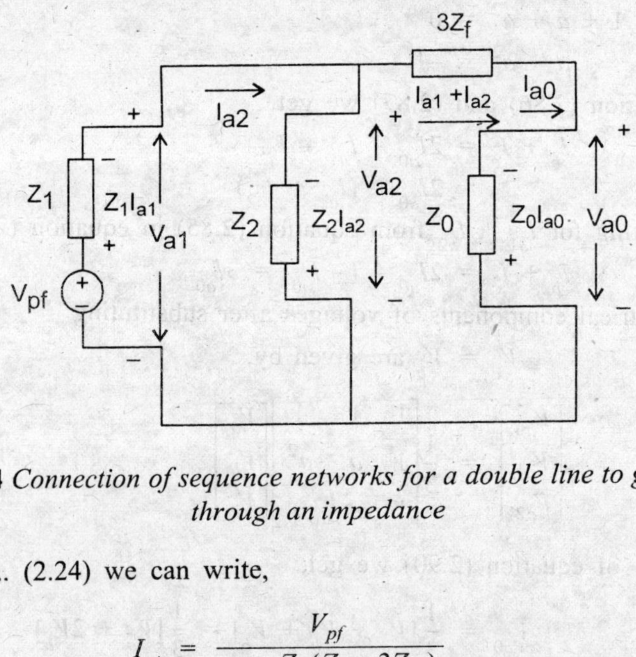

Figure 2.24 *Connection of sequence networks for a double line to ground fault through an impedance*

From Fig.. (2.24) we can write,

$$I_{a1} = \frac{V_{pf}}{Z_1 + \dfrac{Z_2(Z_0 + 3Z_f)}{Z_2 + Z_0 + 3Z_f}} \qquad \text{...(2.96)}$$

$$V_{a1} = V_{pf} - Z_f I_{a1} \qquad \text{...(2.97)}$$

$$V_{a2} = V_{a1};\ I_{a1} = \frac{V_{a2}}{Z_2};\ I_{a0} = -(I_{a1} + I_{a2}) \qquad \text{...(2.98)}$$

From equation (2.89)

$$\boxed{\text{Fault current}, I_f = I_b + I_c = 3I_{a0}}$$

Example 2.1: Two 11*KV*, 20 *MVA*, 3 phase, star connected generators operate in parallel as shown in Fig..(2.1.1). The positive, negative and zero sequence reactances of each being respectively, *j*0.18, *j*0.15, *j*0.10 *p.u*, The star point of one of the generator is Isolated and that of the other is earthed through a 2Ω resistor. *A* single line to ground fault occurs at the terminals of one of tue generators. Estimate (*i*) fault current (*ii*) current in grounded resistor and (*iii*) voltage across grounding resistor.

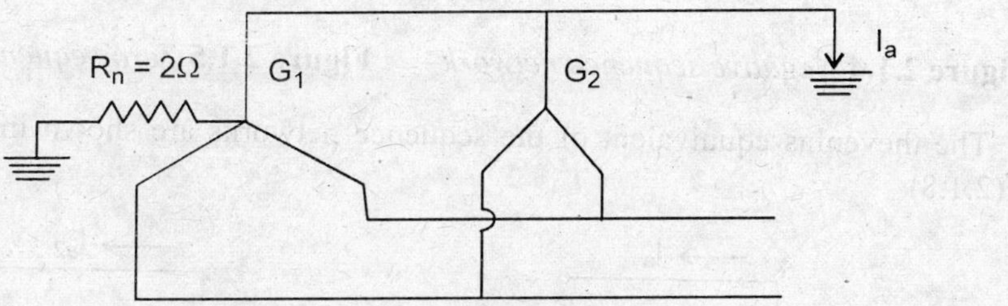

Figure 2.1.1

Solution: Let is choose the generator ratigs as base values.

$$\therefore \quad MVA_b = 20\ MVA;$$

$$KV_b = 11\ KV$$

$$\text{Base impedance,} \quad Z_b = \frac{KV_b^{\,2}}{MVA_b} = \frac{11^2}{10} = 6.05\Omega$$

$$P.u.\ \text{value of neutral resistance} = \frac{\text{Actual value}}{\text{Base impedance}}$$

$$= \frac{2}{6.05} = 0.3306\ P.u$$

The single line diagram and the sequence networks of the given power system are shown in Fig..(2.1.2) to (2.1.5)

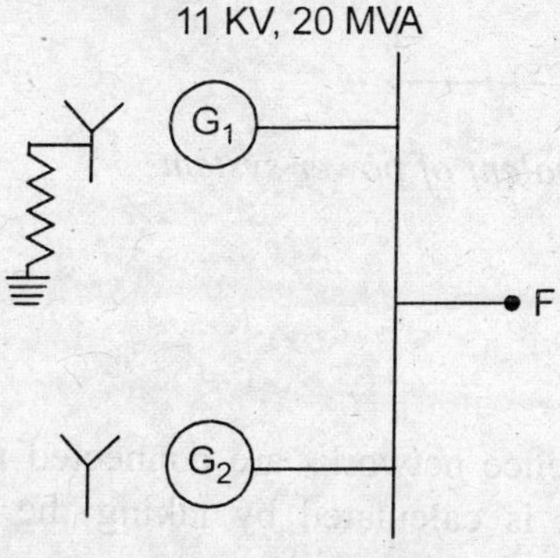

Figure 2.1.2 *Single line diagram*

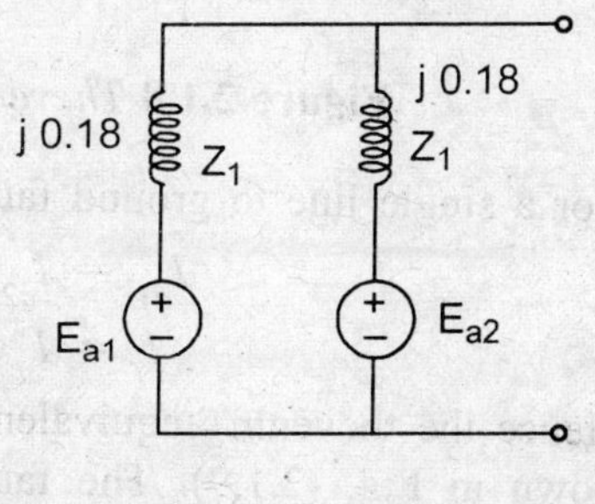

Figure 2.1.3 *Positive sequence network*

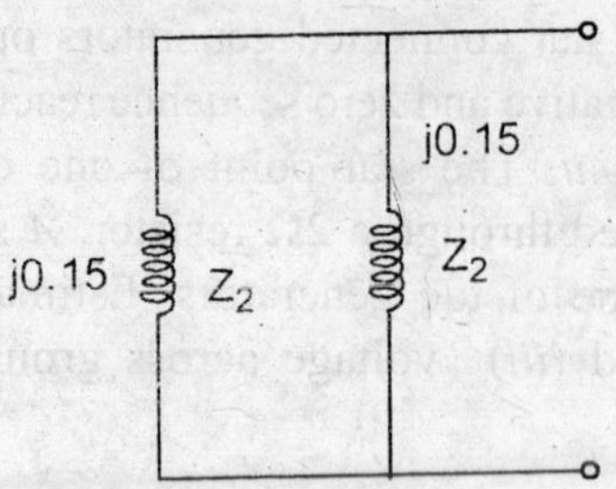

Figure 2.1.4 *Negative sequence network*

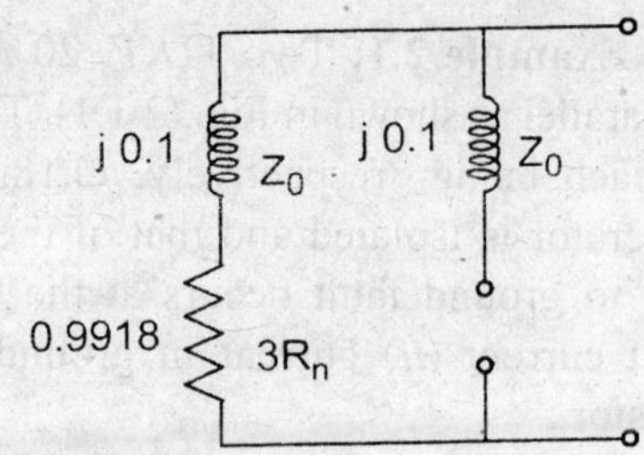

Figure 2.1.5 *Zero sequence network*

The thevenins equuivalent of the sequence networks are shown in Fig. (2.1.6) to (2.1.8)

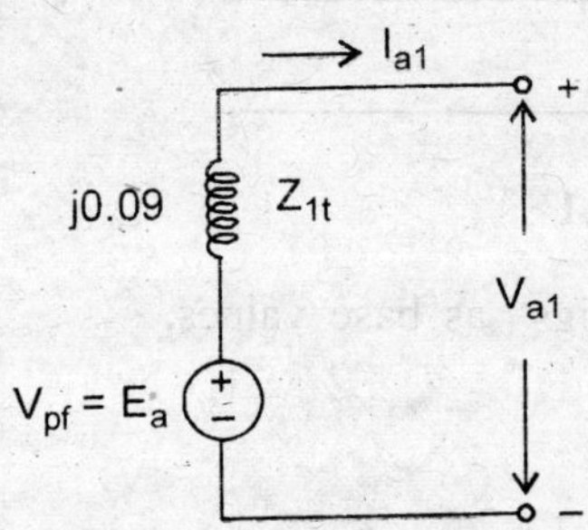

Figure 2.1.6 *Thevenin's equivalent of power system*

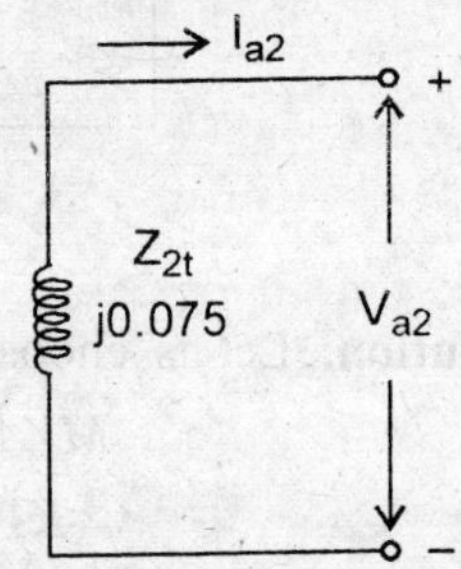

Figure 2.1.7 *Thevenin's equivalent of power system*

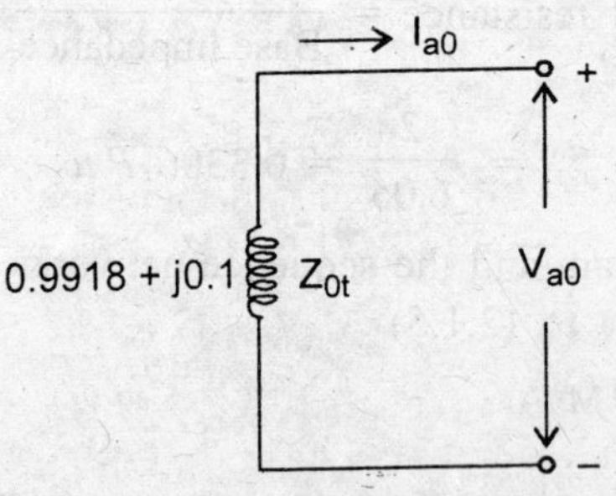

Figure 2.1.8 *Thevenin's equivalent of power system.*

For a single line to ground fault,

$$I_{a1} = I_{a2} = I_{a0}$$

and

$$I_f = I_a = 3I_{a1}$$

Hence the thevenin's equivalent of sequence networks are connected in series as shown in Fig. (2.1.9). The fault current is calculated by taking the prefault voltage $V_{pf} = 1$ *p.u.*

From Fig. (2.1.9) we get,

$$I_{a1} = \frac{1\angle 0°}{j0.009 + j0.075 + 0.9918 + j0.1}$$

$$= \frac{1}{0.9918 + j0.265} = \frac{1}{1.0266\angle 15°} = 0.9741\angle{-15°}\ P.u$$

(i) To find fault current

Fault current, $I_f = I_a = 3I_{a1}$

$= 3 \times 0.9741\ \angle{-15°}$

$= 2.9223\ \angle{-15°}\ P.u$

Base Current, $I_b = \frac{KVA_b}{\sqrt{3}KV_b} = \frac{20 \times 1000}{\sqrt{3} \times 11}$

$$\boxed{I_b = 10.907A}$$

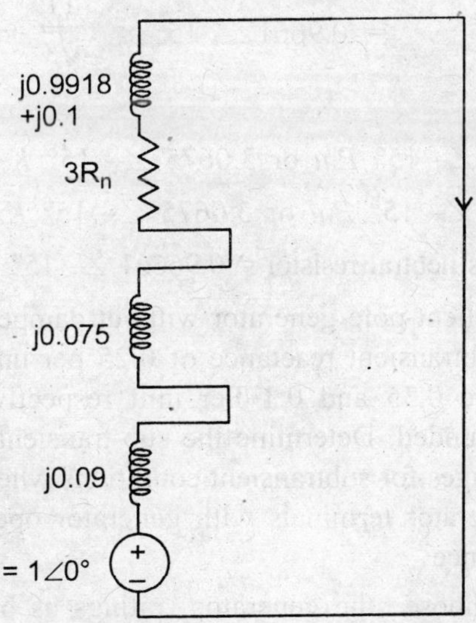

Figure 2.1.9 *Connection of sequence networks*

∴ Actual value of fault current

$= P.u$ value of fault current × Base current

$= 2.9223\angle{-15°}\ \times 1044.7$

$= 3067.5\angle{-15°}$

Actual value of fault current $= 3.0675\angle{-15°}\ KA$

(ii) To Find the current through neutral resistor:

The current through the neutral resitor is same as that of fault current.

$\therefore$ current through neutral resistor = $2.9223\angle{-15°}$ *P.a*

(or) $3.0675\angle{-15°}$ *KA.*

(iii) To find the voltage across grounding resistor:

From the therenin's equivalent of zero sequence network we get,

The voltage across grounding resistor} $= 3R_n I_{a0} = 3R_n I_{a1}$

$$= 3 \times 0.3306 \times 0.9741 \angle{-15°}$$

$$= 0.9661\angle{-15°} \ P.u.$$

Actual value of voltage across grounding resistor

$$= \text{P.u value of voltage} \times \frac{KV_b}{\sqrt{3}} \quad \left[\because KV_b \text{ is line value}\right]$$

$$= 0.9661\angle{-15°} \times \frac{11}{\sqrt{3}} = 6.1356\angle{-15°} \ KV$$

Result:

(*i*) If = 2.9223 $\angle{-15°}$ *P.u* or 3.0675 $\angle{-15°}$ *KA*

(*ii*) In = 2.9223 $\angle{-15°}$ *P.u* or 3.0675 $\angle{-15°}$ *KA*

(*iii*) voltage across neutral resistor = 0.96661 $\angle{-15°}$ *P.u* or 6.1356 $\angle{-15°}$ *KV*

Example 2.2: A salient-pole generator without dampers is rated 20 *MVA*, 13.8 *KV* and has a direct subtransient reactance of 0.25 per unit. the negative and zero sequence reactances are 0.35 and 0.1 Per unit respectively. The neutral of the generator is solidly grounded. Determine the sub-transient current in the generator and the line to line voltages for subtransient conditions when a single line to ground fault occurs at the generator terminals with generator operating unloaded at rated voltage. Neglect resistance

Solution: Let us choose the generator ratings as base values.

$$\therefore \quad MVA_b = 20 \ MVA;$$

$$KV_b = 13.8 \ KV$$

Base current $$I_b = \frac{KVA_b}{\sqrt{3}KV_b} = \frac{20\times1000}{\sqrt{3}\times13.8} = 836.7A$$

Given that, $$Z_1 = j0.25 \ P.u$$

$$Z_2 = j0.35 \ P.u$$

and $$Z_0 = j0.1 \ P.u$$

For a single line to ground fault in an unloaded generator the sequence network are connected in series as shown in Fig. (2.2.1). The following relations exists for a single line to ground fault.

$$I_{a0} = I_{a1} = I_{a2}$$

$$I_{a1} = E_a / Z_1 + Z_2 + Z_0$$

Hence E_a is phase value of induced enf. Let it be 1 $P.u$ (*i.e.* $E_a = 1\angle 0°$ $P.u$)

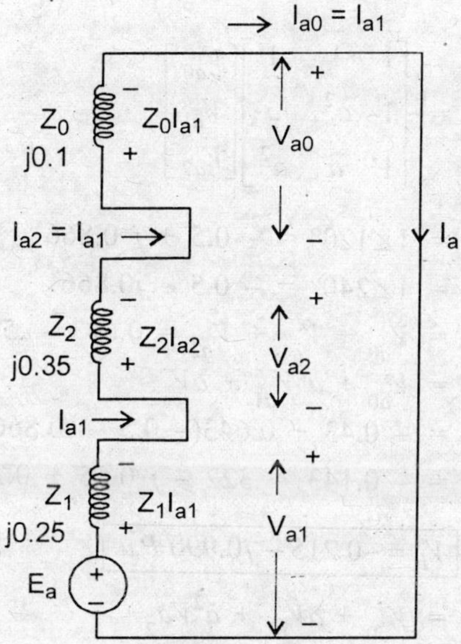

Figure 2.2.1

$$\therefore \quad I_{a1} = \frac{1}{j0.25 + j0.35 + j0.1}$$

$$= \frac{1}{j0.7} = -j1.43 = 1.43\angle -90° \; P.u$$

subtransient fault current in $P.u$, I_f''

$$= |I_a| = 3|I_{a1}| = 3 \times 1.43$$

$$= 4.29 \; P.u$$

Actual value of subtransient fault current = (P.U value of fault) × Base current

$$= 4.29 \times 836.7 = 3589.4 \; A$$

$$= 3.5894 \; KA$$

From the sequence network of the generator the symmetrical components of phase-a voltage are calculated as shown below.

$$V_{a1} = E_a - I_{a1}Z_1 = 1.0 - (-j\ 1.43)\ (j\ 0.25) = 1.0 - 0.357$$

$$V_{a1} = 0.643\ P.u$$

$$V_{a2} = -Z_{a2}Z_2 = -(-j1.43)\ (j\ 0.35) = -0.50\ P.u$$

$$V_{a0} = -I_{a0}Z_0 = -(-j\ 1.43)\ (j\ 0.1)$$

$$\boxed{V_{a0} = -0.143 P.u}$$

The phase voltages are given as,

$$\begin{bmatrix} V_a \\ V_b \\ V_c \end{bmatrix} = \begin{bmatrix} 1 & 1 & 1 \\ 1 & a^2 & a \\ 1 & a & a^2 \end{bmatrix} \begin{bmatrix} V_{a0} \\ V_{a1} \\ V_{a2} \end{bmatrix}$$

where $a = 1\angle 120° = -0.5 + j\ 0.866$

$$a^2 = 1\angle 240° = -0.5 - j0.866$$

$$V_a = V_{a0} + V_{a1} + V_{a2} = 0.643 - .50 - 0.143 = 0$$

$$V_b = V_{a0} + a^2V_{a1} + aV_{a2}$$

$$= -0.43 + 0.643(-0.5 - j0.866) - 0.50(-0.5 + j0.66)$$

$$= -0.143 - .322 - j\ 0.55 + 0.25 - j0.433$$

$$\boxed{V_b = -0.215 - j0.990\ P.u}$$

$$V_c = V_{a0} + aV_{a1} + a^2Va_2$$

$$= -0.143 + .643(-0.5 + j\ 0.866) - 0.5(0.5 - j0.866)$$

$$= -0.143 - 0.322 + j0.557 + 0.25 + j\ 0.433$$

$$\boxed{V_c = -0.215 + j0.99 P.u}$$

Line to line voltage are

$$V_{ab} = V_a - V_b = 0.215 + j0.990 = 1.01\angle 77.7°\ P.u$$

$$V_{bc} = V_b - V_c = -0.215 - j0.99 + 0.215 - j\ 0.99 = -j1.980$$

$$= 1.980\angle\ 270°\ P.u$$

$$V_{ca} = V_c - V_a = -0.215 + j\ 0.99 = 1.01\ \angle 102.3°\ P.u$$

Since the generated voltage to neutral E_a was taken as 1.0 $P.u$, the above line to line voltage are expressed in $P.u$ on the base of Phase voltage. The post fault line voltage expressed in volts are obtained by multiplying the P.u value with base value of phase voltage.

Line value of base voltage = 13.8 KV

Phase value of base voltage $= \dfrac{13.8}{\sqrt{3}} = 7.97\ KV.$

$$V_{ab} = 1.01 \times 7.97 \angle 77.7° = 8.05 \angle 77.7° \ KV$$

$$V_{bc} = 1.98 \times 7.97 \angle 270° = 15.78 \angle 270° \ KV$$

$$V_{ca} = 1.01 \times 7.97 \angle 102.3° = 8.05 \angle 102.3° \ KV$$

Result:

(*i*) subtransiout current = 3.58994 *KA*

(*ii*) $V_{ab} = 8.05 \angle 77.7° \ KV$

$V_{bc} = 15.78 \angle 270° \ KV$

$V_{ca} = 8.05 \angle 102.3° \ KV$

Example 2.3: Find the subtransient currents and line to line voltage at the fault under subtransient conditions when line to line fault between phase b and c occurs at the terminals of the generator described in example (2.2). Assume that the generators unloaded and operating at rated terminal voltage when the fault occurs. Neglect resistance.

Solution: Let us choose the generator ratings as base values.

$$\therefore \quad MVA_b = 20 \ MVA;$$

$$KV_b = 13.8 \ KV$$

Base current, $$I_b = \frac{KVA_b}{\sqrt{3}KV_b} = \frac{200 \times 1000}{\sqrt{3} \times 13.8}$$

$$= 836.7A \approx 837 \ A$$

Given that, $Z_1 = j\,0.25$ P.u, $Z_2 = j\,0.35\ P.u$ and $Z_0 = j\,0.1\ P.u$

For a line to line fault the positive and Negative Sequence network are connected in parallel as shown in Fig. (2.3.1)

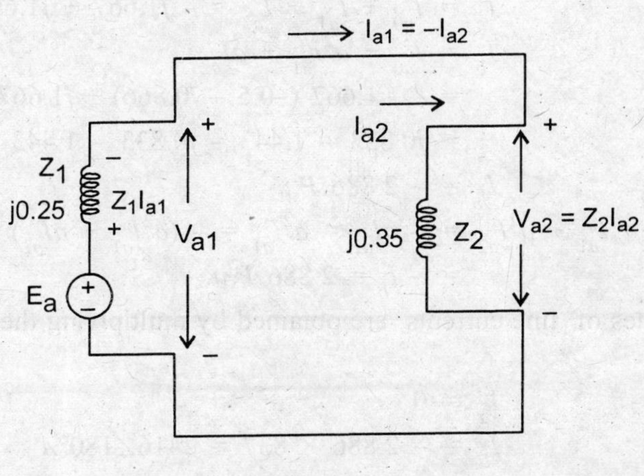

Figure 2.3.1

The following relations exists for a line fault,

$$I_a = 0$$
$$I_b = -I_c$$
$$I_{a0} = 0$$
$$I_{a2} = -I_{a1}$$
$$V_b = V_c$$
$$V_{bc} = 0$$

From Fig (2.3.1) $\quad I_{a1} = \dfrac{E_a}{Z_1 + Z_2}$

Here E_a is phase value of induced emf.

Let it be 1 *p.u* (i.e., $E_a = 1\angle 0°\ P.u$)

$$\therefore \quad I_{a1} = \frac{1}{j0.25 + j0.35} = \frac{1}{j0.6} = -j1.667$$

$$I_{a1} = 1.667\angle -90°\ P.u$$
$$I_{a2} = -Ia_1 = j1.667\ P.u$$
$$I_{a0} = 0$$

The line currents are related to symmetrical components by the following equation.

$$\begin{bmatrix} I_a \\ I_b \\ I_c \end{bmatrix} = \begin{bmatrix} 1 & 1 & 1 \\ 1 & a^2 & a \\ 1 & a & a^2 \end{bmatrix} \begin{bmatrix} I_{a0} \\ I_{a1} \\ I_{a2} \end{bmatrix}$$

where $\quad a = 1\angle 120° = -0.5 + j0.866$

and $\quad a^2 = 1\angle 240° = -0.5 - j0.866$

$$I_a = I_{a0} + I_{a1} + I_{a2} = -j1.667 + j1.667 = 0$$
$$I_b = I_{a0} + a^2 I_{a1} + aI_{a2}$$
$$= -j\,1.667\,(-0.5 - j0.866) + j1.667(-0.5 + j0.866)$$
$$= j0.833 - 1.443 - j0.833 - 1.443$$
$$I_b = -2.886\ P.u$$
$$I_c = I_{a0} + I_{a1} + a^2 I_{a2} = -aI_{a2} - a^2 I_{a1} = -(a^2 I_{a1} + aI_{a2})$$
$$= -I_b = 2.886\ P.u$$

Actual values of line currents are obtained by multiplying the *P.u* values with base current.

$$I_a = 0$$
$$I_b = -2.886 \times 837 = 2416\angle 180° A$$
$$I_c = 2.8865 \times 837 = 2416\angle 0° A$$

The fault current, $I_f = |I_b| = 2416A = 2.416\ KA$.

From the sequence networks of the generator the symmetrical components of phase-a voltage are calculated as shown below.

$$V_{a0} = 0$$
$$V_{a1} = E_a - I_{a1}Z_1;$$
$$V_{a2} = V_{a1}$$

$$\therefore \quad V_{a1} = V_{a2} = 1 - (-j\ 10667)\ (j\ 0.25) = (-\ 0.417) = 0.583\ \ p.u$$

The phase Voltage are given by,

$$\begin{bmatrix} V_a \\ V_b \\ V_c \end{bmatrix} = \begin{bmatrix} 1 & 1 & 1 \\ 1 & a^2 & a \\ 1 & a & a^2 \end{bmatrix} \begin{bmatrix} V_{a0} \\ V_{a1} \\ V_{a2} \end{bmatrix}$$

$$V_a = V_{a0} + V_{a1} + V_{a2} = 0.583 + 0.583 = 1.116\angle 0°\ p.u$$
$$V_b = V_{a0} + a^2V_{a1} + aV_{a2}$$
$$V_b = 0.583(-\ 0.5 - j0.866) + 0.583(-\ 0.5 + j0.8.66)$$
$$= -\ 0.583\ P.u$$
$$V_c = V_b = -\ 0.583\ P.u$$

Line voltage are

$$V_{ab} = V_a - V_b = 1.116 + 0.583 = 1.749\angle 0°\ P.u$$
$$V_{bc} = V_b - V_c = -\ 0.583 + 0.583 = 0\ P.u$$
$$V_{ca} = V_c - V_a = -\ 0.583 - 1.166 = 1.749\angle 180°\ P.u$$

Since the generated phase voltage E_a is taken as $P.u$, the above line voltage are expressed in $P.u.$ on the base of phase voltage.

The post fault line voltage expressed in volts are obtained by multiplying the $p.u$ value with the base value of phase voltage.

Line value of base voltage = $13.8\ KV$

Phase value of base Voltage = $\dfrac{13.8}{\sqrt{3}} = 7.9\ kV.$

$$V_{ab} = 1.749\angle 0° \times 7.97 = 13.94\angle 0° KV$$
$$V_{bc} = 0KV$$
$$V_{ca} = 1.749\angle 180° \times 7.97 = 13.94\ \angle 180° KV.$$

Result:

(*i*) $I_f = 2.416\ KA$

(*ii*) $V_{ab} = 13.94\angle 0°\ KV$

$V_{bc} = 0KV$

$V_{ca} = 13.94\angle 180°\ KV$

Example 2.4: Find the subtransient currents and the line to line voltages at the fault under suberansiant conditions when a double line to ground fault involving phase b and c occurs at the terminals of the generator described in example (2.2). Assume that the generator is unloaded and operating at rated voltage when the fault occurs. Neglect resistance.

Solution: Let us choose the generator rating as base value,

$$\therefore \quad MVA_b = 20\ MVA$$

$$KV_b = 13.8\ KV$$

Base current, $I_b = \dfrac{KVA_b}{\sqrt{3}KV_b} = \dfrac{20\times10^3}{\sqrt{3}\times13.8} = 837A$

Given that, $Z_1 = j0.25\ P.u$

$Z_2 = j0.35\ P.u$

and $Z_0 = j0.1\ p.u$

For a double line to ground fault the positive, negative and zero sequence networks are connected in parallel as shown in Fig. (2.4.1)

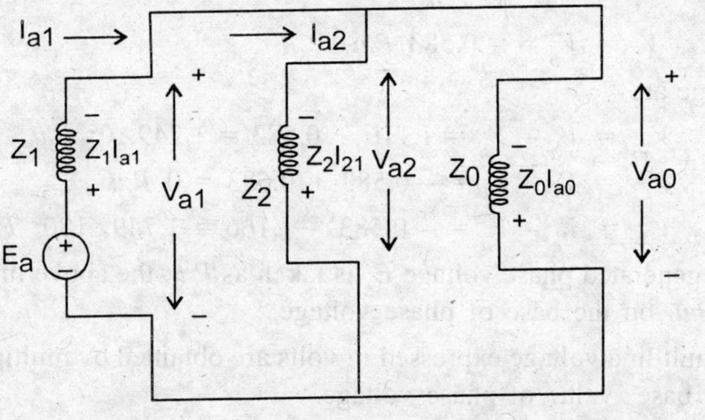

Figure 2.4.1

The following relations exists for a double line to ground fault.

$$I_{a1} = -(I_{a2} + I_{a0})$$

$$V_{a0} = V_{a1} = V_{a2}$$

$$V_b = V_c = 0$$

From fig (2.4.1) we can calculate the symmetrical components of phase-*a* voltage current as shown below.

$$I_{a1} = \frac{E_a}{Z_1 + \dfrac{Z_2 Z_0}{Z_2 + Z_0}} = \frac{1 + j_0}{j0.25 + \dfrac{(j0.35\times j0.10)}{(j0.35 + j0.1)}}$$

$$= \frac{1.0}{j0.25 + j0.0778} = \frac{1.0}{j0.3278} = -j3.05 \ P.u$$

$$V_{a1} = V_{a0} = V_{a2} = E_a - I_{a1}Z_1$$

$$= 1 - (-j3.05)\ (j0.25)$$

$$= 1.0 - 0.763 = 0.237 \ P.u$$

$$I_{a2} = \frac{-V_{a2}}{Z_2} = -\frac{0.237}{j0.35} = j0.68 \ P.u$$

$$I_{a0} = -\frac{V_{a0}}{Z_0} = \frac{-0.237}{j0.10} = j2.37 \ P.u$$

The line currents are related to symmetrical components by the following equation.

$$\begin{bmatrix} I_a \\ I_b \\ I_c \end{bmatrix} = \begin{bmatrix} 1 & 1 & 1 \\ 1 & a^2 & a \\ 1 & a & a^2 \end{bmatrix} \begin{bmatrix} I_{a0} \\ I_{a1} \\ I_{a2} \end{bmatrix}$$

$$I_a = I_{a0} + I_{a1} + I_{a2}$$

$$= j2.37 - j3.05 + j0.68 = 0$$

$$I_b = I_{a0} + a^2 I_{a1} + aI_{a2}$$

$$= j2.37 + (-0.5 - j0.8.66)\ (-j3.05) + (-0.5 + j0.866)\ (j0.68)$$

$$= j1.525 - 2.641 - j0.34 - 0.589 + j2.37.$$

$$= -3.230 + j3.555$$

$$I_b = 4.80\angle 132.3° \ P.u$$

$$I_c = I_{a0} + aI_{a1} + a^2 I_{a2}$$

$$= j2.37 + (-0.5 + j0.866)\ (-j\ 3.05) + (-0.5 - j0.866)\ (j0.68)$$

$$= j1.525 + 2.641 - j0.34 + 0.589 + j2.37$$

$$= 3.230 + j3.555$$

$$I_c = 4.80\angle 47.7° \ P.u$$

Fault current, $I_f = I_b + I_c = -3.230 + j3.555 + 3.230 + j3.555$

$$I_f = j7.11 \ P.u$$

The phase and line voltages are calculated as

$$V_a = V_{a1} + V_{a2} + V_{a0} = 3V_{a1} = 3 \times 0.237 = 0.711\ P.u$$
$$V_b = V_c = 0$$
$$V_{ab} = V_a - V_b = 0.711\ P.u$$
$$V_{bc} = 0$$
$$V_{ca} = V_c - V_a = -\ 0.711\ P.u$$

The actural ralues of line currents and voltages are calculated as shown belowl

$$I_a = 0$$
$$I_b = 837 \times 4.8\angle 132.3° = 4017\angle 132.3° A$$
$$I_c = 837 \times 4.8\angle 47.7° = 4017\angle 4.7.7° A$$
$$I_f = 837 \times 7.11\angle 90° = 5951\angle 90° A$$
$$V_{ab} = 0.711 \times 7.97 = 5.66\angle 0° KV$$
$$V_{bc} = 0$$
$$V_{ca} = -\ 0.711 \times 7.97 = 5.66\angle 180°\ KV.$$

Example 2.5: A generator of negligible resistance having, 1 *P.u* voltage Behind transient reactance is subjected to different types of faults.

Type of fault	Resulting fault current in *P.u*
3 Phase	3.33
L – L	2.23
L – G	3.01

Calculate the per cent value of 3 sequence reactances.

Solution:

Case (i): 3 Phase Fault:

The circuit model used to calculate the transient fault current for *a* 3-phase fault on a g enerator is shown in Fig. (2.5.1).

Given that, $E'_g = 1\ P.u$

$I' = 3.33\ P.u$

When load current is neglected,

$$E_g'' = E_g' = E_g = E_a$$

$$\therefore \quad X_a' = \frac{|E_a|}{|I'|} = \frac{1}{3.33} = 0.3\ P.u$$

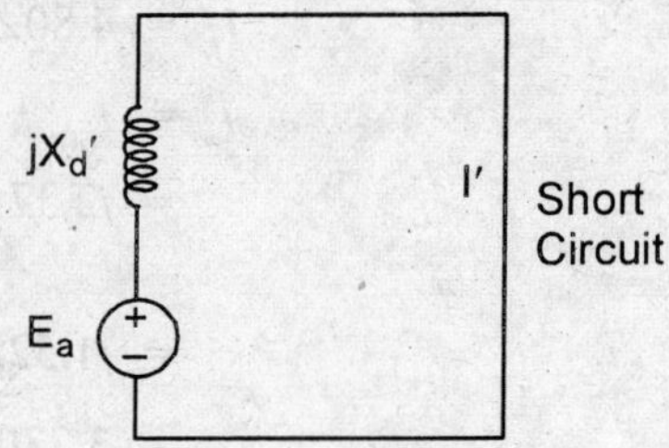

Figure 2.5.1

We know that the reactance during symmetrical fault is positive sequence reactance.

∴ Positive sequence reactance of generator,

$$X_1 = X_d' = 0.3 \ P.u$$

Case (ii): L-L fault:

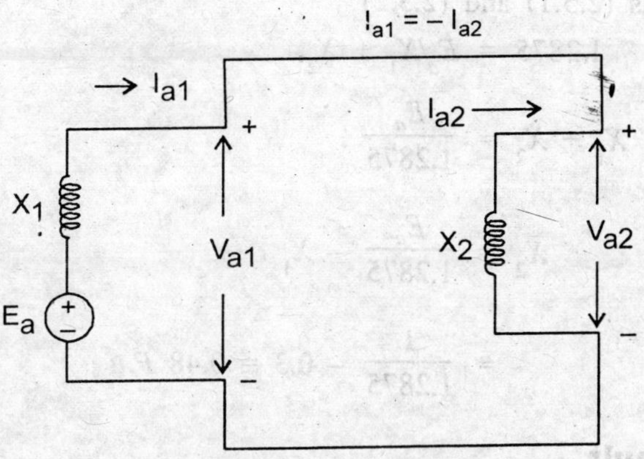

Figure 2.5.2

For a line to line fault on a generator the positive and negative sequence networks can be connected in parallel as shown in Fig. (2.5.2) From Fig. (2.5.2) we get,

$$I_{a1} = Ea/j(X_1 + X_2)$$

$$\therefore \quad |I_{a1}| = \frac{E_a}{X_1 + X_2} \quad \text{...(2.5.1)}$$

Let us assume that the line to line fault is due to short between phase-*b* and phase-*c*. Hence the fault current is I_b.

From the symmetrical compounds of the current we can write,

$$I_b = I_{a0} + a^2 I_{a1} + a I_{a2}$$

For a line to line fault,

$$I_{a0} = 0 \quad \text{and} \quad I_{a2} = -I_{a1}$$

$$\therefore \quad I_b = 0 + a^2 I_{a1} - a I_{a1} = I_{a1}(a^2 - a)$$

$$= I_{a_1}(-0.5 - j0.866 - (-0.5 + j0.866)$$

$$= I_{a_1}(-j1.732) = -j1.733\, I_{a_1}$$

Fault current, $I_f = |I_b| = 1.732\ |I_{a_1}|$

$$\therefore \quad |I_{a_1}| = \frac{I_f}{1.732}$$

Given that $I_f = 2.23$ P.u

$$\therefore \quad |I_{a_1}| = \frac{2.23}{1.732} = 1.2875 \text{ P.u} \qquad ...(2.5.2)$$

From equations (2.5.1) and (2.5.2)

$$1.2875 = E_a/X_1 + X_2$$

$$\therefore \quad X_1 + X_2 = \frac{E_a}{1.2875}$$

$$\therefore \quad X_2 = \frac{E_a}{1.2875} - X_1$$

$$= \frac{1}{1.2875} - 0.3 = 0.48 \ P.u$$

Case (*iii*): *L-G* fault

For a single line to ground fault, on a generator the sequence networks are connected in series as shown in Fig. (2.5.3). For a single line to ground fault, the fault current is $|I_a|$ and it is equal to $3\,|I_{a_1}|$.

Given that, $I_f = 3.01$ P.u

$$\therefore \quad 3\,|I_{a_1}| = I_f = 3.01 \text{ (or) } |I_{a_1}| = \frac{3.01}{3} = 1.0033 \text{ P.u}$$

From Fig. 2.5.3 we can write,

$$I_{a_1} = \frac{E_a}{jX_1 + jX_2 + jX_0}$$

$$\therefore \quad |I_{a_1}| = \frac{E_a}{X_1 + X_2 + X_0}$$

$$X_1 + X_2 + X_0 = \frac{E_a}{|I_{a_1}|}$$

$$X_0 = \frac{E_a}{|I_{a_1}|} - X_1 - X_2$$

$$= \frac{1}{1.0033} - 0.3 - 0.48$$

$$X_0 = 0.22 \text{ P.u}$$

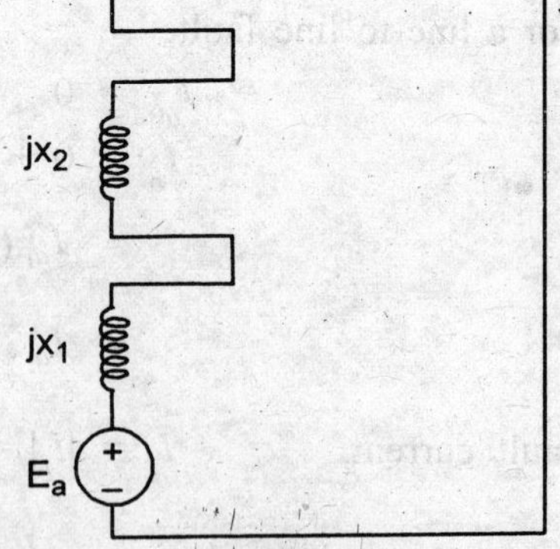

Figure 2.5.3

Result:

+ve sequence reactance = 0.3 P.u

–ve sequence reactance = 0.48 P.u

Zero sequence reactance = 0.22 P.u

EXERCISE

1. A 25 MVA, 11 KV alternator with X_0 = 0.05 P.u, $X_1 = X_2$ = 0.15 P.u is earthed through a reactance of 0.333Ω. Calculate the line current for a single line to ground fault in amperes.
2. A single line to ground fault of 0.05Ω resistance occur in a 3ϕ system supplied by a synchrous generator with a generator emf of 11 KV between lines. The +ve, –ve and zero sequence reactances of the generators and network upto the fault are 0.5Ω, 0.2Ω & 0.1Ω respectively. Find the fault current.
3. Two similar generators each having emfs of 2.2 KV per phase and impedances $Z_1 = 0 + j1$, $Z_2 = 0 + j8$ and $Z_0 = 0 + j0.3\Omega$ per phase are connected in parallel. both generators are solidly grounded. Determine (*i*) total current to ground (*ii*) phase voltage when a double line to ground fault occurs at the terminals.
4. A 20MVA, 11 KV alternator solidly grounded neutral has a subtransient. reactances of 0.25 P.u. the negative and zero sequence reactances are j0.146, j0.66 respectively. A single line to ground fault occurs the terminals of an unloaded generator. Determine the fault currents & line to line voltages. neglect resistances.
5. A synchronous generator is rated 25 *MVA*, 11 *KV*. It is star connected with the neutral points solidly grounded.

   The generator is operating at a rated voltage. It's reactances are $X_1 = X_2$ = 20 and X_0 = 0.08 P.u. calculate the symmetrical subtransient line currents for

   (*i*) *L-G* fault

   (*ii*) *L-L* fault

   (*iii*) *L-L-G* fault

   (*iv*) 3ϕ fault

Short Questions and Answers

Q. 1. Name the fault in which +ve, –ve and zero sequence component currents are equal.

Ans. single line to ground fault.

Q. 2. Name the fault in which +ve and –ve sequence component currents together equal to zero sequence component.

Ans. Double line to ground fault.

Q. 3. Name the various unsymmetrical faults.

Ans. 1. L-G fault

2. L-L fault

3. L-L-G fault

4. Oper conductor fault

Q. 4. Define negative sequence impedance.

Ans. The negative sequence impedance of an equipment is the impedance offered by the equipment to the flow of negative sequence current.

Q. 5. Name the fault which do not have zero sequence current.

Ans. L-L fault.

Q. 6. Name the faults involving ground.

Ans. 1. L-G fault.

2. L-L-G fault.

3. 3ϕ fault

Q. 7. Define +ve sequence impedance.

Ans. The +ve sequence impedance of an equipment is the impedance offered by the equipment to the flow of +ve sequence currents.

Q. 8. In what type of fault the +ve sequence component of current is equal in magnitude but opposite in phase to negative sequence components of current?

Ans. L-L fault.

Q. 9. In which fault the –ve and zero sequence currents are absent?

Ans. 3ϕ fault.

Q. 10. Write the boundary condition in L-G fault.

$$V_a = 0; I_b = I_c = 0$$

Q. 11. Write the boundary condition for L-L fault.

$$I_a = 0; I_b + I_c = 0; V_b = V_c$$

Q. 12. Write the boundary condition for Double line to ground fault and 3ϕ fault..

(*i*) **Double line to ground fault**

$$I_b = 0; V_b = 0; V_c = 0$$

(*ii*) **3ϕ fault**

$$I_a + I_b = I_c = 0; V_a = V_b = V_c = 0$$

2.7 FORMATION OF ZBUS USING SINGULAR TRANSFORMATION

Consider the node basis matrix equation representing the power system,

$$Y_{bus}\ V = I \quad ...(2.99)$$

On premultiplying the equation (2.99) by Y^{-1}_{bus} we get,

$$V = Y^{-1}_{bus}\ I \quad ...(2.100)$$

Now the elements of Y^{-1}_{bus} will be impedances and so the matrix Y^{-1}_{bus} **can be** represented by a impedance matrix called bus impedance matrix, Z_{bus}.

$$Z_{bus} = Y^{-1}_{bus} \quad ...(2.101)$$

and

$$V = Z_{bus}\ I \quad ...(2.102)$$

Since the bus admittance matrix is symmetrical, the bus impedance matrix is also symmetrical around the principal diagonal. In bus impedance matrix, the elements on the main diagonal are called driving point impedance of the buses or nodes and the off-diagonal elements are called the transfer impedances of the buses or nodes. The bus impedance matrix is very useful in fault analysis.

The bus impedance matrix can be determined by two methods. In one method we can form the bus admittance and then taking its inverse to get the bus impedance matrix. In another method the bus impedance matrix can be directly formed from the reactance diagram and this method requires the knowledge of the modifications of existing bus impedances matrix due to addition of new bus or addition of a new line between existing buses.

Modification of an existing bus impedance matrix

Let us denote the original Z_{bus} of a system with n-number of independent buses as Z_{orig}. When a branch of impedance Z_b is added to the system the Z_{orig} gets modified. The branch impedance Z_b can be added to the original system in the following four different ways.

Case 1: Adding a branch of impedance Z_b from a new bus-*p* to the reference bus.

Case 2: Adding a branch of impedance Z_b from a new bus-*p* to an existing bus-*q*.

Case 3: Adding a branch of impedance Z_b from an existing bus-*q* to the reference bus.

Case 4: Adding a branch of impedance Z_b between two existing buses *h* and *q*

The modification of Z_{oring} for the above 4 cases have been discussed here,

Case 1: Adding Z_b from a new bus-*p* to the reference bus.

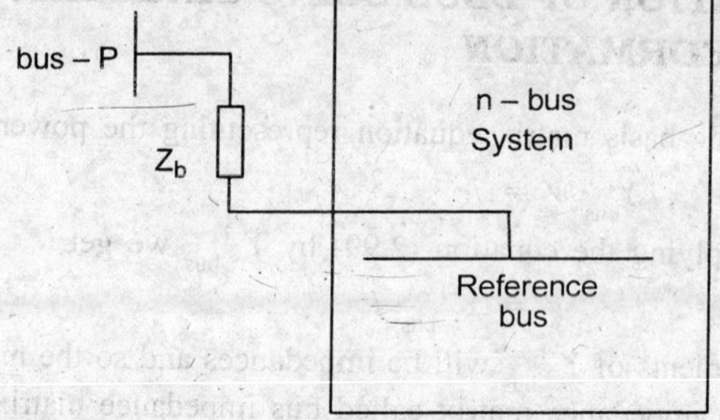

Figure 2.25 *Adding a new bus through an impedance to reference bus*

Consider a *n*-bus system as shown in Fig. 2.25. Let us add a bus-*p* through an impedance Z_b to the reference bus. the addition of a bus will increase the order of the bus impedance matrix by one.

In this case the dements of $(n + 1)^{th}$ column and row are all zeros except the diagonal.

The diagonal element is the added branch impedance Z_b. The elements of original Z_{bus} matrix are not altered. The new bus impedance matrix will be as shown in equation (2.103).

$$Z_{\text{bus, new}} = \left[\begin{array}{ccc|c} & & & 0 \\ & & & 0 \\ & Z_{\text{orig}} & & \vdots \\ \hline 0 & 0 & \cdots & Z_b \end{array}\right] \qquad \text{...(2.103)}$$

Case 2: Adding Z_b from a new bus-*p* to existing bus-q

Consider a *n*-bus system as shown in Fig. 2.26. in which a new bus-*p* is added through an impedance Z_b to an existing bus-*q*. The addition of a bus will increase the order of the bus impedance matrix by one.

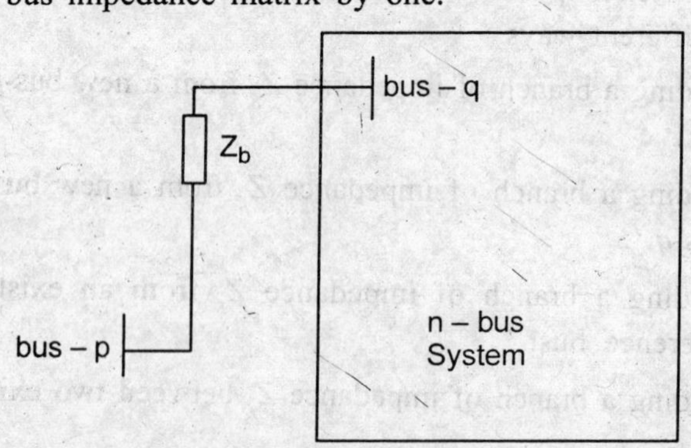

Figure 2.26 *Adding a new bus through an impedance to an existing bus*

In this the elements of $(n + 1)^{th}$ column are the elements of qth column and elements of $(n + 1)^{th}$ row are the elements q^{th} now. The diagoral element is givenby sum of Z_{qq} and Z_b. The elements of original Z_{bus} matrix are not altered. The new bus impedance matrix will be as shown in equation (2.104),

$$Z_{\text{bus, new}} = \left[\begin{array}{ccc|c} & & & Z_{1q} \\ & & & Z_{2q} \\ & Z_{orig} & & \vdots \\ & & & Z_{nq} \\ \hline Z_{q1} & Z_{q2} & \cdots \; Z_{qn} & Z_{qq} + Z_b \end{array}\right] \quad ...(2.104)$$

Case 3: Adding Z_b from an existing bus-q to the reactance bus:

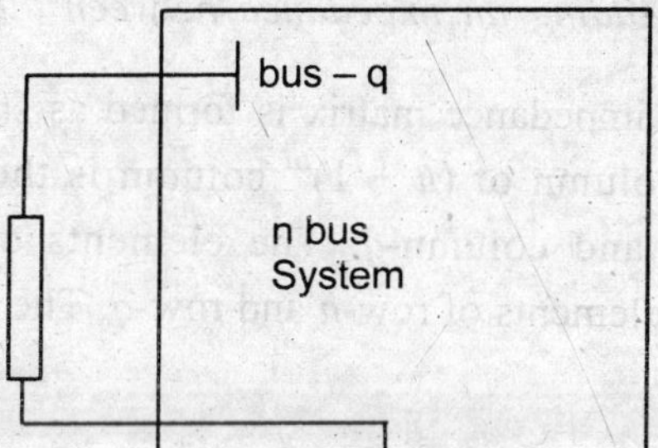

Fig. 2.27 *Adding an impedance between existing bus and reference bus*

Consider a n-bus system shown in Fig. 2.27 in which an impedance Z_b is added from an existing bus-q to the reference bus. Let us consider as if the impedance Z_b is connected from a new bus-p and existing bus-q. Now it will be an addition as that of case-2. The new bus impedance matrix order $(n + 1)$ can be formed as that of case-2. Then we can short circuit the bus-q to reference bus. This equivalent to eliminating $(n + 1)^{\text{th}}$ row and $(n + 1)^{\text{th}}$ column. This reduced bus impedance matrix is the actual new bus impedance matrix. Every element of actual new bus impedance matrix can be determined using the equation (2.105)

$$Z_{jk,act} = Z_{jk} - \frac{Z_{j(n+1)} Z_{(n+1)k}}{Z_{(n+1)(n+1)}} \quad ...(2.105)$$

Note:

1. $Z_{\text{jk,act}}$ is the impedance corresponding to row-j and column k of actual new bus impedance matrix.
2. Z_{jk}, $Z_{\text{(n+1)k}}$, $Z_{\text{j(n+1)}}$, $Z_{\text{(n+1)(n+1)}}$ are impedances of new bus impedance matrix of order $(n + 1)$.
3. Since bus impedance matrix is symmetrical

$$Z_{\text{jk,act}} = Z_{\text{kj,act}}$$

Case 4: Adding Z_b between two existing buses *h* and *q*.

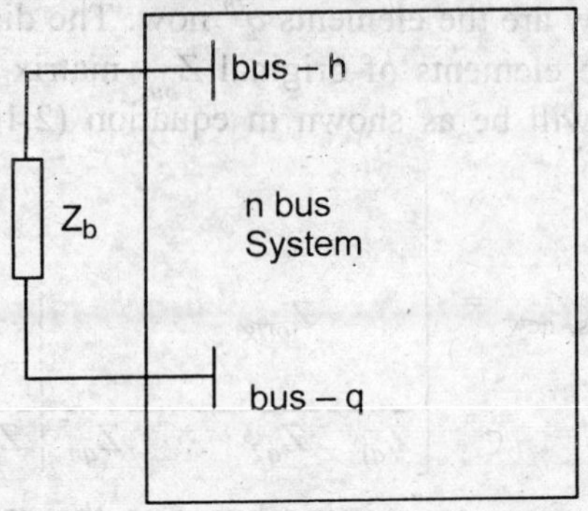

Figure 2.28 *Adding an impedance between bus-h and bus-q.*

In this case the bus impedance matrix is formed as shown in equation (2.106). Here the elements of column of $(n + 1)^{th}$ column is the difference between the elements of column-h and column-*q*. The elements of $(n + 1)$th row is two difference between the elements of row-*h* and row-*q*. The diagonal element is given by equation (2.107).

$$Z_{bus,\,new} = \left[\begin{array}{cccc|c} & & & & Z_{1h} - Z_{1q} \\ & Z_{orig} & & & Z_{2h} - Z_{2q} \\ & & & & \vdots \\ & & & & Z_{nh} - Z_{nq} \\ \hline Z_{h1-q1} & Z_{h2-q2} & \cdots & Z_{hn} - Z_{qn} & Z_{(n+1)(n+1)} \end{array}\right] \quad ...(2.106)$$

$$Z_{(n+1)(n+1)} = Z_b' + Z_{hh} + Z_{qq} - 2\,Z_{hq} \quad ...(2.107)$$

Since the modification does not involve addition of new bus, the order of new bus impedance matrix has to be reduced to n×n by eliminating the $(n + 1)^{th}$ column and $(n + 1)^{th}$ now.

This reduced bus impedance matrix is the actual new bus impedance matrix. Every element of this actual new bus impedance matrix can be determined using equation (2.105).

Direct determination of a bus impedance matrix

The bus impedance matrix can be directly obtained from impedance or reactance diagram instead of forming Y_{bus} and then inverting it. In direct determination of Z_{bus} first consider an impedance Z_a connected form bus-1 to reference bus. Now the bus impedance matrix will have a single element as shown below.

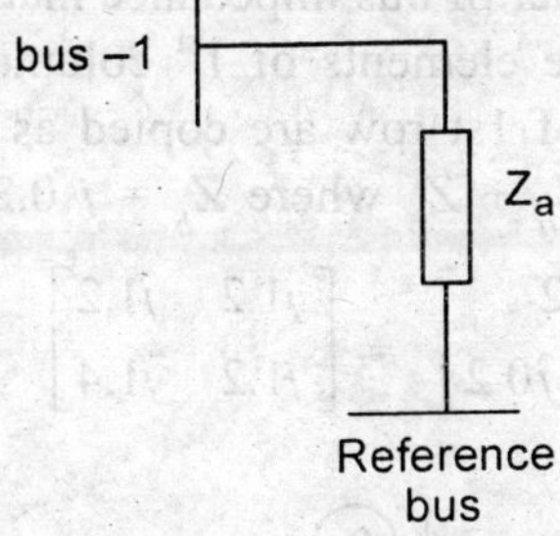

$Z_{bus} = [Z_a]$

Then add each element of impedance or reacctance diagram one by one modify the Z_{bus} in each step. Each modification of Z_{bus} involve any one of the four cases discussed above.

Example 2.6: Defermine Z_{bus} for system whose reactance diagram is shown in Fig. 2.6.1. Where the impedance is given in p.u. preserve all the three nodes.

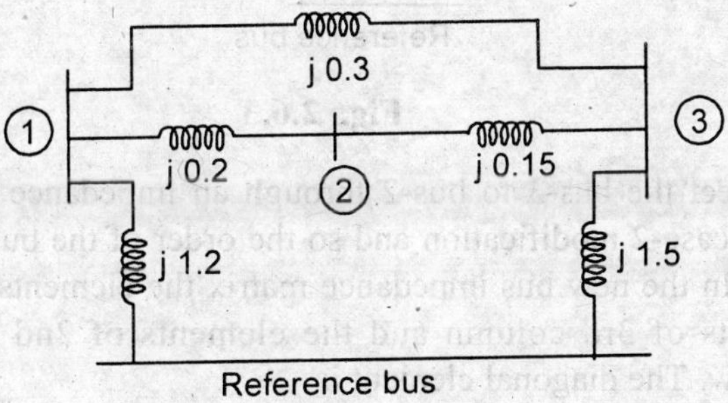

Figure 2.6.1

Note: Preserve all the three nodes means the final matrix should have 3 × 3 matrix.

Solution:

Step 1: Consider the branch with impedance j 1.2 p.u. connected between bus-1 and reference bus as shown in Fig. 2.6.2. The system shown in Fig. 2.6.2 has a single bus and so the order of bus impedance matrix is one, as shown below,

$Z_{bus} = [j1.2]$

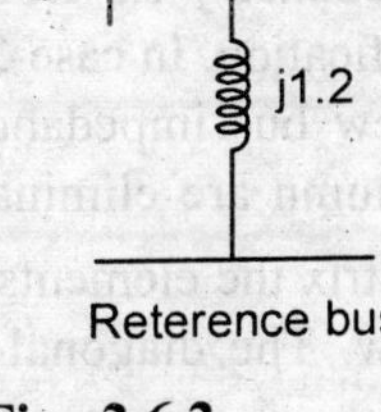

Fig. 2.6.2

Step 2: Connect bus-2 to bus-1 through an impedance $j\,0.2$ as shown in Fig. 2.6.3. This is case-2 modification and so the order of bus impedance matrix increases by one. In the new bus impedance matrix, the elements of 1st column are copied as elements of 2nd column and the elements of 1st row are copied as elements of 2nd row. The diagonal element is given by $Z_a + Z_b$ where $Z_b = j\,0.2$.

$$\therefore \quad Z_{bus} = \begin{bmatrix} j1.2 & j1.2 \\ j1.2 & j1.2 + j0.2 \end{bmatrix} = \begin{bmatrix} j1.2 & j1.2 \\ j1.2 & j1.4 \end{bmatrix}$$

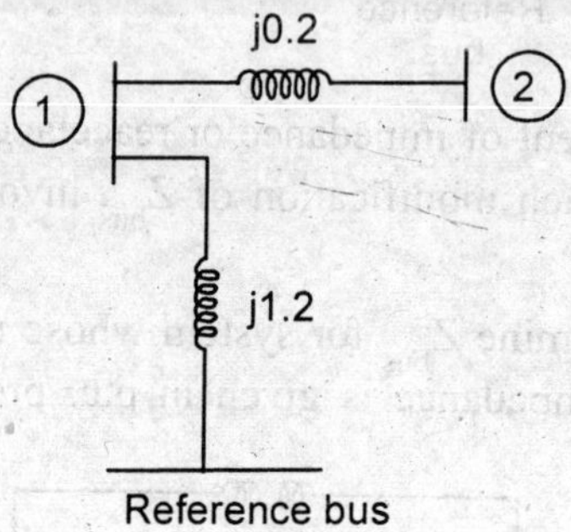

Fig. 2.6.3

Step 3: Connect the bus-3 to bus-2 through an impedance j 0.15 as shown in Fig. 2.6.4. This is case-2 modification and so the order of the bus impedance matrix increased by one. In the new bus impedance matrix the elements of 2nd column are copied as elements of 3rd column and the elements of 2nd row are copied as elements of 3rd row. The diagonal element is given by $Z_{22} + Z_b$ where $Z_b = j0.15$

$$\therefore \quad Z_{bus} = \begin{bmatrix} j1.2 & j1.2 & j1.2 \\ j1.2 & j1.4 & j1.4 \\ j1.2 & j1.4 & j1.4 + j0.15 \end{bmatrix}$$

$$= \begin{bmatrix} j1.2 & j1.2 & j1.2 \\ j1.2 & j1.4 & j1.4 \\ j1.2 & j1.4 & j1.55 \end{bmatrix}$$

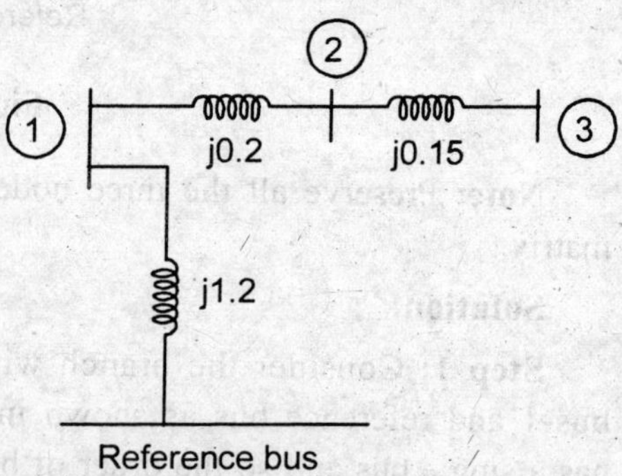

Figure 2.6.4

Step 4: Connect the impedance j 1.5 from bus-3 to reference bus as shown in Fig. 2.6.5. This is case-3 modification. In case-3 modification the new bus impedance. In case-3 modification the new bus impedance matrix is formed as that of case-2 and then the last row and column are eliminated by node elimination techniques.

In new bus impedance matrix the elements of 3rd column are copied as elements of 4th column and of 4th row. The diagonal element is given by $Z_{33} + Z_b$ where $Z_b = j1.5$.

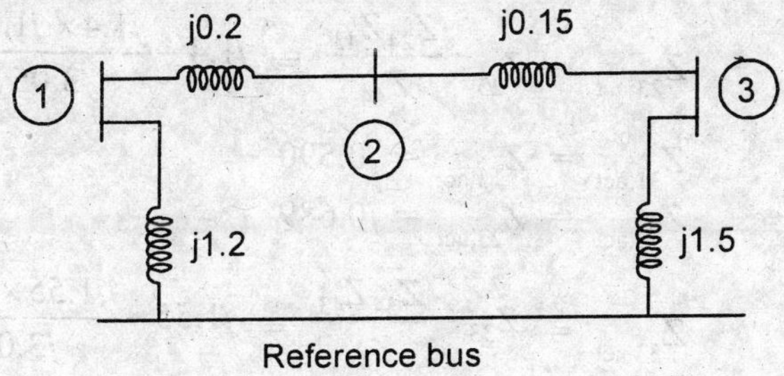

Fig. 2.6.5

$$\therefore \quad Z_{bus} = \begin{bmatrix} j1.2 & j1.2 & j1.2 & j1.2 \\ j1.2 & j1.4 & j1.4 & j1.4 \\ j1.2 & j1.4 & j1.55 & j1.55 \\ j1.2 & j1.4 & j1.55 & j1.55+j1.5 \end{bmatrix}$$

$$= \begin{bmatrix} j1.2 & j1.2 & j1.2 & j1.2 \\ j1.2 & j1.4 & j1.4 & j1.4 \\ j1.2 & j1.4 & j1.55 & j1.55 \\ j1.2 & j1.4 & j1.55 & j3.05 \end{bmatrix}$$

The actual new bus impedance matrix is obtained by eliminating the 4th row and 4th column. The element Z_{jk} of the actual new bus impedance matrix is given by,

$$\therefore \quad Z_{jk,act} = Z_{jk} - \frac{Z_{j(n+1)}Z_{(n+1)}K}{Z_{(n+1)}Z_{(n+1)}}$$

where $n = 3$; $j = 1, 2, 3$ and $k = 1, 2, 3$

$$Z_{11,act} = Z_{11} - \frac{Z_{14}Z_{41}}{Z_{44}} = j\ 1.2 - \frac{j1.2 \times j1.2}{j3.05} = j0.728$$

$$Z_{12,act} = Z_{12} - \frac{Z_{14}Z_{42}}{Z_{44}} = j1.2 - \frac{j1.2 \times j1.4}{j3.05} = j\ 0.649$$

$$Z_{13,act} = Z_{13} - \frac{Z_{14}Z_{43}}{Z_{44}} = j1.2 - \frac{j1.2 \times j1.55}{j3.05} = j\ 0.590$$

$$Z_{21,act} = Z_{12,act} = j\ 0.649$$

$$Z_{22,act} = Z_{22} - \frac{Z_{24}Z_{42}}{Z_{44}} = j1.4 - \frac{j1.4 \times j1.4}{j3.05} = j0.757$$

$$Z_{23,act} = Z_{23} - \frac{Z_{24}Z_{43}}{Z_{44}} = j1.4 - \frac{j1.4 \times j1.55}{j3.05} = j0.689$$

$$Z_{31,act} = Z_{13,act} = j0.590$$

$$Z_{32,act} = Z_{23,act} = j0.689$$

$$Z_{33,act} = Z_{33} - \frac{Z_{34}Z_{43}}{Z_{44}} = j1.55 - \frac{j1.55 \times 1.55}{j3.05} = j0.762$$

$$\therefore \quad Z_{bus} = \begin{bmatrix} j0.728 & j0.649 & j0.590 \\ j0.649 & j0.757 & j0.689 \\ j0.590 & j0.689 & j0.762 \end{bmatrix}$$

Steps: Connect the impedance $j0.3$ between bus-1 and bus-3 as shown in Fig. 2.6.6. This is case-4 modification.

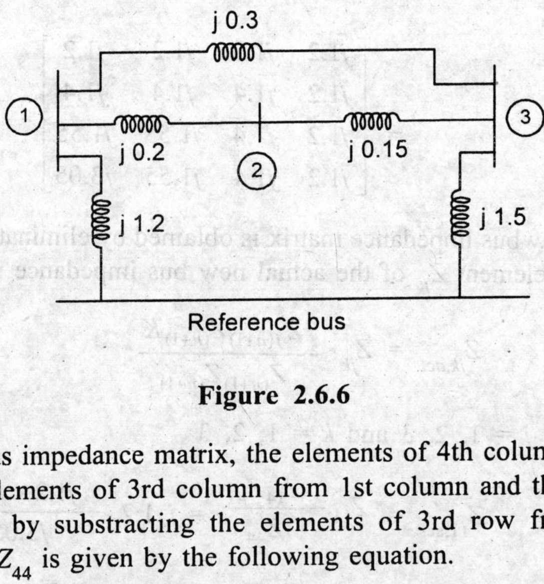

Figure 2.6.6

In the new bus impedance matrix, the elements of 4th column are obtained by substracting the elements of 3rd column from 1st column and the elements of 4th row are obtained by substracting the elements of 3rd row from 1st row. The diagonal element Z_{44} is given by the following equation.

$$Z_{44} = Z_b + Z_{11} + Z_{33} - 2Z_{13}$$

where $Z_b = j0.3$

$$\therefore \quad Z_{44} = j0.3 + j0.728 + j0.762 - 2(j0.59) = j0.61$$

$$Z_{bus} = \begin{bmatrix} j0.728 & j0.649 & j0.590 & j0.728 - j0.59 \\ j0.649 & j0.757 & j0.689 & j0.649 - j0.689 \\ j0.59 & j0.689 & 0.762 & j0.59 - j0.762 \\ j0.728 - j0.59 & j0.649 - j0.689 & j0.59 - j0.762 & j0.61 \end{bmatrix}$$

$$Z_{bus} = \begin{bmatrix} j0.728 & j0.649 & j0.590 & j0.138 \\ j0.649 & j0.757 & j0.689 & -0.04 \\ j0.59 & j0.689 & j0.762 & -j0.172 \\ j0.138 & -0.04 & -j0.172 & j0.61 \end{bmatrix}$$

Since this modification does not add a new bus, the 4th row and column has to be eliminated using node elimination technique, to determine the actual new bus impedance matrix.

The element Z_{jk} of actual new bus impedance matrix is given by,

$$\therefore \quad Z_{jk,act} = Z_{jk} - \frac{Z_{j(n+1)}Z_{(n+1)k}}{Z_{(n+1)}Z_{(n+1)}}$$

where $n = 3,\ j = 1, 2, 3$ and $k = 1, 2, 3$

$$Z_{11,act} = Z_{11} - \frac{Z_{14}Z_{41}}{Z_{44}}$$

$$= j0.728 - \frac{j0.138 \times j0.138}{j0.61} = j0.697$$

$$Z_{12,act} = Z_{12} - \frac{Z_{14}Z_{42}}{Z_{44}}$$

$$= j0.649 - \frac{j0.138 \times (-j0.04)}{j0.61} = j0.658$$

$$Z_{21,act} = Z_{12,act} = j0.658$$

$$Z_{22,act} = Z_{22} - \frac{Z_{24}Z_{42}}{Z_{44}}$$

$$= j0.757 - \frac{(-j0.04)(-j0.04)}{j0.61} = j0.754$$

$$Z_{23,act} = Z_{23} - \frac{Z_{24}Z_{43}}{Z_{44}}$$

$$= j0.689 - \frac{(-j0.04)(-j0.172)}{j0.61} = j0.678$$

$$Z_{31,act} = Z_{13,act} = j0.629$$

$$Z_{32,act} = Z_{23,act} = j0.678$$

$$Z_{33,act} = Z_{33} - \frac{Z_{34}Z_{43}}{Z_{44}} = j0.762 - \frac{(-j0.172)(-j0.172)}{j0.61}$$

$$Z_{33,act} = j0.714$$

$$\therefore \quad Z_{bus} = \begin{bmatrix} j0.697 & j0.658 & j0.629 \\ j0.658 & j0.754 & j0.678 \\ j0.629 & j0.678 & j0.714 \end{bmatrix}$$

Example 2.7: Find the bus impedance matrix for the system whose reactance diagram is shown in Fig. 2.7.1. All the impedances are in p.u.

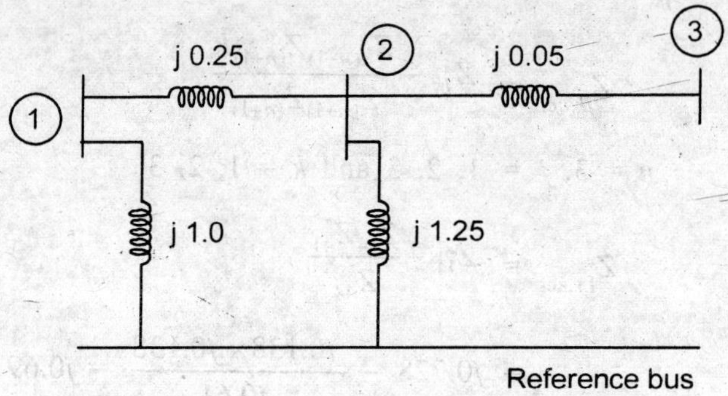

Fig. 2.7.1

Solution:

Step 1: Consider the branch with impedance $j1$ P.u connected between bus-1 and reference as shown in Fig. 2.7.2. The system shown in Fig. 2.7.2 has single bus and so the order of bus impedance matrix is one, as shown below.

$$Z_{bus} = [j1.0]$$

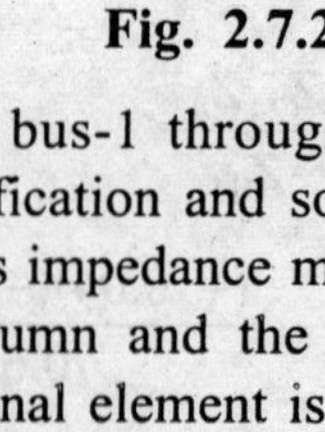

Fig. 2.7.2

Step 2: Connect bus-2 to bus-1 through an impedance $j0.25$ as shown in Fig. 2.7.3. This is case-2 modification and so the order of bus impedance matrix increases by one. In the new bus impedance matrix, the elements of 1st column are copied as elements of 2nd column and the elements of 1st row are copied as elements of 2nd row. The diagonal element is given by $Z_{11} + Z_b$ where $Z_b = j0.25$

$$\therefore \quad Z_{bus} = \begin{bmatrix} j1.0 & j1.0 \\ j1.0 & j1.0 + j0.25 \end{bmatrix} = \begin{bmatrix} j1.0 & j1.0 \\ j1.0 & j1.25 \end{bmatrix}$$

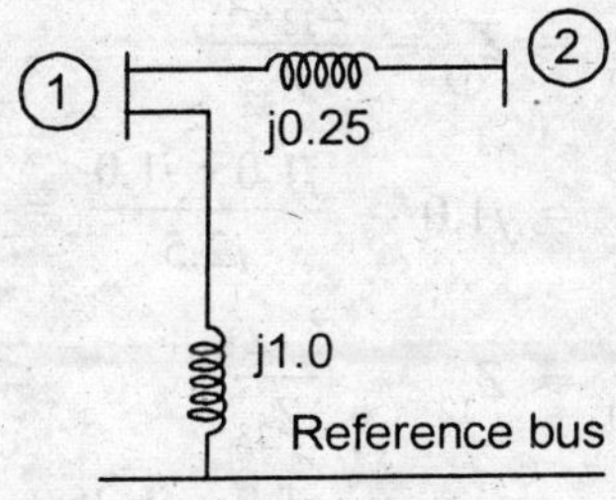

Fig. 2.7.3

Step 3: Connect the impedancce $j1.25$

From bus-2 to reference bus as shown in Fig. 2.7.4. This is case-3 modification the new bus impedance matrix is formed as that of case-2 and then the last row and column are eliminated by node elimination technique.

In the new bus impedance matrix the elements of 2nd column are copied as elements of 3rd column and the elements of 2nd row are copied as elements of 3rd row. The diagonal element is given by $Z_{22} + Z_b$ where $Z_b = j1.25$

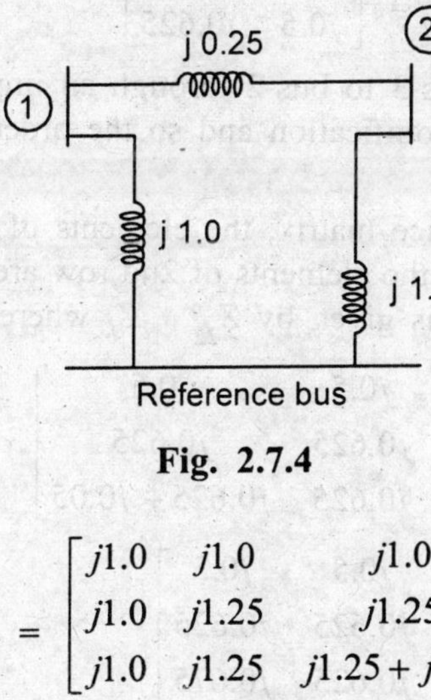

Fig. 2.7.4

$$\therefore \qquad Z_{bus} = \begin{bmatrix} j1.0 & j1.0 & j1.0 \\ j1.0 & j1.25 & j1.25 \\ j1.0 & j1.25 & j1.25 + j1.25 \end{bmatrix}$$

$$Z_{bus} = \begin{bmatrix} j1.0 & j1.0 & j1.0 \\ j1.0 & j1.25 & j1.25 \\ j1.0 & j1.25 & j2.5 \end{bmatrix}$$

The actual new bus impedance matrix is obtained by eliminating the 3rd row and 3rd column. The element Z_{j12} of the actual new bus impedance matrix is given by,

$$\therefore \qquad Z_{jk,act} = Z_{jk} - \frac{Z_{j(n+1)(n+1)}k}{Z_{(n+1)(n+1)}}$$

where $\quad n = 2; j = 1, 2$ and $k = 1, 2$

$$Z_{11,act} = Z_{11} - \frac{Z_{13}Z_{31}}{Z_{33}}$$

$$= j1.0 - \frac{j1.0 \times j1.0}{j2.5} = j0.6$$

$$Z_{12,act} = Z_{12} - \frac{Z_{13}Z_{32}}{Z_{33}}$$

$$= j1.0 - \frac{j1.0 \times j1.25}{j2.5} = j0.5$$

$$Z_{22,act} = Z_{12,act} = j0.5$$

$$Z_{22,act} = Z_{22} - \frac{Z_{23}Z_{32}}{Z_{33}}$$

$$= j1.25 - \frac{j1.25 \times j1.25}{j2.5} = j0.626$$

$$\therefore \quad Z_{bus} = \begin{bmatrix} j0.6 & j0.5 \\ j0.5 & j0.625 \end{bmatrix}$$

Step 4: Connect the bus-3 to bus-2 through an impedance $j0.05$ as shown in Fig. 2.7.5. This is case-2 modification and so the order of bus impedance matrix increased by one.

In the new bus impedance matrix, the elements of 2nd column are copied as elements of 3rd column and the elements of 2nd row are copied as elements of 3rd row. The diagonal element is given by $Z_{22} + Z_b$ where $Z_b = j0.05$

$$Z_{bus} = \begin{bmatrix} j0.6 & j0.5 & j0.5 \\ j0.5 & j0.625 & j0.625 \\ j0.5 & j0.625 & j0.625 + j0.05 \end{bmatrix}$$

$$Z_{bus} = \begin{bmatrix} j0.6 & j0.5 & j0.5 \\ j0.5 & j0.625 & j0.625 \\ j0.5 & j0.625 & j0.675 \end{bmatrix}$$

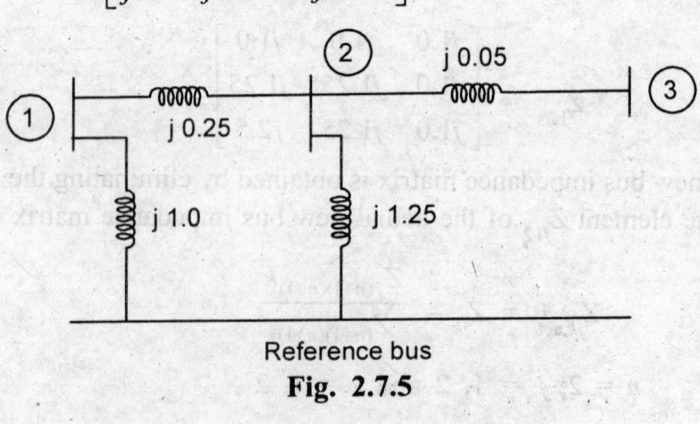

Fig. 2.7.5

EXERCISE

6. For the system shown in Fig. E.2.6.1, determine Z_{bus}

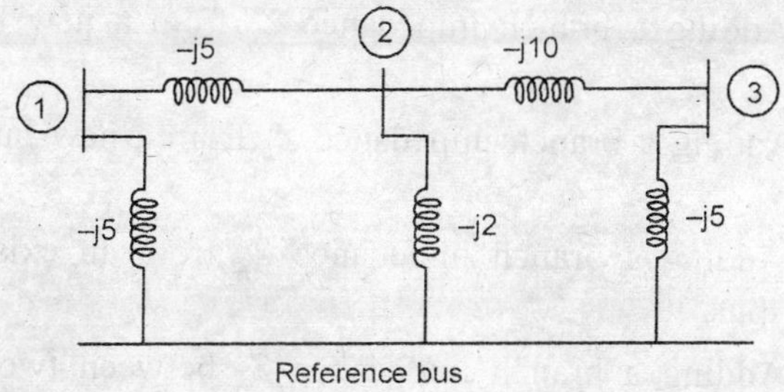

7. Determine Z_{bus} for the Fig. E-2.7.1 where the impedances are given in p.u. preserve all the 3 nodes.

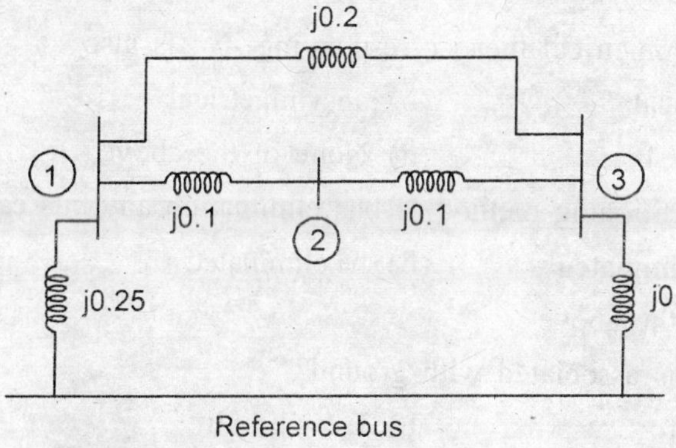

Short Questions & Answers

Q. 13. What is bus impedance matrix?

Ans. The matrix consisting of driving point impedances and transter impedances of the network of a power system is called impedance matrix. It is given by the inverse of bus admittance matrix and it is denoted as Z_{bus}. The bus impedance matrix is symmetrical.

Q. 14. Name the diagonal elements and off diagonal elements of bus impedance matrix.

Ans. The diagonal elements of bus impedance matrix are called driving point impedances of the buses and off diagonal elements of bus impedance matrix are called transter impedances of the buses.

Q. 15. What are the methods available for forming bus impedance matrix?

Method 1: From the bus admittance matrix and then take its inverse to get bus impedance matrix.

Method 2: Directly form the bus impedance matrix from the reactance diagram.

Q. 16. Write the four ways of adding an impedance to an existing system so as to modify bus impedance matrix.

Case 1: Adding a branch impedance Z_b from a new bus-p to reference bus.

Case 2: Adding a branch impedance Z_b from a new bus-p to an existing bus-q.

Case 3: Adding a branch impedance Z_b from an existing bus-q to the reference bus.

Case 4: Adding a branch impedance Z_b between two existing buses h and q.

Choose The Best Answer

1. If Y_{bus} is symmetrical then corresponding Z_{bus} is also

(a) symmetrical (b) ansymmetrical

(c) both a & b (d) None of the above

2. In matrix partitioning method of bus elimination any bus can

(a) not be eliminated (b) be eliminated

(c) none of the above

3. Which fault is associated with ground?

(a) L-G fault (b) L-L-G fault

(c) 3ϕ fault (d) all of these

4. Choose which is not an unsymmetrical fault

(a) L-G fault (b) L-L fault

(c) 3ϕ fault (d) an conductor fault

5. Choose the correct boundary condition for single line to ground fault.

(a) $V_a = 0; I_b = I_c = 0$ (b) $I_a = 0; I_b + I_c = 0; V_b = V_c$

(c) $I_a = 0; V_b = 0; V_c = 0$ (d) $I_a + I_b + I_c = 0; V_a = V_b = V_c = 0.$

6. In which fault +ve, –ve and zero sequence currents are equal?

(a) L-G fault (b) L-L fault

(c) L-L-G fault (d) all of the above

7. Name the falut which has only +ve sequence component alone.

(a) 3ϕ fault (b) L-G fault

(c) L-L fault (d) None of the above

ANSWERS

1. a 2. a 3. d 4. c 5. a 6. a 7. a

Fill in the Blanks

1. ________ fault is most common in a power system.
2. If the currents are ________ then it is known as unsymmetrical fault.
3. The positive sequence voltage is ________ for unloaded generator.
4. Except ________ the positive and negative sequence networks are same.
5. ________ sequence component does not have phase displacement.
6. The diagonal point of Y_{bus} are called ________ and off diagonal elements are called ________
7. The diagonal elements of Z_{bus} are called ________ and off diagonal elements are called ________

ANSWERS

1. L-G fault
2. Unequal
3. $V_{a1} = Ea - I_a z_1$
4. Emf source
5. Zero
6. Self admittances, mutual admittances.
7. driving point impedances, transfer impedances.

Chapter-3

LOAD FLOWS

3.1 INTRODUCTION

The load flow studies involves the solution of the power system network under steady state conditions, The solution is obtained by considering certain inequality constraints imposed on node voltages and reactive power of generators.

The main informations obtained from a load flow studv are the magnitude and phase angle of the voltage at each bus is the real and reactive power flowing in each line.

The load flow solution also gives initial conditions of the system when the transient behavior of the system is to be studied.

The load flow study of a power system is essential to decide the best operation of existing system and for planning the future expansion of the system. It is also essential for designing a new power system. The load flow solution of power system under balanced load condition have been presented in this chapter. For analysis of balanced system single phase equivalent circuit of three phase system is used to make the calculation as easy as possible.

A load flow study of a power system generally require the following steps

1. Representation of the system by single line diagram
2. Determining the impedance diagram using the information in single line diagram.
3. Formulation of network equations.
4. Solution of network equation.

Under steady state conditions the network equations will be in the form of simple algebraic equation. In practical system the laod is changing and hence generation of power level also continuously changing, but for calculation purpose we assume that the loads and generation are fixed at a particular value for a suitable period of time. *e.g.* 10 minutes (or) 30 minutes etc.;

3.2. NODAL ADMITTANCE MATRIX (Y_{BUS})

The meeting point of various components in a power system is called bus. The bus bar is a conductor made of copper or aluminium having negligible resistance. Hence the bus bar will have zero voltage drop when it conducts the rated current. Therefore the buses ar considered as point of constant voltage in a power system.

When the power system is represented by impedance (or) reactance diagram, it can be considered as a circuit or network. The buses can be treated as nodes and the voltages of all buses (nodes) can be solved by conventional node analysis technique.

Let N be the number of major (or) Principal nodes in a circuit or network. Since the voltage of a node can be measured only with respect to reference point, one of the node is considered as reference node. Now the network will have (N–1) independent voltages. In nodal analysis, the independent voltages are solved by writing kirchoff's current law (*KCL*) equaton for (N–1) nodes in the circuit. For writing KCL equations, the voltage sources in the circuit should be converted to equivalent current sources.

Let, V_1, V_2, V_3............V_n = Node voltages of 1, 2, 3, ...n respectively.

I_{11}, I_{22}, I_{33}.........I_{mn} = sum of current sources connected to nodes 1, 2, 3.....n respectively.

Y_{jj} ⇒ sum of admittance connected to node j.

y_{jk} ⇒ Negative of sum of admittance connected between node-j and node-k.

Now the n-number of node equations for N-bus system will be in the form shown below (here, n = N–1)

$$Y_{11}V_1 + Y_{12}V_2 + Y_{13}V_3 + \ldots + Y_{1n}V_n = I_{11}$$
$$Y_{21}V_1 + Y_{22}V_2 + Y_{23}V_3 + \ldots + 1/2_nV_n = I_{22}$$
$$\vdots \qquad\qquad \vdots \quad \vdots$$
$$Y_{n1}V_1 + Y_{n2}V_2 + Y_{n3}V_3 + \ldots + Y_{nn}V_n = I_{nn}$$

The above n number of equation can be arranged in the matrix form as shown in equation (3.1).

$$\begin{bmatrix} Y_{11} & Y_{12} & Y_{13} & \cdots & Y_{1n} \\ Y_{21} & Y_{22} & Y_{23} & \cdots & Y_{2n} \\ Y_{31} & Y_{32} & Y_{33} & \cdots & Y_{3n} \\ \vdots & \vdots & \vdots & & \vdots \\ Y_{n1} & Y_{n2} & Y_{n3} & \cdots & Y_{nn} \end{bmatrix} \begin{bmatrix} V_1 \\ V_2 \\ V_3 \\ \vdots \\ V_n \end{bmatrix} = \begin{bmatrix} I_{11} \\ I_{22} \\ I_{33} \\ \vdots \\ I_{nn} \end{bmatrix} \quad ...(3.1)$$

In matrix notation of the equation (3.1) can be written as,

$$YV = I$$

In Power system the Y matrix is designated as Y_{bus}, and called bus admittance matrix. The node voltage are called bus voltages.

Therefore the matrix notation of equation (3.1) can be written as,

$$Y_{bus}V = I \quad ...(3.2)$$

Where,

Y_{bus} = Bus admittance matrix of order $(n \times n)$

V = Bus voltage matrix of order $(n \times 1)$

I = Current Source matrix of order $(n \times 1)$

n = Number of independent buses in the system.

$$\text{Bus admittance matrix, } Y_{bus} = \begin{bmatrix} Y_{11} & Y_{12} & Y_{13} & \cdots & Y_{1n} \\ Y_{21} & Y_{22} & Y_{23} & \cdots & Y_{2n} \\ \vdots & \vdots & \vdots & \vdots & \vdots \\ Y_{n1} & Y_{n2} & Y_{n3} & \cdots & Y_{nn} \end{bmatrix} \quad ...(3.3)$$

The bus admittance matrix, Y_{bus} is symmetrical around the principal diagonal. The admittances Y_{11}, Y_{22}, Y_{33}, ...Y_{nn} are called self admittances at the buses and the other admittances are called mutual admittances.

In general,

Y_{ji} = sum of all admittances connected to bus-j

Y_{jk} = Negative sum of all a admittances connected between bus-j and bus-k.

Also, $$Y_{jk} = Y_{kj}$$

> ➡ If the power system is represented by reactance diagram, then all the elements of the network are inductive susceptances. In this case Y_{ji} will be negative and Y_{jk} will be positive.

Solution of bus voltage

Consider the node basis matrix equation (equation 3.3) of N-bus system.

$$Y_{bus} V = I \quad \text{...(3.4)}$$

On Premultiplying the equation (3.4) by Y_{bus}^{-1} we get,

$$V = Y_{bus}^{-1} I \quad \text{...(3.5)}$$

We known that, $Y_{bus}^{-1} = \dfrac{\text{Adjoint of } Y_{\text{bus}}}{\text{Determinant of } Y_{\text{bus}}}$

Let $\Delta \rightarrow$ Determinant of Y_{bus}

and Δ_{ij} = Cofactor of Y_{jk}

Now,

$$\text{Adjoint of } Y_{bus} = \begin{bmatrix} \Delta_{11} & \Delta_{12} & \cdots & \Delta_{1n} \\ \Delta_{21} & \Delta_{22} & \cdots & \Delta_{2n} \\ \vdots & \vdots & & \vdots \\ \Delta_{n1} & \Delta_{n2} & \cdots & \Delta_{nn} \end{bmatrix} = \begin{bmatrix} \Delta_{11} & \Delta_{21} & \cdots & \Delta_{n1} \\ \Delta_{21} & \Delta_{22} & \cdots & \Delta_{n2} \\ \vdots & \vdots & & \vdots \\ \Delta_{1n} & \Delta_{2n} & \cdots & \Delta_{nn} \end{bmatrix}$$

...(3.6)

Also,
$$V = \begin{bmatrix} V_1 \\ V_2 \\ \vdots \\ V_n \end{bmatrix} \quad \text{and} \quad I = \begin{bmatrix} I_{11} \\ I_{22} \\ \vdots \\ I_{nn} \end{bmatrix} \quad \text{...(3.7)}$$

Using equations (3.6) and (3.7), the equation (3.5) can be written as shown in equation (3.8)

$$\begin{bmatrix} V_1 \\ V_2 \\ V_3 \\ \vdots \\ V_n \end{bmatrix} = \frac{1}{\Delta} \begin{bmatrix} \Delta_{11} & \Delta_{21} & \Delta_{31} & \cdots & \Delta_{n1} \\ \Delta_{12} & \Delta_{22} & \Delta_{32} & \cdots & \Delta_{n2} \\ \Delta_{13} & \Delta_{23} & \Delta_{33} & \cdots & \Delta_{n3} \\ \vdots & \vdots & \vdots & & \vdots \\ \Delta_{1n} & \Delta_{2n} & \Delta_{3n} & \cdots & \Delta_{nn} \end{bmatrix} \begin{bmatrix} I_{11} \\ I_{22} \\ I_{33} \\ \vdots \\ I_{nn} \end{bmatrix} \quad \text{...(3.8)}$$

By matrix multiplication the equation (3.8) can be expressed as n-number of linear independent equation shown below.

$$V_1 = \frac{1}{\Delta}\left[\Delta_{11} I_{11} + \Delta_{21} I_{22} + \Delta_{31} I_{33} + \ldots\ldots \Delta_{n1} I_{nn}\right]$$

$$V_2 = \frac{1}{\Delta}\left[\Delta_{12} I_{11} + \Delta_{22} I_{22} + \Delta_{32} I_{33} + \ldots\ldots \Delta_{n2} I_{nn}\right]$$

$$\vdots \qquad \vdots$$

$$V_n = \frac{1}{\Delta}\left[\Delta_{1n} I_{11} + \Delta_{2n} I_{22} + \Delta_{3n} I_{33} + \ldots\ldots \Delta_{nk} I_{nn}\right]$$

In general the Kth bus voltage is given by

$$V_k = \frac{1}{\Delta}\left[\Delta_{1k}I_{11} + \Delta_{2k}I_{22} + \Delta_{3k}I_{33} + ...\Delta_{nk}I_{nn}\right]$$

$$\therefore \quad V_k = \frac{1}{\Delta}\sum_{j=1}^{n}\Delta_{jk}I_{jj} \qquad ...(3.9)$$

Where, $\Delta \rightarrow$ Determinant of Y_{bus} matrix

$I_{jj} \rightarrow$ Sum of current sources injecting current to node-j

$\Delta_{jk} \rightarrow$ cofactor of the element Y_{jk} of bus amittance matrix.

The equation (3.9) is crammer's rule and this equation can be expressed in another simpler form as

K^{th} bus voltage, $K = \dfrac{\Delta k}{\Delta}$

Where, $\Delta \rightarrow$ Determinate of Y_{bus} matrix

$D_k \rightarrow$ Determinate of Y_{bus} matrix after replacing K^{th} column by current source vector I.

Bus (or) node diminution by matrix algebra

The buses or nodes which does not have any current sources can be eliminated by matrix manipulation of the standard node equation.

Consider the general form of node basis matrix equation,

$$Y_{bus}\ V = I \qquad ...(3.10)$$

Now tue matrices in the equation (3.10) can be partitioned using the guideline given below.

- The column matrices V and I are rearranged such that the elements associated with buses to be eliminated are in the lower rows of the matrices.
- The square matrix Y_{bus} is rearanged such that the elements associated with buses to be eliminated are in the last rows and columns of the matrices.
- The bus admittance matrix is partitioned so that elements identified only with nodes to be eliminated are separated form the other elements by horizontal and vertical lines.

Consider a bus admittance matrix of order (5 × 5). Now the matrix equation (3.10) for n = 5 can be written as shown in equation (3.11). Let us assume that the buses 4 and 5 does not have any current source and so they can be eliminated.

$$\begin{bmatrix} Y_{11} & Y_{12} & Y_{13} & \vdots & Y_{14} & Y_{15} \\ Y_{21} & Y_{22} & Y_{23} & \vdots & Y_{24} & Y_{25} \\ Y_{31} & Y_{32} & Y_{33} & \vdots & Y_{34} & Y_{35} \\ \cdots & \cdots & \cdots & \cdots & \cdots & \cdots \\ Y_{41} & Y_{42} & Y_{43} & \vdots & Y_{44} & Y_{45} \\ Y_{51} & Y_{52} & Y_{53} & \vdots & Y_{54} & Y_{55} \end{bmatrix} \begin{bmatrix} V_1 \\ V_2 \\ V_3 \\ V_4 \\ V_5 \end{bmatrix} = \begin{bmatrix} I_{11} \\ I_{22} \\ I_{33} \\ I_{44} \\ I_{55} \end{bmatrix} \quad \text{...(3.11)}$$

Let us partition the matrix (3.11) and define the following sub matrices.

$$K = \begin{bmatrix} Y_{11} & Y_{12} & Y_{13} \\ Y_{21} & Y_{22} & Y_{23} \\ Y_{21} & Y_{32} & Y_{33} \end{bmatrix}; \qquad L = \begin{bmatrix} Y_{14} & Y_{15} \\ Y_{24} & Y_{25} \\ Y_{34} & Y_{35} \end{bmatrix}$$

$$L^T = \begin{bmatrix} Y_{41} & Y_{42} & Y_{43} \\ Y_{51} & Y_{52} & Y_{53} \end{bmatrix}; \qquad M = \begin{bmatrix} Y_{44} & Y_{45} \\ Y_{54} & Y_{55} \end{bmatrix}$$

$$V_A = \begin{bmatrix} V_1 \\ V_2 \\ V_3 \end{bmatrix} \qquad V_x = \begin{bmatrix} V_4 \\ V_5 \end{bmatrix}$$

$$I_A = \begin{bmatrix} I_{11} \\ I_{22} \\ I_{33} \end{bmatrix} \qquad I_x = \begin{bmatrix} I_{44} \\ I_{55} \end{bmatrix}$$

Where,

$I_x \rightarrow$ sub matrix composed of the currents entering the buses to be eliminated.

$V_x \rightarrow$ sub matrix composed of the voltage of the buses to be eliminated.

$K \rightarrow$ sub matrix composed of self and mutual admittances identified only with buses to be retained.

$M \rightarrow$ sub matrix composed of self and mutual admittancs identified only with buses to be eliminated.

$L \rightarrow$ Sub matrix composed of mutural admittances between buses to be retained and eliminated.

$L^T \rightarrow$ Transpose of L.

Using the submetrices as defined above, the matric equation (3.11) can be written as shown in equation (3.12)

$$\begin{bmatrix} K & L \\ L^T & M \end{bmatrix}\begin{bmatrix} V_A \\ V_X \end{bmatrix} = \begin{bmatrix} I_A \\ I_X \end{bmatrix} \qquad ...(3.12)$$

By matrix multiplication the equation (3.12) can be written as,

$$KV_A + LV_X = I_A \qquad ...(3.13)$$

$$L^T V_A + MV_x = I_x \qquad ...(3.14)$$

From equation (3.14) we get,

$$MV_X = I_A - L^T V_A \qquad ...(3.15)$$

Here all the elements of the submatrix I_x are zero, because the buses to be eliminated does not have any current source.

$$\therefore \quad MV_X = -L^T V_A \qquad ...(3.16)$$

On premultiplying equaiton (3.16) by M^{-1} we get,

$$V_X = -M^{-1}L^T V_A \qquad ...(3.17)$$

On substituting for V_X from equation (3.17) in equation (3.13) we get,

$$KV_A - LM^{-1}L^T V_A = I_A$$

$$\therefore \quad [K - Lm^{-1}L^T]\, V_A = I_A$$

$$Y_{bus,\ new}\, V_A = I_A \qquad ...(3.18)$$

Where, $$Y_{bus,\ new} = K - LM^{-1}L^T \qquad ...(3.19)$$

The new bus admittance matrix $Y_{bus,\ new}$ used to reconstruct the circuit with the unwanted buses eliminated.

The matrix partitioning method discussed above is a general procedure and it is suitable for computer solutions. When large number of buses are to be eliminated, then the size of matrix M will be large and finding M^{-1} will be difficult.

The process of elimination can be simplified if one bus is eliminated at a time. In this case the matrix M will have only one element and M^{-1} is the reciprocal of this element. Here, the bus to be eliminated must be the highest numbered bus and renumbering may be required. This can be achieved by row interchange and column interchange. When bus-P has to be eliminated in a system with n-independent buses, the Pth row is interchanged with n^{th} row and P^{th} column is interchanged with n^{th} column in the bus admittance matrix.

Consider a bus admittance matrix of order $(n \times n)$, in which the nth bus has to be eliminated.

$$Y_{bus} = \left[\begin{array}{cccc:c} \multicolumn{4}{c}{\overbrace{\hphantom{Y_{11} \quad Y_{12} \quad \cdots \quad Y_{1(n-1)}}}^{K}} & \overbrace{\hphantom{Y_{1n}}}^{L} \\ Y_{11} & Y_{12} & \cdots & Y_{1(n-1)} & Y_{1n} \\ Y_{21} & Y_{22} & \cdots & Y_{2(n-1)} & Y_{2n} \\ \vdots & & & & \vdots \\ Y_{(n-1)1} & Y_{(n-2)2} & \cdots & Y_{(n-1)(n-1)} & Y_{(n-1)n} \\ \hdashline Y_{n1} & Y_{n2} & \cdots & Y_{n(n-1)} & Y_{nn} \\ \multicolumn{4}{c}{\underbrace{\hphantom{Y_{11} \quad Y_{12} \quad \cdots \quad Y_{1(n-1)}}}_{L^T}} & \underbrace{\hphantom{Y_{nn}}}_{M} \end{array}\right] \quad ...(3.20)$$

Let us Partition the bus admittance matrix as shown in equaiton (3.20). Submatrices are given below.

$$K = \begin{bmatrix} Y_{11} & Y_{12} & \cdots & Y_{1(n-1)} \\ Y_{21} & Y_{22} & \cdots & Y_{2(n-1)} \\ \vdots & & & \\ Y_{(n-1)1} & Y_{(n-2)2} & \cdots & Y_{(n-1)(n-1)} \end{bmatrix}$$

$$L = \begin{bmatrix} Y_{1n} \\ Y_{2n} \\ \vdots \\ Y_{(n-1)n} \end{bmatrix}$$

$$L^T = \begin{bmatrix} Y_{n1} & Y_{n2} & \cdots & Y_{n(n-1)} \end{bmatrix}$$

$$M = Y_{nn} \quad ; \quad M^{-1} = \frac{1}{Y_{nn}}$$

From equation (3.19) we get,

$$Y_{bus,\ new} = K - LM^{-1}L^T \quad ...(3.21)$$

On substituting the submatrices in equation (3.21) we get,

$$Y_{bus,\ new} = \begin{bmatrix} Y_{11} & Y_{12} & \cdots & Y_{1(n-1)} \\ Y_{21} & Y_{22} & \cdots & Y_{2(n-1)} \\ \vdots & \vdots & & \vdots \\ Y_{(n-1)1} & Y_{(n-2)2} & \cdots & Y_{(n-1)(n-1)} \end{bmatrix} - \begin{bmatrix} Y_{1n} \\ Y_{2n} \\ \vdots \\ Y_{(n-1)n} \end{bmatrix} \left[\frac{1}{Y_{nn}}\right] \left[Y_{n1} Y_{n2} .... Y_{n(n-1)}\right]$$

$$Y_{bus,\ new} = \begin{bmatrix} Y_{11} & Y_{12} & \cdots & Y_{1(n-1)} \\ Y_{21} & Y_{22} & \cdots & Y_{2(n-1)} \\ \vdots & \vdots & & \vdots \\ Y_{(n-1)1} & Y_{(n-2)2} & \cdots & Y_{(n-1)(n-1)} \end{bmatrix} - \frac{1}{Y_{nn}} \begin{bmatrix} Y_{1n}Y_{n1} & Y_{1n}Y_{n2} & \cdots & Y_{1n}Y_{(n-1)} \\ Y_{2n}Y_{n1} & Y_{2n}Y_{n2} & \cdots & Y_{2n}Y_{n(n-1)} \\ \vdots & \vdots & & \vdots \\ Y_{(n-1)n}Y_{n1} & Y_{(n-1)n}Y_{n2} & \cdots & Y_{(n-1)n}Y_{n(n-1)} \end{bmatrix}$$

...(3.22)

From equation (3.22) the element Y_{jk} of the resulting $(n - 1) \times (n - 1)$ new bus admittance matrix is given by

$$Y_{jk,\ new} = Y_{jk} - \frac{Y_{jn} Y_{nk}}{Y_{nn}} \qquad ...(3.23)$$

For $j = 1, 2, 3....(n - 1)$ and $K = 1, 2, 3....(n - 1)$

Where Y_{jk}, Y_{jn}, Y_{nk} and Y_{nn} are elements of original or given bus admittance matrix.

The equation (3.23) can be used to compute an element of bus admittance matrix using equaiton (3.23).

Step 1: Consider an element Y_{jk} of original bus admittance matrix.

Step 2: Get the product of last element of row-j and Last dement of column-K.

Step 3: Divide the Product obtained in Step-2 by Y_{nn} of original bus admittance matrix.

Step 4: On substracting the resultant value obtained in Step-3 from Y_{jk} of original bus admittance matrix we get, the new value of Y_{jk}.

> ➡ Since bus admittance matrix is symmetrical we can take $Y_{kj,new} = Y_{jk,\ new}$ and can avoid $m(m - 1)/2$ calculationss for a new bus admittance matrix of order $(m \times m)$, where $m = (n - 1)$.

Example 3.1: Solve the node voltage V_1 and V_2 in the network shown in Shown in Fig (3.1.1). The voltage and impedances are in *P.u.*

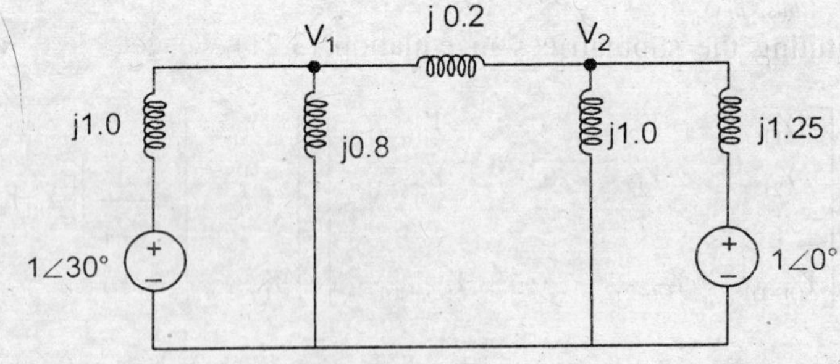

Figure 3.1.1

Solution: The voltage sources in the network can be converted to current sources by source transformation as shown in Fig. (3.1.2)

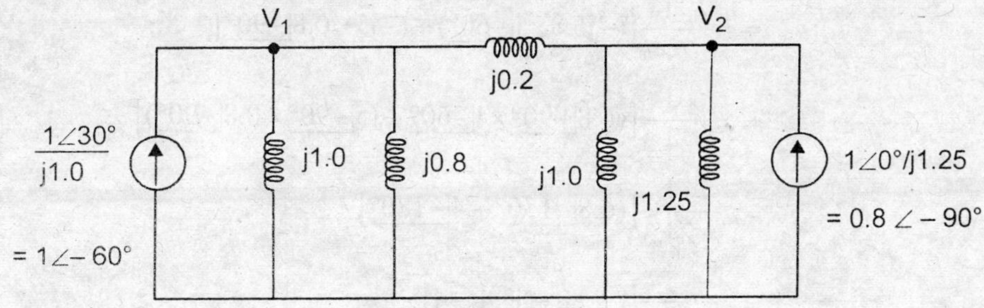

Figure 3.1.2

The node basis matrix equation is formed as shown below.

$$\begin{bmatrix} \frac{1}{j1}+\frac{1}{j0.8}+\frac{1}{j0.2} & -\frac{1}{j0.2} \\ -\frac{1}{j0.2} & \frac{1}{j0.2}+\frac{1}{j1}+\frac{1}{j1.25} \end{bmatrix} \begin{bmatrix} V_1 \\ V_2 \end{bmatrix} = \begin{bmatrix} 1\underline{|30^\circ}/j1 \\ 1\underline{|0^\circ}/j1.25 \end{bmatrix}$$

$$\begin{bmatrix} -j1-j1.25-j5 & j5 \\ j5 & -j5-j-j0.8 \end{bmatrix} \begin{bmatrix} V_1 \\ V_2 \end{bmatrix} = \begin{bmatrix} 1\underline{|30^\circ}/1\underline{|90^\circ} \\ 1\underline{|0^\circ}/1.25\underline{|90^\circ} \end{bmatrix}$$

$$\begin{bmatrix} -j7.25 & j5 \\ j5 & -j6.8 \end{bmatrix} \begin{bmatrix} V_1 \\ V_2 \end{bmatrix} = \begin{bmatrix} 1\underline{|-60^\circ} \\ 0.8\underline{|-90^\circ} \end{bmatrix}$$

$$\Delta = \begin{bmatrix} -j7.25 & j5 \\ j5 & -j6.8 \end{bmatrix}$$

$$= (-j7.25)(-j6.8) - (j5)(j5)$$

$$= j^2(7.25 \times 6.8) - j^2 25 = -(7.25 \times 6.8) + 25 = -243$$

In node basis analysis, K^{th} node voltage V_k, is given by,

$$V_k = \frac{1}{\Delta}\sum_{j=1}^{n} \Delta_{jk} I_{jj}$$

Where, $\Delta_{jk} \rightarrow$ Cofaotor of Y_{jk}

$\Delta \rightarrow$ Determinine of Y

$I_{jj} \rightarrow$ Sum of current sources connected to nodej

Here $n = 2$

$$\therefore \quad V_1 = \frac{1}{\Delta}[\Delta_{11}I_{11} + \Delta_{21}I_{22}]$$

$$= -\frac{1}{24.3}[(-j6.8 \times 1\angle -60°) + (-j5 \times 0.8\angle -90°]$$

$$= -\frac{1}{24.3}[(6.8\angle -90° \times 1\angle -60° + (5\angle -90° \times 0.8\angle -90°)]$$

$$= -\frac{1}{24.3}[(6.8\angle -150° + 4\angle -180°]$$

$$= -\frac{1}{24.3}[-5.88 - j3.4 - 4]$$

$$= \frac{-9.88 - j3.4}{-24.3} = \frac{9.88}{24.3} + j\frac{3.4}{24.3} = 0.4066 + j0.1399$$

$$\boxed{V_1 = 0.43\angle 19° P.u}$$

$$V_2 = \frac{1}{\Delta}[\Delta_{12} I_{11} + \Delta_{22} I_{22}]$$

$$= \frac{-1}{24.3}[(-j5 \times \angle -60°) + (-j7.25 \times 0.8\angle -90°)]$$

$$= \frac{5\angle -150° + 5.8\angle -180°}{-24.3} = \frac{-4.33 - j2.5 - 5.8}{-24.3}$$

$$= \frac{-10.13 - j2.5}{-24.3} = \frac{10.13}{24.3} + j\frac{2.05}{24.3}$$

$$= 0.4169 + j1.29 = 0.43\angle 14° \; p.u$$

Ans: $V_1 = 0.43\angle 19° \; P.u$

$V_2 = 0.43\angle 14° \; P.u$

Example 3.2: For the network shown in Fig (3.2.1). Form the bus admittance matrix. Determine the reduced admittance matrix by eliminating node A. The values are marked in P.u.

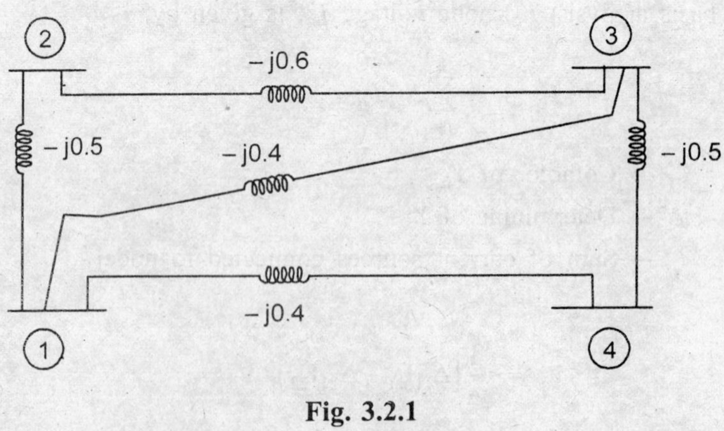

Fig. 3.2.1

Solution: The Y_{bus} matrix of the network is,

$$Y_{bus} = \begin{bmatrix} -(j0.5+j0.4+j0.4) & j0.5 & j0.4 & j0.4 \\ j0.5 & -(j0.5+j0.6) & j0.6 & 0 \\ j0.4 & j0.6 & -(j0.6+j0.5+j0.4) & j0.5 \\ j0.4 & 0 & j0.5 & -(j0.5+j0.4) \end{bmatrix}$$

$$Y_{bus} = \begin{bmatrix} -j1.3 & j0.5 & j0.4 & j0.4 \\ j0.5 & -j1.1 & j0.6 & 0 \\ j0.4 & j0.6 & -j1.5 & j0.5 \\ j0.4 & 0 & j0.5 & -j0.9 \end{bmatrix}$$

The elements of new bus admittance matrix after eliminating the 4th row and 4th column is given by

$$Y_{kj,\ new} = y_{jk} - \frac{Y_{jn}Y_{nk}}{Y_{nn}}$$

where, $n = 4$

$j = 1, 2, 3$

and $k = 1, 2, 3$

The bus admittance matrix is symmetrical

$$\therefore \quad Y_{kj,\ new} = Y_{jk,\ new}$$

$$Y_{11,\ new} = Y_{11} - \frac{Y_{14}Y_{41}}{Y_{44}} = -j1.3 - \frac{(j0.4)(j0.4)}{-j0.9} = j0.5$$

$$Y_{12,\ new} = Y_{12} - \frac{Y_{14}Y_{42}}{Y_{44}} = +j0.5 - \frac{(j0.4)0}{-j0.9} = j0.5$$

$$Y_{13,\ new} = Y_{13} - \frac{Y_{14}Y_{43}}{Y_{44}} = +j0.4 - \frac{(j0.4)(j0.5)}{-j0.9}$$

$$Y_{13,\ new} = j\ 0.622$$

$$Y_{21,\ new} = Y_{12,\ new} = j0.5$$

$$Y_{22,\ new} = Y_{22} - \frac{Y_{24}Y_{42}}{Y_{44}} = -j1.1 - \frac{(0)(0)}{-j0.9} = -\ j1.1$$

$$Y_{23,\ new} = Y_{23} - \frac{Y_{24}Y_{43}}{Y_{44}} = -j0.6 - \frac{(0)(j0.5)}{-j0.9} = j0.6$$

$$Y_{31,\ new} = Y_{13,\ new} = j\ 0.622$$

$$Y_{32,\ new} = Y_{23,\ new} = j0.06$$

$$Y_{33,\ new} = Y_{33} - \frac{Y_{34}Y_{43}}{Y_{44}} = -j1.5 - \frac{(j0.5)(j0.5)}{-j0.9} = -\ j1.222$$

The reduced bus admittance matrix after eliminating 4th row is shown below.

$$Y_{bus} = \begin{bmatrix} -j1.12 & j0.5 & j0.622 \\ j0.5 & -j1.1 & j0.6 \\ j0.622 & j0.6 & -j1.222 \end{bmatrix}$$

Ans:

$$Y_{bus} = \begin{bmatrix} -j1.12 & j0.5 & j0.622 \\ j0.5 & -j1.1 & j0.6 \\ j0.622 & j0.66 & -j1.222 \end{bmatrix}$$

Example 3.3: Eliminate buses 3 and 4 in the given bus admittance matrix and form bus admittance matrix.

$$Y_{bus} = \begin{bmatrix} -j9.8 & 0.0 & j4.0 & j5.0 \\ 0.0 & -j8.3 & j2.5 & j5.0 \\ j4.0 & j2.5 & -j14 & j8.0 \\ j5.0 & j5.0 & j8.0 & -j18.0 \end{bmatrix}$$

Solution: The elements of new bus admittance after eliminating 4th rowad 4th colum is given by

$$Y_{jk,\ new} = Y_{jk} - \frac{Y_{jn}Y_{nk}}{Y_{nn}}$$

where $n = 4$

$j = 1, 2, 3$

and $K = 1, 2, 3$

$$Y_{11,\ new} = Y_{11} - \frac{Y_{14}Y_{41}}{Y_{44}} = -j9.8 - \frac{j5.0 \times j5.0}{-18.0} = -j8.411$$

$$Y_{12,\ new} = Y_{12} - \frac{Y_{14}Y_{42}}{Y_{44}} = 0.0 - \frac{j5.0 \times j5.0}{-j18.0} = j1.388$$

$$Y_{13,\ new} = Y_{13} - \frac{Y_{14}Y_{43}}{Y_{44}} = j4.0 \frac{-j5 \times j8}{-j18.0} = j6.222$$

$$Y_{21,\ new} = Y_{12,\ new} = j1.3888$$

$$Y_{22,\ new} = Y_{22} - \frac{Y_{24}Y_{42}}{Y_{44}} = -j8.3 - \frac{j5 \times j5}{-j18.0} = -j6.911$$

$$Y_{23,\ new} = Y_{22} - \frac{Y_{24}Y_{43}}{Y_{44}} = j2.5 - \frac{j5 \times j8}{-j18} = j\ 4.722$$

$$Y_{31,\,new} = Y_{13,\,new} = j6.22$$

$$Y_{32,\,new} = Y_{23,\,new} = j4.722$$

$$Y_{33,\,new} = Y_{33} - \frac{Y_{34}Y_{43}}{Y_{44}} = -j14 - \frac{j8 \times j8}{-j18} = -j10.444$$

The reduced bus admittance matrix after eliminating 4th node is given by

$$Y_{bus} = \begin{bmatrix} -j8.411 & j1.3889 & j6.222 \\ j1.388 & -j6.911 & j4.722 \\ j6.222 & j4.722 & -j10.444 \end{bmatrix}$$

Elimination of node 3: $Y_{nn} = Y_{33} = -j10.444$

The other elemaents ofthe reduced bus admittance matrix can be formed from the equation.

$$Y_{jk,\,new} = Y_{jk} \frac{-Y_{jn}Y_{nk}}{Y_{nn}}$$

where,

$$n = 3$$
$$j = 1, 2, 3$$
$$K = 1, 2, 3$$

$$Y_{11,\,new} = Y_{11} - \frac{Y_{13}Y_{31}}{Y_{33}}$$

$$= -j8.411 - \frac{j6.222 \times j6.222}{-j10.444} = -j4.7043$$

$$Y_{12,\,new} = Y_{12} - \frac{Y_{13}Y_{32}}{Y_{33}}$$

$$= -j1.388 - \frac{j6.222 \times j4.722}{-j10.444} = j4.2010$$

$$Y_{21,\,new} = Y_{12,\,new} = j4.2011$$

$$Y_{22,\,new} = Y_{22} - \frac{Y_{23}Y_{32}}{Y_{33}} = -j6.911 - \frac{j4.722 \times j4.722}{-j10.444}$$

$$= -j4.7761$$

The reduced bus admittance matrix after eliminating node 3 and 4 is

$$Y_{bus} = \begin{bmatrix} -j4.7043 & j4.2011 \\ j4.2011 & -j4.7761 \end{bmatrix}$$

EXERCISE

1. Solve the node voltage.

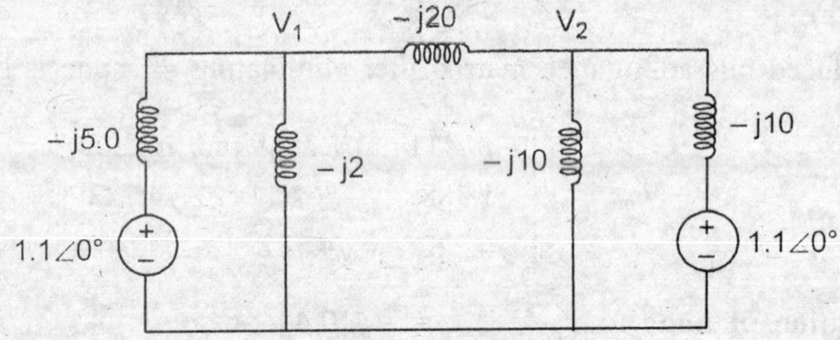

Figure E 3.1.1

2. For the network shown in Figure E. 3.2.1. Give the total number of elements, nodes, buses and braches.write the element of Y_{bus} matrix directly by inspection.

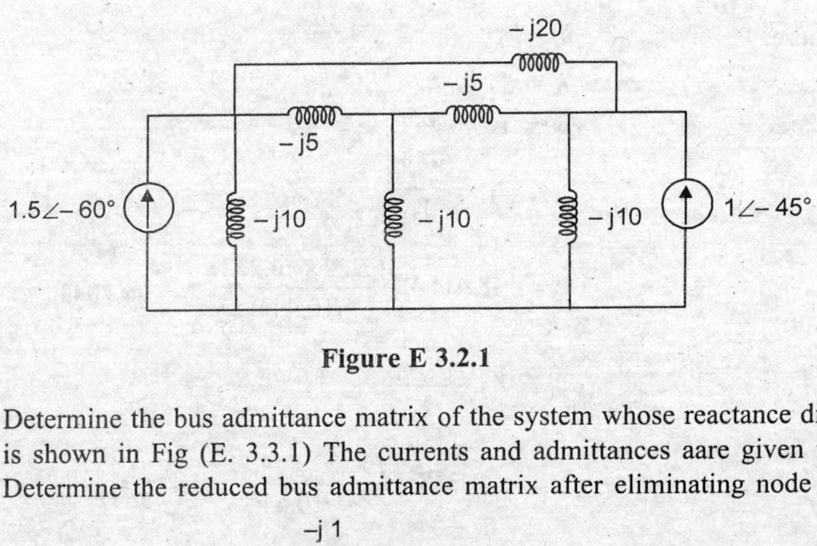

Figure E 3.2.1

3. Determine the bus admittance matrix of the system whose reactance diagram is shown in Fig (E. 3.3.1) The currents and admittances aare given in P.u. Determine the reduced bus admittance matrix after eliminating node = 3.

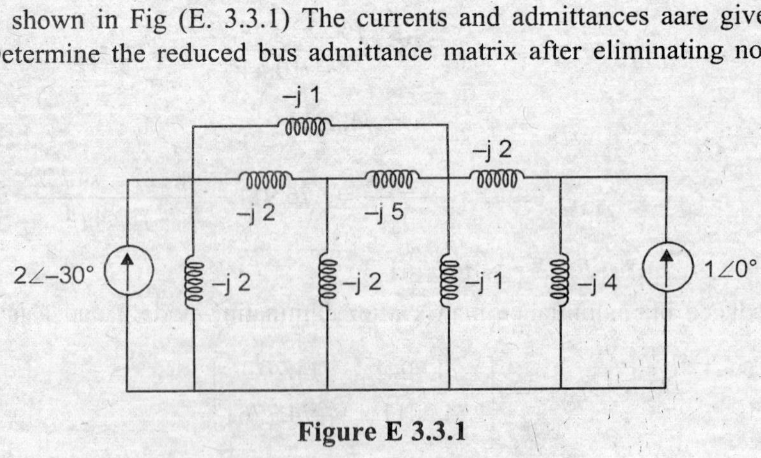

Figure E 3.3.1

3.3 TYPES OF BUSES

In a Power system the buses are meeting points of various components. The generators will feed energy to buses and loads will draw energy from buses. In the network of a power system the buses becomes nodes and so a voltage can be specified for each bus.

Therefore each bus in a power system, is associated with four quantities and they are real power, reactive power, magnitude of voltage and phase angle of voltag. In a load flow porblem two quantities (out of four) are specified for each bus and remaining two quantities are objained by solving the load flow equations.

The buses of a power system can be classified into following three types based on the quantities being specified for the buses. The below table shows the quantities specified and to be obtained for each type of bus.

(i) Load bus (or PQ bus)

(ii) Generator bus (or) voltage controlled or PV bus)

(iii) Slack bus (or swing bus or referance bus)

Bus Type	Quantities specified	Quantities to be obtained
Load bus	P, Q	$\|V\|$, Q
Generator bus	P, $\|V\|$	Q, δ
Slack bus	$\|V\|$, δ	P, Q

$$P = P_G - P_D$$

$|V|$ = Magnitude of bus voltage

$$Q = Q_G - Q_D$$

δ = phase of bus voltage

P_G, Q_G = Real and reactive power generated by generators connected to the bus respectively.

P_D, Q_D= Real and reactive power dorwin bythe loads connected tothe bus respectively.

Load bus: The bus is called load bus, when real and reactive components of power are specified for the bus. The load flow equations can be solved to find the magnitude and phase of bus voltage.In a load bus the voltage is allowed to vary within permissible limits, ofr example I5%

Generator bus: The bus is called generator bus, when real power and magnitude of bus voltage are specified for the bus. The load flow equation can be solved to find the reactive power and phase of bus voltage. Usually for generator buses reactive power limits will be specified.

Slack bus: The bus is called slack bus if the magnitude and phase of bus voltage are specified for the bus. The slack bus is the reference bus for load flow equation and usually one of the generator bus is selected as the slack bus.

Need for slack bus: Basicaly the power system has only two types of buses and they are load and generator buses. In these only power injected by gnerators and power drawn by load are specified, but the Power loss in transmission lines are not accounted. In a power system the total power generated will be equal to sum of power consumed by load and losses.

Therefore in a power system, Total complex Power

$$\left.\begin{matrix}\text{Sum of complex}\\\text{Power of generator}\end{matrix}\right\} = \begin{bmatrix}\text{Sum of complex}\\\text{Power of loads}\end{bmatrix} + \begin{bmatrix}\text{Total complex Power}\\\text{loss in transmission lines}\end{bmatrix}$$

or

$$\begin{bmatrix}\text{Total complex Power}\\\text{Loss in transmission lines}\end{bmatrix} = \begin{bmatrix}\text{Sum of complex}\\\text{Power of generators}\end{bmatrix} - \begin{bmatrix}\text{Sum of complex}\\\text{Power of loads}\end{bmatrix}$$

The transmission line losses can be eliminated only if the real and reactive Power of all buses are known. The Powers in the buses will be known only after solving the load flow equations.

For these reasons, the real and reactive power of one of the generator bus is not specified and this bus is called slack bus. It is assumed that the slack bus generates the real and reactive power required for transmission line losses. Hence for slack bus, the magnitude and phase of bus, are specified and real and ractive powers are obtained through the load flow solution.

3.4 FORMULATION OFLOAD FLOW EQUATIONS USING Y_{BUS} MATRIX

The load flow equations can be formed using either the mesh or node basis equataions of a power system. Howerver, from the Point of computer time and memory, the nodal admittance formulation using the nodal voltages as the independent variables is the most economic.

The node basis matrix equation of a n-bus system is given by

$$Y_{bus}V = I \qquad ...(3.24)$$

Where, $Y_{bus} \rightarrow$ Bus admittance matrix of order $(n \times n)$

$V \rightarrow$ Bus(node) voltage matrix of order $(n \times 1)$

$I \rightarrow$ Source current matrix of order $(n \times 1)$

The equation (3.24) can be written in the form shown below.

$$\begin{bmatrix} I_1 \\ I_2 \\ \vdots \\ I_p \\ \vdots \\ I_n \end{bmatrix} = \begin{bmatrix} Y_{11} & Y_{12} & \cdots & Y_{1P} & \cdots & Y_{in} \\ Y_{21} & Y_{22} & \cdots & Y_{2P} & \cdots & Y_{2n} \\ \vdots & & & & & \\ Y_{P1} & Y_{P2} & \cdots & Y_{PP} & \cdots & Y_{pn} \\ \vdots & \vdots & & & & \\ Y_{n1} & Y_{n2} & \cdots & Y_{nP} & \cdots & Y_{nn} \end{bmatrix} \begin{bmatrix} V_1 \\ V_2 \\ \vdots \\ V_P \\ \vdots \\ V_n \end{bmatrix} \quad ....(3.25)$$

Let I_P = Current injected to bus-P

V_P = Voltage of bus-P

From equation (3.25) the current I_p can be expressed in equation (3.26)

$$I_P = Y_{P1}V_1 + Y_{P2}V_2 + ...Y_{PP}V_P + ...Y_{Pn}V_n$$

$$\therefore \quad I_P = \sum_{q=1}^{P-1} Y_{Pq}V_q + Y_{pp}V_p + \sum_{q=P+1}^{n} Y_{pq}V_q \quad (3.26)$$

Let S_P = Complex power of bus-P

P_p = Real power of bus-P

Q_P = Reactive power of bus-P.

Now, $S_P = P_p + jQ_P$

Also, $S_P = V_p I_p^*$

$$\therefore \quad V_p I_p^* = P_p + jQ_p \quad ...(3.27)$$

The load flow problem can be handled more conveniently by use of I_p rather than I_p^*.

$\therefore$ Take conjugate of equation (3.27)

$$(V_p I_p^*)^* = (P_p + j\,Q_p)^*$$

$$V_p^* I_p = P_p - j\,Q_p$$

$$\therefore \quad I_p = \frac{P_p - jQ_p}{V_p^*} \quad ...(3.28)$$

On equating the equations (2.26) and (3.28)

$$\sum_{q=1}^{p-1} Y_{pq}V_q + Y_{pp}V_p + \sum_{q=p+1}^{n} Y_{pq}V_q = \frac{P_p - jQ_p}{V_p^*} \quad ...(3.29)$$

Method 1 : Gauss (or) Gauss-Seidel Method

The equation (3.29) can be written as shown

$$Y_{pp}V_p = \frac{P_p - jQ_p}{V_p} - \sum_{q=1}^{p-1} Y_{pq}V_q - \sum_{q=p+1}^{n} Y_{pq}V_q$$

$$\therefore \qquad V_p = \frac{1}{Y_{pp}}\left[\frac{P_p - jQ_p}{V_p^*} - \sum_{q=1}^{p-1} Y_{pq}V_q - \sum_{q=p+1}^{n} Y_{pq}V_q\right] \qquad ...(3.30)$$

$$V_p = \frac{1}{Y_{pp}}\left[\frac{P_p - jQ_p}{V_p^*} - \sum_{\substack{q=1\\q\neq p}}^{n} Y_{pq}V_q\right] \qquad ...(3.31)$$

The equation (3.30) is called load flow equation for Gausss (or) Gauss-seidel method.

Method 2 : Newton-Raphson Method

$$P_p - jQ_p = V_p^* \sum_{q=1}^{n} Y_{pq}V_q \qquad ...(3.32)$$

Let,

$$V_p \rightarrow e_p + j f_p \qquad ...(3.33)$$

$$V_q = e_q + j f_q \qquad ...(3.34)$$

$$Y_{pq} = G_{pq} - j B_{pq} \qquad ...(3.35)$$

Where $e_p, f_p \Rightarrow$ Real and imaginary part of V_p respectively

$e_p, f_q \Rightarrow$ Real and imaginary part of V_q respectively

$G_{pq}, B_{pq} \Rightarrow$ Conductance and susceptance of the admittance Y_{pq} respectively

On substituting for V_p, V_q and Y_{pq} from equation (3.33), (3.34) and (3.35) in equation (3.32) we get,

$$P_p - j\,Q_p = (e_p + j f_p)^* \sum_{q=1}^{n}(G_{pq} - j B_{pq})(e_q + j f_q)$$

$$= (e_p - j f_p)\sum_{q=1}^{n}(e_p G_{pq} - j e_q B_{pq} + j f_q G_{pq} - j^2 f_q B_{pq})$$

$$= (e_p - j f_q)\sum_{q=1}^{n}[(e_q G_{pq} + f_q B_{pq}) + j(f_q G_{pq} - e_q B_{pq})]$$

$$\sum_{q=1}^{n}(e_p - jf_q)\left[(e_q G_{pq} + f_q B_{pq} + j(f_q G_{pq} - e_q B_{pq})\right]$$

$$= \sum_{q=1}^{n}\left[\begin{matrix} e_p(e_p G_{pq} + f_q B_{pq}) + je_p(f_q G_{pq} - e_p B_{pq} \\ -jf_q(e_p G_{pq} + f_q B_{pq}) - j^2 f_p(f_q G_{pq} - e_p B_{pq}) \end{matrix}\right]$$

$$= \sum_{q=1}^{n}\left[e_p(e_q G_{pq} + f_q B_{pq}) + f_p(f_q G_{pq} - e_q B_{pq})\right]$$

$$-j\sum_{q=1}^{n}\left[f_p(e_q G_{pq} + f_q B_{pq}) - e_p(f_q G_{pq} - e_q B_{pq})\right] \qquad ...(3.36)$$

On sep[illegible]ng the real and imaginary parts of equation (3.36) we get,

$$P_p = \sum_{q=1}^{n}[e_p(e_q G_{pq} + f_q B_{pq}) + f_p(f_p G_{pq} - e_q B_{pq}) \quad \text{...(3.37)}$$

$$Q_q = \sum_{q=1}^{n}[f_p(e_p G_{pq} + f_q B_{pq}) - e_p(f_q G_{pq} - e_q B_{pq})] \quad \text{...(3.38)}$$

From equation (3.33) we get,

$$|V_p|^2 = e_p^2 + f_p^2 \quad \text{...(3.39)}$$

The equations (3.36), (3.37) and (3.38) are called load-flow equations of Newton-Rapshon method.

3.5 LOAD FLOW SOLUTION BY GAUSS-SEIDEL METHOD

The Gauss-seidel method is an iterative algorithm for solving a set of non-linear load flow equations. The non-linear load flow equations are given by equation (3.30), when $P = 1, 2, \ldots n$ and this equation is,

$$V_p = \frac{1}{Y_{pp}}\left[\frac{P_p - jQ_p}{V_p^*} - \sum_{q=1}^{p-1} Y_{pq}V_q - \sum_{q=p+1}^{n} Y_{pq}V_q\right] \quad \text{...(3.40)}$$

Where, $P = 1, 2, 3 \ldots n$

The variables in the equations obtained from equation (3.40) for $P = 1, 2, 3 \ldots n$ are the node voltages $V_1, V_2, V_3 \ldots V_n$. In Gauss-Seidel method an initial value of voltages are assumed and they are denoted as $V_1^0, V_2^0, V_3^0 \ldots V_n^0$. On substituting these initial values in equation (3.40) and by taking $p = 1$, the revised value of bus-1 voltage V_1' is computed. Now replace the V_1' for V_1^0 and V_2' for V_2^0 and perform the calculation for bus-3 and so on.

The process of computing all the bus voltages as explained above is called one iteration. The iterative process is then repeated till the bus voltage converges within prescribed accuracy. the convergence of bus voltage is quite. Sensitive to the initial values assumed. Based on practical experience it is easier to get a set of initial voltage very close to final solution. In view of the above discussions the load flow equation (3.40) can be written in the modified form as shown below.

$$V_p^{k+1} = \frac{1}{Y_{pp}}\left[\frac{P_p - jQ_p}{(V_p^k)^*} - \sum_{q=1}^{p-1} Y_{pq}V_q^{k+1} - \sum_{q=p+1}^{n} V_q^k\right] \quad \text{...(3.41)}$$

Where, V_i^k = kth iteration value of bus voltage V_i

V_i^{k+1} = $(k + 1)^{\text{th}}$ iteration value of bus voltage V_i

In equation (3.41), to compute the $(k + 1)$th iteration value of bus-p voltage the $(k + 1)$th iteration of voltages are used for all buses less than p and k^{th} iteration values of voltages are used for all buses greater than or equal to p. The equation (2.41) is applicable for load bus, since in load bus changes in both magnitude and phase of voltages are allowed. But in generator bus the magnitude of voltage remains constant and so the equation (3.41) is used to calculate the phase of the voltage.

It is important to note that the slack bus is a reference bus and so its voltage will not change. Therefore, in each iteration the slack bus voltage is not modified.

For a generator bus, the reactive power is not specified. Therefore in order to calculate the phase of bus voltage of a generator bus using equation (3.41), we have to estimate the reactive power from the bus voltages and admittance as shown below.

From equation (3.29) we get,

$$\frac{P_p - jQ_p}{V_p^*} = \sum_{q=1}^{p-1} Y_{pq}V_q + V_{pp}V_p + \sum_{q=p+1}^{n} Y_{pq}V_q$$

$$\therefore \quad P_p - j\,Q_p = V_p^*\left[\sum_{q=1}^{p-1} Y_{pq}V_q + \sum_{q=p}^{n} Y_{pq}V_q\right] \quad \text{...(3.42)}$$

From equation (3.42) the equation for complex power in bus-p during $(k + 1)$th iteration can be obtained as shown in equation (3.43).

$$P_p^{k+1} - jQ_p^{k+1} = (V_p^k)^*\left[\sum_{q=1}^{p-1} Y_{pq}V_q^{k+1} + \sum_{q=p}^{n} Y_{pq}V_q^k\right] \quad \text{...(3.43)}$$

The reactive power of bus-p during $(k + 1)^{th}$ iteration is given by imaginary part equation (3.43).

$$\left.\begin{matrix}\text{Reactive powr of bus-}P\\ \text{during }(k+1)^{th}\,\textit{iteration}\end{matrix}\right\}$$

$$Q^{k+1} = (-1)\times I_m\left\{(V_p^k)^*\left[\sum_{q=1}^{p-1} Y_{pq}V_q^{k+1} + \sum_{q=p}^{n} Y_{pq}V_q^k\right]\right\} \quad \text{...(3.44)}$$

Also, for generator buses a lower and upper limit for reactive powers will be specified. In each iteration, the reactive power of generator bus is calculated using equation (3.44) and then checked with specified limits. If it violates the specified limits then the reactive power of the bus is equated to the limit violated and it is treated as load bus. It is does not violates the limits then the bus is treated as generator bus.

Computation of Slack Bus Power and Line Flows

The slack bus power can be calculated after computing the bus voltage upto the specified accuracy. The equation (3.42) can be used to calculate the slack bus power. Here, bus-p is slack bus.

$$P_p - j\,Q_p = V_p^* \sum_{q=1}^{n} Y_{pq} V_q^k$$

The line flows are the power fed by the buses into varies lines and they are calculated as shown below.

Consider a line connecting bus-p and bus-q as shown in Fig. (3.1) usually the transmission line is connected to buses using transformers at its ends. The π-equivalent of the transmission line with transformers at its end consist of a series admittance Y_{pq} and shunt admittances $\frac{Y'_{pa}}{2}$ as shwon in Fig. 3.1.

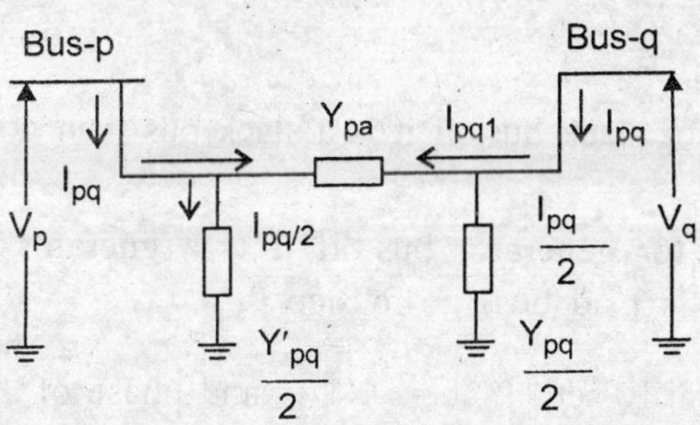

Figure 3.1 *II-equivalent of transmission line for evaluting line flows*

In Figure (3.1)

$$I_{pq} = I_{pq1} + I_{pq2} = (V_p - V_q) Y_{pq} + V_p \frac{Y_{pq}}{2} \qquad ...(3.46)$$

Complex power injected by bus–P n line–pq

$$S_{pq} = P_{pq} - j\,Q_{pq} = V_p^*\, I_{pq}$$

$$= V_p^* \left[(V_p - V_q) Y_{pq} + V_p \frac{Y'_{pq}}{2} \right] \qquad ...(3.47)$$

In Fig. (3.1), $$I_{qp} = I_{qp1} + I_{qp2} = (V_q - V_p)\, Y_{pq} + V_q \frac{Y'_{pq}}{2} \qquad ...(3.48)$$

Complex power injected by bus–q n line–pq

$$S_{pq} = P_{qp} - j\,Q_{qp} = V_p{*}\, I_{qp}$$

$$= V_q^* \left[(v_q - V_p)Y_{pq} + V_q \frac{Y'_{pq}}{2} \right] \qquad ...(3.49)$$

Power loss in the transmission line – pq

$$S_{pq}, \text{ loss} = S_{pq} + S_{qp} \qquad ...(3.50)$$

Procedure for Load Flow Solution by Gauss-Seidel Method

Step 1: Assume a flat voltage profile $1 + j\,0$ for all buses (nodes) except the slack bus. The voltage of slack bus is the specified voltage it is not modified in any iteration.

Step 2: Assume a suitable value of $\in$ called convergence criterian. Here $\in$ is a specified change in bus voltage that is used to compare the actual change in bus voltage between k^{th} and $(k + 1)^{th}$ iteration.

Step 3: Set iteration count, $k = 0$ and assumed voltage profile of the buses are denoted as $V_1^0, V_2^0, V_3^0, ...V_n^0$ except slack bus.

Step 4: Set bus, $P = 1$.

Step 5: Check for slack bus. If it is a slack bus then go to step-12, otherwise go to next step.

Step 6: Check for generator bus. If it is a gnerator bus go to next step, otherwise (*i.e.*, if it is load bus) go to step 9.

Step 7: Temporarily set $|V_p^k| = |V|_{spec}$ and phase of V_p^k as the kth iteration value if the bus-p is a generator bus where $|V_p^p|_{spec}$ is the specified magnitude of voltage for bus-p. Then calculate the reactive power of the generator bus using the following equation.

$$Q_{p,\,cal}^{k+1} = (-1) \times I_m \left\{ (v_p^k)^* \left[\sum_{q=1}^{p-1} Y_{pq} V_q^{k+1} + \sum_{q=p}^{n} Y_{pq} V_q^k \right] \right\}$$

(Here "I_m" stands for "Imaginary part of")

The calculated reactive power may be within specified limits or it may violate the limits.

If the calculated reactive power is within the specified limits than consider this bus as generator bus set $Q_p = Q_{p,\,cal}^{k+1}$ for this iteration and go to step 8. If the calculated reactive power vioiates the specified Limit for reactive power then treat this bus as load bus. the magnitude of the reactive power at this bus will correspond the limit it has violated.

i.e., if $Q_{p,\,cal}^{k+1} < Q_{p,\,\min}$ then $Q_p = Q_{p,\,\min}$

or $Q_{p,\,cal}^{k+1} > Q_{p,\,\max}$ then $Q_p = Q_{p,\,\max}$

Since the bus is treated as load bus, take actual value of V_p^k for $(k + 1)^{th}$, iteration *i.e.*, (V_p^k) need not be replaced by $|V_p|_{spec}$ when the generator bus is treated as load bus. Go to step 9.

Step 8: For generator bus the magnitude of voltage does not change and so for all iterations the magnitude of bus voltage is the specified value. The phase of the bus voltage can be calculated as shown below.

$$V_{P,\text{temp}}^{k+1} = \frac{1}{Y_{pp}}\left[\frac{P_p - jQ_p}{(V_p^k)^*} - \sum_{q=1}^{p-1} Y_{pq}V_q^{k+1} - \sum_{q=p+1}^{n} Y_{pq}V_q^k\right]$$

$$\delta_p^{k+1} = \tan^{-1}\left[\frac{\text{Imaginary part of } V_{p,temp}^{k+1}}{\text{Real part of } V_{p,temp}^{k+1}}\right]$$

Now, the $(k + 1)^{th}$ iteration voltage of the generator bus is given by

$$V_p^{k+1} = |V_p|_{\text{spec}} \angle \delta_p^{k+1}$$

Where $|V_p|_{spec}$ = magnitude of specified bus voltage.

After calculating V_p^{k+1} for generator bus go to step 11.

Step 9: For the load bus the $(k + 1)^{th}$ iteration value of load bus-*p* volttage, V_p^{k+1} can be calculated using the following equation.

$$V_p^{k+1} = \frac{1}{Y_{pp}}\left[\frac{P_p - jQ_p}{(V_p^k)^*} - \sum_{q=1}^{p-1} Y_{pq}V_q^{k+1} - \sum_{q=p+1}^{n} Y_{pq}V_q^k\right]$$

Step 10: An acceleration factor, α can be used for faster convergence. If acceleration factor is specified then modify the $(k + 1)$th iteration value of bus-*p* voltage using the following equation.

$$V_{p,\,acc}^{k+1} = V_p^k + \alpha(V_p^{k+1} - V_p^k)$$

Then set, $V_p^{k+1} = V_{p,\,acc}^{k+1}$

> A suitable value of α for a particular system can be obtained by running trial load flows. For most of the systems the recommended value of α is 1.6.

Step 11: Calculate the change in the bus-*p* voltage, using the relation,

$$\Delta V_p^{k+1} = V_p^{k+1} - V_p^k$$

Step 12: Repeat the step 5 to 11 until all the bus voltages have been calculated. For this increment the bus count by 1 and go to step 5, until the bus count is *n*.

Step 13: Find out the largest of the absolute value of the change in voltage.

i.e., find the largest among $|\Delta V_1^{k+1}|, |\Delta V_2^{k+1}|, |\Delta V_3^{k+1}| ..., |\Delta V_n^{k+1}|$. Let this change be $|\Delta V_{max}|$. Check whether this largest change $|\Delta V_{max}|$ is less than the prespecified tolerance $\in$. If $|\Delta V_{max}|$ is less than $\in$ then more to the next step. Otherwise increment the iteration count and go to step 4.

Step 14: Calculate the line flows and slack bus power using the bus (node) voltages.

A computer program can be developed using the procedure given above. A flow chart for the computational procedure have been pressented in Fig. 3.2.

> For pure load bus the neal and reactive power are considered as negative. For pure generator buses the real and reactive power are always considered as positive, even if the bus is treated as load bus. In buses having generators and loads connected to it, either the net power will be specified or the generated and load powers will be individually specified. The specified wet power is the difference between generator and load power and so it can be used as such in calculation without changing the sign. when the generator and load power are individually specified, then the power used for calculation is the value obtained by subtracting load power from generator power
>
> (*i.e.*, $\delta_p = \Sigma S_G - \Sigma S_L$ and $S_p = P_p + j\,Q_p$).

Now we will see the flow chart,

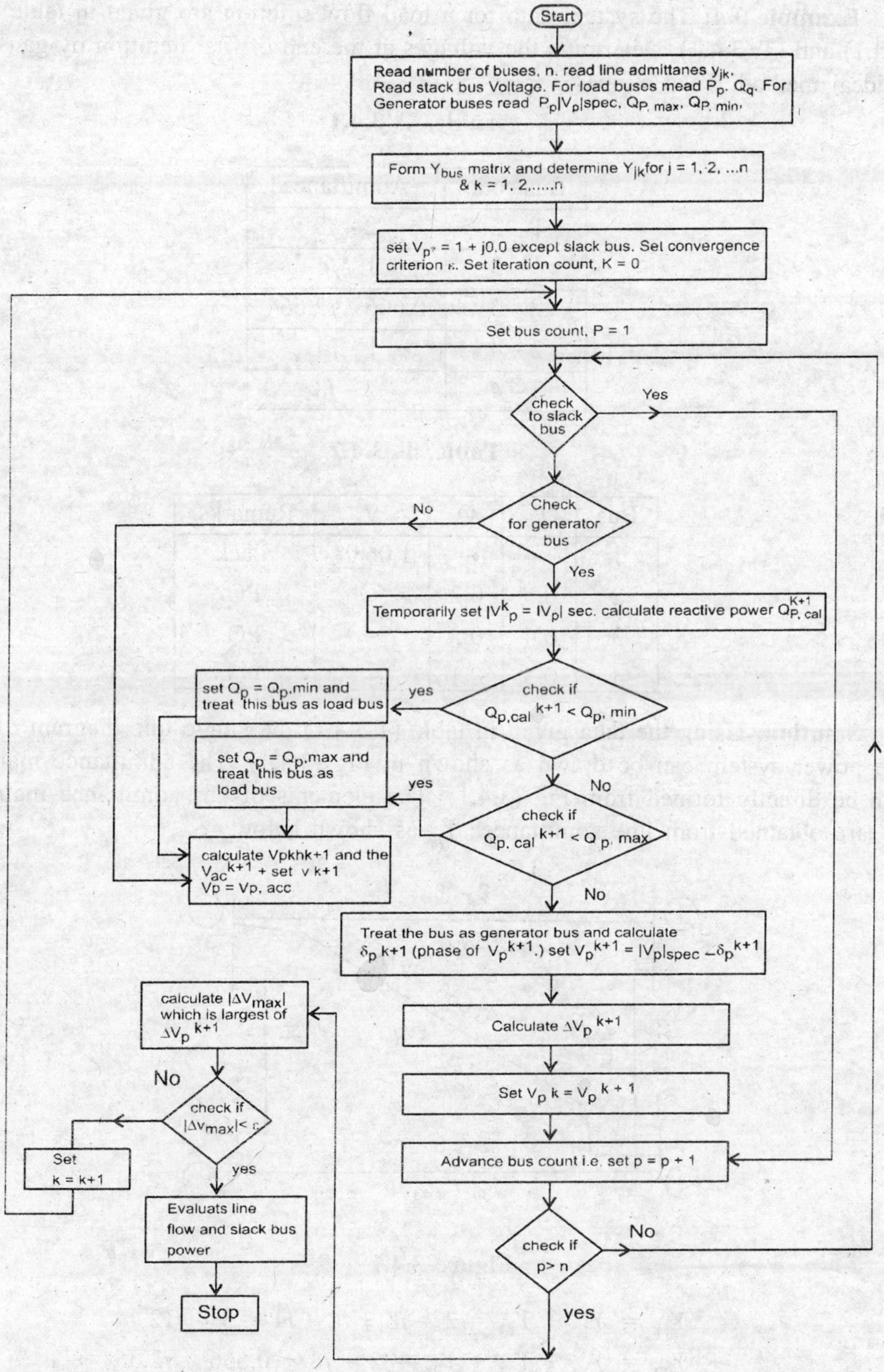

Figure 3.2 *Flow chart for load flow solution by Gauss-seidel method*

Example 3.4: The system data for a load flow solution are given in table (T. 3.4.1) and (T. 3.4.1). Determine the voltages at the end of first iteration by gausss-seideal method. Take $\alpha = 1.6$.

Table. T. 3.4.1

Bus code	Admittance
1 – 2	$2 - j8$
1 – 3	$1 - j4$
2 – 3	$0.666 - j2.664$
2 – 4	$1 - j4$
3 – 4	$2 - j8$

Table. T. 3.4.2

Bus	P	Q	V	Remarks.
1	–	–	$1.06\angle 0°$	Slack
2	0.5	0.2	–	PQ
3	0.4	0.3	–	PQ
4	0.3	0.1	–	PQ

Solution: Using the data given in table (T.3.4.1) the single line diagram of the power system can be drawn as shown in Fig (3.4.1). The admittance matrix can be directly formed from Fig (3.4.1). The elements of bus admittance matrix Y_{jk} are obtained from line admittances Y_{jk} as shown below.

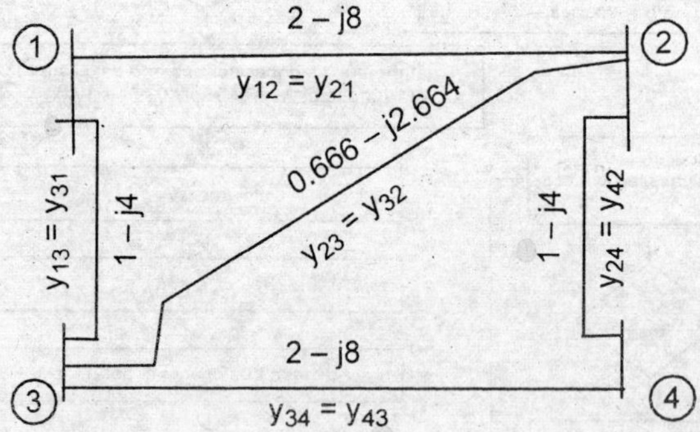

Figure 3.4.1

$$Y_{11} = Y_{12} + Y_{13} = 2 - j8 + 1 - j4 = 3 - j12$$

$$Y_{22} = Y_{12} + Y_{23} + Y_{24} = 2 - j8 + 0.666 - j2.664 + 1 \quad - j4$$

$$= 3.666 - j14.664$$

$$Y_{33} = Y_{13} + Y_{23} + Y_{34} = 1 - j4 + 0.666 - j2.664 + 2 - j8$$
$$= 3.666 - j14.664$$
$$Y_{44} = Y_{24} + Y_{34} = 1 - j4 + 2 - j8 = 3 - j12$$
$$Y_{12} = Y_{21} = -Y_{12} = -(-2 - j8) = -2 + j8$$
$$Y_{13} = Y_{31} = -Y_{13} = -(1 - j4) = -1 + j4$$
$$Y_{14} = Y_{41} = 0$$
$$Y_{23} = Y_{32} = -Y_{23} = -(0.666 - j2.664) = -0.666 + j2.664$$
$$Y_{24} = Y_{42} = -Y_{24} = -(1 - j4) = -1 + j4$$
$$Y_{34} = Y_{43} = -Y_{34} = -(2 - j8) = -2 + j8$$

$$Y_{bus} = \begin{bmatrix} Y_{11} & Y_{12} & Y_{13} & Y_{14} \\ Y_{21} & Y_{22} & Y_{23} & Y_{24} \\ Y_{31} & Y_{32} & Y_{33} & Y_{34} \\ Y_{41} & Y_{42} & Y_{43} & Y_{44} \end{bmatrix}$$

$$= \begin{bmatrix} 3 - j12 & -2 + j8 & -1 + j4 & 0 \\ -2 + 28 & 3.666 - j14.664 & -0.666 + j2.664 & -1 + j4 \\ -1 + j4 & -0.66 + j2.664 & 3.66 - j14.664 & -2 + j8 \\ 0 & -1 + j4 & -2 + j8 & 3 - j12 \end{bmatrix}$$

The initial values of bus voltages are considered as 1 *P.u.* except the slack bus.

$$\therefore \quad V_2^0 = 1 + j0$$
$$V_3^0 = 1 + j0$$
$$V_4^0 = 1 + j0$$

The bus–1 is a slack bus and so its voltage remain at the specified value for all iterations.

i.e., $\quad V_1^0 = V^1 = \ldots\ldots\ldots\ldots V_1^k = 1.06 + j0 \ p.u$

Since the buses are *PQ* buses the specified real and reactive Powers are considered as load powers. Threfore a negative sign is attached to the specified power.

The $(K + 1)^{th}$ iteration voltage of a *PQ*(Load) bus–*P* is given by

$$V_p^{k+1} = \frac{1}{Y_{PP}}\left[\frac{P_p - jQ_P}{\left(V_P^K\right)^*} - \sum_{q=1}^{P-1} Y_{Pq}V_q^{K+1} - \sum_{q=p+1}^{n} Y_{Pq}V_q^K\right]$$

For first iteration, $K = 0$. The system has four buses and so *P* will take values from 1 to 4. Here all the buses are lood buses except bus-1. The calculation of bus voltages for first iteration are shown below.

$$V_1^1 = V_1^0 = 1.06 + j0 \ P.u \qquad (\therefore \text{ Bus-1 is a slack bus})$$

$$V_2^{\ 1} = \frac{1}{Y_{22}}\left[\frac{p_2 - jQ_2}{\left(V_2^0\right)*} - Y_{21}V_1^1 - Y_{23}V_3^0 - Y_{24}V_4^0\right]$$

$$= \frac{1}{3.666 - j14.664}\left[\frac{-0.5 + j0.2}{1 - j0} - (-2 + j8)(1.06) - (-0.666 + j2.6664)(1 + j0) - (-1 + j4)(1 + j0)\right]$$

$$= \frac{-0.5 + j0.2 + 2.12 - j8.48 + 0.666 - j2.664 + 1 - j4}{3.666 - j14.664}$$

$$= \frac{3.28 - j14.974}{3.666 - j14.664} = \frac{15.3010\angle{-77.6^\circ}}{15.1153\angle{-75.96^\circ}}$$

$$= 1.0123\angle{-1.64^\circ}$$

$$= 1.019 - j0.0290 \ P.u$$

$$V_2^{\ 1}{}_{,\ acc} = V_2^{\ 0} + \alpha(V_2^{\ 1} - V_2^{\ 0})$$

$$= 1 + 1.6(1.0119 - j0.0290 - 1)$$

$$= 1 + 1.6(0.0119 - j0.0290) = 1.0190 - j0.0464$$

Now, $V_2^{\ 1} = V_2^{\ 1}{}_{,acc} = 1.090 - j0.0464 \ P.u = 1.0201\angle{-2.61^\circ}$ P.u

$$V_3^{\ 1} = \frac{1}{Y_{33}}\left[\frac{P_3 - jQ_3}{\left(V_3^0\right)*} - Y_{31}V_1^1 - Y_{32}V_2^1 - Y_{34}V_4^0\right]$$

$$= \frac{1}{3.6660 - j14.664}\left[\frac{-0.4 + j0.3}{1 - j0} - (-1 + j4)(1.06) - (-.666 + j2.664)(1.0190 - j0.0464 - (-2 + j8)(1 + j0)\right]$$

$$= \frac{-0.4 + j0.3 + 1.06 - j4.24 - (-.5550 + j2.7455) + 2 - j8}{3.666 - j14.664}$$

$$= \frac{3.215 - j14.6855}{3.666 - j14.664} = \frac{15.0333\angle{-77.65^\circ}}{15.1153\angle{-75.96^\circ}} = 0.9946\angle - 1.69^\circ$$

$$V_3^{\ 1} = 0.9942 - j0.0293 \ P.u$$

$$V_3^{\ 1}{}_{,\ acc} = V_3^{\ 0} + \alpha(V_3^{\ 1} - V_3^{\ 0})$$

$$= 1 + 1.6(0.9942 - j0.0293 - 1)$$

$$= 1 + 1.6(-0.0058 - j0.0293)$$

$$V^1_{3,\,acc} = 0.9907 - j0.0469$$

Now, $V^1_3 = V^1_{3,\,acc} = 0.9907 - j0.0469\ p.u$

$$V^1_3 = 0.9918\angle -2.71°\ P.u$$

$$V_4^1 = \frac{1}{Y_{44}}\left[\frac{P_4 - jQ_4}{(V_4^0)^*} - Y_{41}V_1^1 - Y_{42}V_2^1 - Y_{43}V_3^1\right]$$

$$= \frac{1}{3-j12}\left[\frac{-0.3+j0.1}{1-j0} - (0\times 1.06) - (-1+j4)(1.0190 - j0.0464)\right.$$

$$\left. -(-2+j8)(0.9907 - j0.0469)\right]$$

$$= \frac{-0.3 + j0.1 - (-.833 + j4.1224) - (-1.6062 + j8.0194)}{3-j12}$$

$$= \frac{2.1396 - j12.0418}{3-j12} = \frac{12.2304\angle -79.92°}{12.3693\angle -75.96°}$$

$$= 0.9888\angle -3.96°$$

$$= 0.9864 - j0.683\ P.u$$

$$V^1_{4,\,acc} = V_4^0 + \alpha(V_4^1 - V_4^0)$$

$$= 1 + 1.6(0.9864 - j0.0683 - 1)$$

$$= 1 + 1.6(-0.0136 - j0.0683)$$

$$= 0.9782 - j0.1093$$

Now, $V_4^1 = V^1_{4,\,acc} = 0.9782 - j0.1093\ P.u$

$$V_4^1 = 0.9843\angle -6.38°\ P.u$$

Result:

The bus voltages at the end of first iteration are

$$V_1^1 = 1.06 + j0 = 1.06\angle 0°\ P.u$$

$$V_2^1 = 1.019 - j0.0461 = 1.0201\angle -2.61°\ P.u$$

$$V_3^1 = 0.9907 - j0.0469 = 0.9918\angle -2.7°\ P.u$$

$$V_4^1 = 0.9782 - j0.1093 = 0.9843\angle -6.38°\ P.u$$

Example 3.5: For the system shown in Fig. (3.5.1), determine the voltage at the end of first derivation by gauss-seides method. take $\alpha = 1$ and bus specifications are given in table T.3.51.

Table T. 3.5.1

Bus code	P	Q	V	Remarks.
1	–	–	$1.06\angle 0°$	Slack
2	0.5	$0.1 \le Q_2 \le 1$	1.04	PV
3	0.4	0.3	–	PQ
4	0.2	0.1	–	PQ

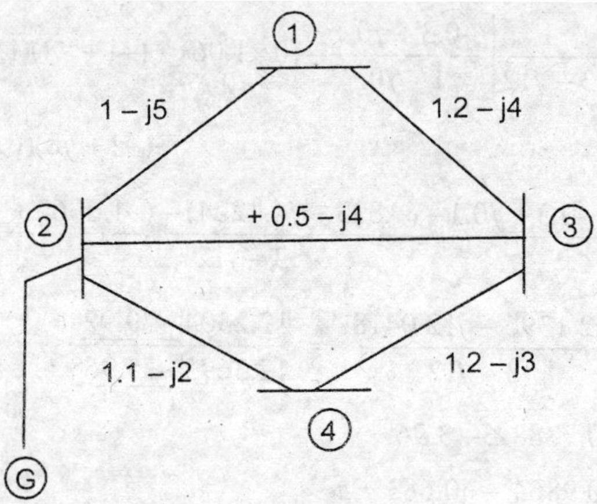

Figure 3.5.1

Solution: The Line admittances are, (refer Fig.3.5.1)

$$Y_{12} = Y_{21} = 1 - j5$$
$$Y_{13} = Y_{31} = 1.2 - j4$$
$$Y_{23} = Y_{32} = 0.5 - j4$$
$$Y_{24} = Y_{42} = 1.1 - j2$$
$$Y_{34} = Y_{43} = 1.2 - j3$$

The elements of bus admittance matrix Y_{jk} are obtained from line admittance Y_{jk} as shown below.

$$Y_{11} = Y_{12} + Y_{13} = 1 - j5 + 1.2 - j4 = 2.2 - j9$$
$$Y_{22} = Y_{12} + Y_{23} + Y_{24} = 1 - j5 + 0.5 - j4 + 1.1 - j2 = 2.6 - j11$$
$$Y_{33} = Y_{13} + Y_{23} + Y_{34} = 1.2 - j4 + 0.5 - j4 + 1.2 - j3$$
$$= 2.9 - j11$$
$$Y_{44} = Y_{24} + Y_{34} = 1.1 - j2 + 1.2 - j3 = 2.3 - j5$$
$$Y_{12} = Y_{21} = -Y_{12} = -(1 - j5) = -1 + j5$$

$$
\begin{aligned}
Y_{13} &= Y_{31} = -Y_{13} = -(1.2 - j4) = -1.2 + j4 \\
Y_{14} &= Y_{41} = 0 \\
Y_{23} &= Y_{32} = -Y_{23} = -(0.5 - j4) = -0.5 + j4 \\
Y_{24} &= Y_{42} = -Y_{24} = -(1.1 - j2) = -1.1 + j2 \\
Y_{34} &= Y_{43} = -Y_{34} = -(1.2 - j3) = -1.2 + j3
\end{aligned}
$$

$$
Y_{bus} = \begin{bmatrix} Y_{11} & Y_{12} & Y_{13} & Y_{14} \\ Y_{21} & Y_{22} & Y_{23} & Y_{24} \\ Y_{31} & Y_{32} & Y_{33} & Y_{34} \\ Y_{41} & Y_{42} & Y_{43} & Y_{44} \end{bmatrix}
$$

$$
= \begin{bmatrix} 2.2 - j9 & -1 + j5 & -1.2 + j4 & 0 \\ -1 + j5 & 2.6 - j11 & -0.5 + j4 & -1.1 + j2 \\ -1.2 + j4 & -0.5 + j4 & 2.9 - j11 & -1.2 + j3 \\ 0 & -1.1 + j2 & -1.2 + j3 & 2.3 - j5 \end{bmatrix}
$$

In the given system bus-1 is slack bus, bus-2 is generator bus and bus-3 and bus-4 are load buses The initial voltages of load buses are assumed as $1 + j0$ $P.u.$ Ior slack and generator buses the specified voltages are used as initial values.

$$
\therefore \quad V_1^0 = V_1^1 = \ldots\ldots = V_1^k = V_1
$$

$$
= 1.06 + j0 \ P.u \ \text{(slack bus)}
$$

$$
V_2^0 = 1.04 + j0 \ P.u
$$

(Generator bus) (Initial phase is assumed as zero)

$$
V_3^0 = 1 + j0 \ P.u \ \text{(Load bus)}
$$

$$
V_4^0 = 1 + j0 \ P.u \ \text{(Load bus)}
$$

For generator bus the specified powers are considered as positive Powers but for load buses the specified powers are considered as negative powers.

For first iteration, $k = 0$. In each iteration the slack bus voltage need not be recalculated. In each iteration the reactive power for generator bus has to be calculated and checked for violation of the specified limits. If the limits are violated then it is treated as load bus.

The calculations of bus voltages for first iteration are shown below.

$$
V_1^1 = V_1^0 = 1.06 + j0 \ P.u \qquad \text{(Bus-1 is a slack bus)}
$$

The bus-2 is a generator bus and so calculate its reactive power, Q_2.

$$
Q_{p,cal}^{k+1} = -1 \times I_m \left\{ \left(V_p^k\right)^* \left[\sum_{q=1}^{n-1} Y_{pq} V_q^{k+1} + \sum_{q=p}^{n} Y_{pq} V_q^k \right] \right\}
$$

Here $P = 2$, $K = 0$, $n = 4$

$$\therefore \quad Q_{2,\,cal}^{1} = -1 \times I_m\left\{\left(V_2^0\right)^*\left[Y_{21}V_1^1 + Y_{22}V_2^0 + Y_{23}V_3^0 + Y_{24}V_4^0\right]\right\}$$

Note: Here $|V_2^0|$ is Same as $|V_2|_{spec}$ and so V_2^0 is used for calculation as such, if it is not same then we have to replace $|V_2^0|$ with $|V_2|_{spec}$.

$$\therefore \quad Q_{2,cal}^{1} = -1 \times I_m\left\{1.04\begin{bmatrix}(-1 + j5(1.06 + j0) + (2.6 - j11)(1.04 + j0)\\ +(-0.5 + j4)(1 + j0) + (-1.1 + j2)(1 + j0)\end{bmatrix}\right\}$$

$$= - \times I_m\{1.04[-1.06 + j5.3 + 2.704 - j11.44 - 0.5 + j4 - 1.1 + j2]\}$$

$$= -1 \times I_m\{0.0458 - j0.1456\}$$

$$Q_{2,\,cal}^{1} = 0.1456\ P.u$$

The specified range for Q_2 is. $0.1 \leq Q_2 \leq 1.0$. The calculated value of Q_2 is within this range and so the reactive power limit is not violated. Therefore the bus can be treated as generator bus.

Now

$$P_2 = 0.5$$
$$A_2 = 0.1456$$
$$V_2^0 = 1.04 + j0$$

Since the bus-2, is treated as generator bus, the $|V_2^1| = |V_2|_{spec}$ and phase of V_2^1 is given by the phase of $V_{2,temp}^{1}$

$$V_{p,temp}^{k+1} = \frac{1}{Y_{pp}}\left[\frac{P_p - jQp}{\left(V_p^k\right)*} - \sum_{q=1}^{p-1} Y_{pq}V_q^{k+1} - \sum_{q=p+1}^{n} Y_{pq}V_q^k\right]$$

$$V_{2,temp}^{1} = \frac{1}{Y_{22}}\left[\frac{P_2 - jQ_2}{\left(V_2^0\right)*} - Y_{21}V_1^1 - Y_{23}V_3^0 - Y_{24}V_4^0\right]$$

$$= \frac{1}{2.6 - j11}\left[\frac{0.5 - j0.1456}{1.04 - j0} - (-1 + j5)(1.06 + j0) - (-0.5 + j4)(1 + j0) - (-1.1 + j2)(1 + j0)\right]$$

$$= \frac{1}{2.6 - j11}[0.4808 - j0.14 + 1.06 - j5.3 + 0.5 - j4 + 1.1 - j2]$$

$$= \frac{3.1408 - j11.44}{2.6 - j11} = \frac{11.8633\lfloor -74.65^\circ}{11.3031\lfloor -76.70^\circ} = 1.0496\angle 2.05^\circ\ P.u$$

$$\therefore \quad \delta_2^1 = \angle V_{2,\,temp}^1 = 2.05°$$

$$V_2^1 = |V|_{2spec} \angle \delta_2^1 = 1.04\angle 2.05° \ P.u$$

$$= 1.0393 + j0.0372 \text{ P.u}$$

The bus-3 and bus-4 are load buses. The voltage of load bus are calculated using the following equation.

$$V_p^{k+1} = \frac{1}{Y_{pp}}\left[\frac{P_p - jQ_p}{\left(V_p^k\right)^*} - \sum_{q=1}^{p-1} Y_{pq}V_q^{k+1} - \sum_{q=p+1}^{n} Y_{pq}V_q^k\right]$$

$$V_3^1 = \frac{1}{Y_{33}}\left[\frac{P_3 - jQ_3}{\left(V_3^0\right)^*} - Y_{3,}V_1^1 - Y_{32}V_2^1 - Y_{34}V_4^0\right]$$

$$= \frac{1}{2.9 - j11}\left[\frac{-0.4 + j0.3}{1 - j0} - (-1.2 + j4)(1.06 + j0) - \right.$$

$$\left. -(-0.5 + j4)(1.6393 + j0.0372) - (-1.2 + j3)(1 + j0)\right]$$

$$= \frac{1}{2.9 - j11}[-\ 0.4\ +\ j0.3\ +\ 1.272\ -\ j4.24\ -\ (-\ 0.6685\ +$$

$$j4.1386)\ +\ 1.2\ -\ j3]$$

$$= \frac{2.7405 - j11.0786}{2.9 - j11} = \frac{11.4125\angle -76.11°}{11.3759\angle -75.23°}$$

$$= 1.0032\angle -0.88° \ P.u$$

$$V_3^1 = 1.0031 - j0.0154 \ P.u$$

$$V_4^1 = \frac{1}{Y_{44}}\left[\frac{P_4 - jQ_4}{\left(V_4^0\right)^*} - V_{41}V_1^1 - Y_{42}V_2^1 - Y_{43}V_3^1\right]$$

$$= \frac{1}{2.3 - j5}\left[\frac{-0.2 + j0.01}{1 - j0} - (0 \times 1.06) - (-1.1 + j2)(1.0393\right.$$

$$\left. + j0.0372 - (-1.2 + j3)(1.0031 - j0.0154)\right]$$

$$= \frac{1}{2.3 - j5}[-0.2 + j0.1 - (-1.2176 + j2.0377) - (-1.1575 + j3.6278)]$$

$$= \frac{2.1751 - j4.9655}{2.3 - j5} = \frac{5.4210\angle -66.34°}{5.5036\angle -65.30}$$

$$V_4^{\,1} = 0.9850\angle -1.04°\ P.u = 0.9848 - j0.0179\ P.u$$

Result: The bus voltage at the and of first iteration are,

$$V_1^{\,1} = 1.06 + j0 = 1.06\angle 0°\ P.u$$

$$V_2^{\,1} = 1.0393 + j0.0371 = 1.04\angle\ 2.05°\ P.u$$

$$V_3^{\,1} = 1.0031 - j0.0154 = 1.0032\angle -0.88\ P.u$$

$$V_4^{\,1} = 0.9848 - j0.0179 = 0.9880\angle -1.04°\ P.u$$

Example 3.6: Let the reactive power constraint on generator bus-2 in example 3.5 be changed to $0.2 \le Q_2 \le 1.0$. With other data of example 3.5 remaining same, solve the voltages at the end of first iteration by Gauss-seidel method.

Solution: The formation of bus impedance matrix and calculation of $Q_{2\,,cal}^{\,1}$ are same as in example (3.5). The $Q_{2\,,cal}^{\,1}$ corresponding to initial gauess $V_2^{\,0} = 1.04 + j0$ is 0.1456 P.u. This value of Q_2 Violates the lower limit of the specified range for Q_2. Therefore the reactive power generation for bus-2 is fixed at 0.2 (lower limit) and the bus-2 is treated as load bus for first iteration. Now $V_2^{\,0} = 1 + j0.0$, similar to other load buses for first iteration. But P and Q are considered as positive for bus-2 and P and Q are negative for other load buses. The voltages are calculated using the following equation.

$$V_p^{k+1} = \frac{1}{Y_{pp}}\left[\frac{P_p - jQ_p}{\left(V_p^k\right)^*} - \sum_{q=1}^{p-1} Y_{pq}V_q^{k+1} - \sum_{q=p+1}^{n} Y_{pq}V_q^k\right]$$

$$V_1^{\,1} = V_1 = 1.06 + j0\ P.u \qquad (\therefore \text{ Bus-1 is a slack bus})$$

$$V^1{}_2 = \frac{1}{y_{22}}\left[\frac{P_2 - jQ_2}{(V_2^0)} - Y_{2,}V_1^1 - Y_{23}V_3^0 - Y_{24}V_4^0\right]$$

$$V_2^{\,1} = \frac{1}{2.6 - j11}\left[\frac{0.5 - j0.2}{1 - j0} - (-1 + j5)(1.06 + j0) - (-0.5 + j4)\right.$$
$$\left.(1 + j0) - (-1.1 + j2)(1 + j6)\right]$$

$$= \frac{1}{2.6 - j11}[0.5 - j0.02 + 1.06 - j5.3 + 0.5 - j4 + 1.1 - j2]$$

$$= \frac{3.16 - j11.5}{2.6 - j11} = \frac{11.9263\angle -74.64°}{11.3031\angle -76.70°}$$

$$= 1.0551\angle 2.06°\ P.u$$

$$V_2^{\,1} = 1.0544 + j0.0379\ P.u$$

$$V_3^{\,1} = \frac{1}{Y_{33}}\left[\frac{P_3 - jQ_3}{\left(V_3^0\right)^*} - Y_{3,}V_1^1 - Y_{32}V_2^1 - Y_{34}V_4^0\right]$$

$$= \frac{1}{2.9 - j11}\left[\frac{-0.4 + j0.3}{1 - j0} - (-1.2 + j4)(1.06 + j0) - (-0.5 + j4)\right.$$

$$\left.(1.0544 + j0.0379) - (-1.2 + j3)(1 + j0)\right]$$

$$= \frac{1}{2.9 - j11}[-0.4 + j0.3 + 1.272 - j4.24 - (-0.6788 + j4.1987) + 1.2 - j3]$$

$$= \frac{2.7508 - j8.4387}{2.9 - j11} = \frac{8.8757\angle -71.95^\circ}{11.3759\angle -75.23^\circ} = 0.7802\angle 3.28^\circ \ P.u$$

$$V_3^1 = 0.7789 + j0.0446 \ P.u$$

$$V_4^1 = \frac{1}{Y_{44}}\left[\frac{P_4 - jQ_4}{\left(V_4^0\right)^*} - Y_{41}V_1 - Y_{42}V_2^1 - Y_{43}V_3^1\right]$$

$$= \frac{1}{2.3 - j5}\left[\frac{-0.2 + j0.1}{1 - j0} - (0 \times 1.06) - (-1.1 + j2)(1.0544 + j0.0379)\right.$$

$$\left.-(-1.2 + j3)(0.7789 + j0.0446)\right]$$

$$= \frac{1}{2.3 - j5}[-0.2 + j0.1 - (-1.2356 + 2.0671) - (-1.0685 + j2.2832)]$$

$$= \frac{2.1041 - j4.2503}{2.3 - j5} = \frac{4.7426\angle -63.66^\circ}{5.5036\angle -65.30^\circ}$$

$$V_4^1 = 0.8617\angle -1.64^\circ \ P.u$$

$$= 0.8613 + j0.0247 \ P.u$$

Result:

$$V_1^1 = 1 + .06 + j0 = 1.06\angle 0^\circ \ P.u$$

$$V_2^1 = 1.0544 + j0.0379 = 1.0551\angle 2.06^\circ \ P.u$$

$$V_3^1 = 0.7789 + j0.0446 = 0.7802 \ \angle 3.28^\circ \ P.u$$

$$V_4^1 = 0.8613 + j0.0247 = 0.8617 \ 0\text{-}1.64^\circ \ P.u$$

Example 3.7: In the system shown in Fig (3.7.1), generators are connected to all the four buses, while loads are connected to buses 2 and 3. The specifications of the buses are given in table (*T.* 3.7.1) and line impedance in table (*T.* 3.7.2). Assume that all the buses other than slack bus are *PQ* type. By taking a flat voltage profile, determine the bus voltages at the and of first Gauss-Seidel iteration.

Table-T. 3.7.1

Bus code	P	Q	V	Remarks
1	–	–	$1.05\angle 0°$	slack bus
2	0.5	–0.2	–	PQ bus
3	–1.0	0.5	–	PQ bus
4	0.3	–0.1	–	PQ bus

Table T. 3.7.2

Line	R in P.u	X in P.u
1–2	0.05	0.15
1–3	0.10	0.30
1–4	0.20	0.40
2–4	0.10	0.30
3–4	0.05	0.15

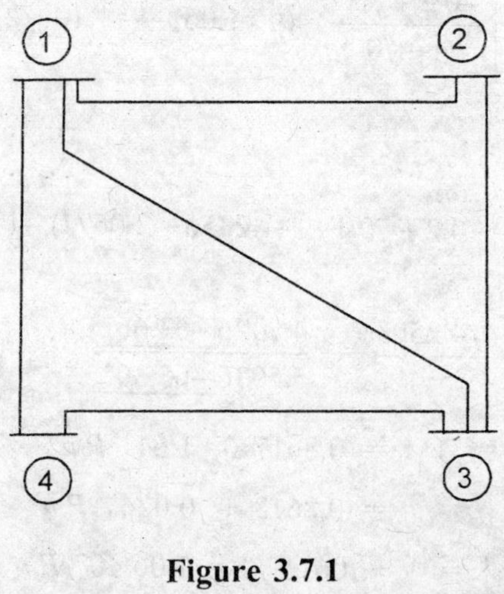

Figure 3.7.1

Solution: First convert the line impedances to line admittance. The line admittances are given by the inverse of line impedances.

From table (*T.* 3.7.2) we get the following line impedances.

$$Z_{12} = 0.05 + j0.15\ P.u$$
$$Z_{13} = 0.10 + j\ 0.30\ P.u$$
$$Z_{14} = 0.20 + j0.40\ P.u$$
$$Z_{24} = 0.10 + j0.30\ P.u$$
$$Z_{34} = 0.05 + j0.15\ P.u$$

$$\therefore \quad Y_{12} = \frac{1}{Z_{12}} = \frac{1}{0.05 + j0.15} = 2 - j6$$

$$Y_{14} = \frac{1}{Z_{14}} = \frac{1}{0.20 + j0.40} = 1 - j2$$

$$Y_{34} = \frac{1}{Z_{34}} = \frac{1}{0.05 + j0.15} = 2 - j6$$

$$Y_{13} = \frac{1}{Z_{13}} = \frac{1}{0.10 + j0.30} = 1 - j3$$

$$Y_{24} = \frac{1}{Z_{24}} = \frac{1}{0.10 + j0.30} = 1 - j3$$

The elements of bus admittance matrix can be obtained from the line admittances as shown below.

$$Y_{11} = Y_{12} + Y_{13} + Y_{14} = 2 - j6 + 1 - j3 + 1 - j2 = 4 - j11$$

$$Y_{22} = Y_{12} + Y_{24} = 2 - j6 + 1 - j3 = 3 - j9$$

$$Y_{33} = Y_{13} + Y_{34} = 1 - j3 + 2 - j6 = 3 - j9$$

$$Y_{44} = Y_{14} + Y_{24} + Y_{34} = 1 - j2 + 1 - j3 + 2 - j6 = 4 - j11$$

$$Y_{12} = Y_{21} = -Y_{12} = -(2 - j6) = -2 + j6$$

$$Y_{13} = Y_{31} = -Y_{13} = -(1 - j3) = -1 + j3$$

$$Y_{14} = Y_{41} = -Y_{14} = -(1 - j2) = -1 + j2$$

$$Y_{23} = Y_{32} = 0$$

$$Y_{24} = Y_{42} = -Y_{24} = -(1 - j3) = -1 + j3$$

$$Y_{34} = Y_{43} = -Y_{34} = -(2 - j6) = -2 + j6$$

$$Y_{bus} = \begin{bmatrix} Y_{11} & Y_{12} & Y_{13} & Y_{14} \\ Y_{21} & Y_{22} & Y_{23} & Y_{24} \\ Y_{31} & Y_{32} & Y_{33} & Y_{34} \\ Y_{41} & Y_{42} & Y_{43} & Y_{44} \end{bmatrix}$$

$$= \begin{bmatrix} 4-j11 & -2+j6 & -1+j3 & -1+j2 \\ -2+j6 & 3-j9 & 0 & -1+j3 \\ -1+j3 & 0 & 3-j9 & -2+j6 \\ -1+j2 & -1+j3 & -2+j6 & 4-j11 \end{bmatrix}$$

The initial value of load bus voltages can be assumed as $1 + j0$ $P.u$

$$\therefore \quad V_2^0 = 1 + j0$$
$$V_3^0 = 1 + j0$$
$$V_4 = 1 + j0$$

The slack bus (bus–1) voltage is not modified in any iteration.

$$\therefore \quad V_1^0 = V_1^1 = \cdots\cdots = V_1^n = V_1 = 1.05 + j0 \; P.u$$

The load bus voltages can be calculated using the following equation. For the first iteration $K = 0$. Here The table (T.3.7.1) Provides the net power and so the volues of P and Q are used as such in the following equation.

$$V_P^{k+1} = \frac{1}{Y_{pp}}\left[\frac{P_p - jQ_p}{\left(V_p^k\right)*} - \sum_{q=1}^{p-1} Y_{pq}V_q^{k+1} - \sum_{q=p+1}^{n} Y_{pq}V_q^k\right]$$

$$V_1^1 = V_1 = 1.05 + j0 = 1.05\angle 0° \qquad (\because \text{ Bus–1 isslack bus})$$

$$V_2^1 = \frac{1}{Y_{22}}\left[\frac{P_2 - jQ_2}{\left(V_2^0\right)*} - Y_{21}V_1^1 - Y_{23}V_3^0 - Y_{24}V_4^0\right]$$

$$= \frac{1}{3.j9}\left[\frac{0.5 + j0.2}{1 - j0} - (-2 + j6)(1.05 + j0) - 0 \times (1 + j6) \\ -(-1 + j3)(1 + j0)\right]$$

$$= \frac{1}{3.j9}[0.5 + j0.2 + 2.1 - j6.3 + 1 - j3]$$

$$= \frac{3.6 - j9.1}{3 - j9} = \frac{9.7862\lfloor -68.42°}{9.4868\lfloor -71.57°}$$

$$= 1.0316\angle 3.15°$$

$$= 1.0306 + j0.0567 \; P.u$$

$$V_3^1 = \frac{1}{Y_{33}}\left[\frac{P_3 - jQ_3}{\left(V_3^0\right)*} - Y_{31}, V_1^1 - Y_{32}V_2^1 - Y_{34}V_4^0\right]$$

$$= \frac{1}{3-j9}\left[\frac{-1-j0.5}{1-j0} - (-1+j3)(1.05+j0) - (0+V_2^1) - (-2+j6)(1+j0)\right]$$

$$= \frac{1}{3-j9}[-1-j0.5+1.05-j3.15+2-j6]$$

$$= \frac{2.05-j9.65}{3-j9} = \frac{9.8653\angle -78.01°}{9.4868\angle -71.57°}$$

$$= 1.0399\angle -6.44°$$

$$= 1.0333 - j0.1166 \; P.u$$

$$V_4^{\,1} = \frac{1}{Y_{44}}\left[\frac{P_4 - jQ_4}{\left(V_4^0\right)*} - Y_{41}V_1^1 - Y_{42}V_2^1 - Y_{43}V_3^1\right]$$

$$= \frac{1}{4-j11}\left[\frac{0.3+j0.1}{1-j0} - (-1+j2)(1.05+j0) - (-1+j3)(1.03+j0.0567) - (-2+j6)(1.0333-j0.1166)\right]$$

$$= \frac{1}{4-j11}\left[0.3+j0.1+1.05-j2.1-(-1.2001+j3.0333) - (1.367+j6.433)\right]$$

$$= \frac{3.9171-j11.4663}{4-j11} = \frac{12.1169\angle -71.14°}{11.7047\angle -70.02°}$$

$$V_4^{\,1} = 1.0352\angle -1.12° = 1.0350 - j0.0202 \; P.u$$

Result:

$$V_1^{\,1} = 1 + .05 + j0 = 1.05\angle 0° \; P.u$$

$$V_2^{\,1} = 1.0300 + j0.0567 = 1.0316\angle 3.15° \; P.u$$

$$V_3^{\,1} = 1.0333 - j0.1166 = 1.0399\angle -6.44° \; P.u$$

$$V_4^{\,1} = 1.0350 - j0.0202 = 1.0352\angle -1.12° \; P.u$$

Example 3.8: In example 3.7 Let the bus-2 be a *PV* bus (Generator bus) with $|V_2| = 1.07 \; P.u.$ The reactive power constraint of the generator bus is $0.3 \le Q_2 \le 1.0$. with other data of example 3.7 remaining same (except Q_2), calculate the bus voltages at the and of first iteration using G-s method.

Solution: The bus admittance matrix is same as that of example 3.7 the initial value of *PQ* bus (Load bus) voltages can be assumed as $1 + j0 \; P.u.$

$$\therefore \quad V_3^{\,0} = 1 + j0 \qquad V_4^{\,0} = 1 + j0$$

The bus-1 is a slack bus,

$$\therefore \quad V_1^{\,0} = V_1^{\,1} = \dots\dots\dots = V_1^{\,n} = V_1 = 1.05 + j0$$

The bus-2 is a generator bus,

$\therefore \quad V_2^0 = 1.07 + j0$

Since the bus-2 is a generator bus its reactive power has to be calculated and checked for violation of specified limits.

(In first iteration $K = 0$)

$$Q_{2,cal}^1 = (-1) \times I_m \left\{ (V_2^0)^* \left[Y_{21}V_1^1 + Y_{22}V_2^0 + Y_{23}V_3^0 + Y_{24}V_4^0 \right] \right\}$$

$$= -1 \times I_m \left\{ (1.07 - j0) \begin{bmatrix} (-2 + j6)(1.05 + j0) + (3 - j9)(1.07 + j0) \\ +(0 + V_3^0) + (-1 + j3)(1 + j0) \end{bmatrix} \right\}$$

$$= -1 \times I_m \{1.07[-2.1 + j6.3 + 3.21 - j9.63 - 1 + j3]\}$$

$$= -1 \times I_m \{1.07 [0.11 - j0.33]\} = 0.3531 \ P.u$$

The calculated value of reactive power, $Q_{2,\ cal}^1$ is within the specified limits. Hence the bus-2 can be treated as generator bus. Now, $Q_2 = Q_{2,\ cal}^1 = .3531 \ P.u$, $P_2 = 0.5$ and $|V_2| = 1.07$ Since the bus-2 is a generator bus, magnitude of the bus-2 voltage remains at 1.07 $P.u$. The phase of bus-2 voltage can be calculated using the following equation.

$$V_{p,temp}^{k+1} = \frac{1}{Y_{pp}} \left[\frac{P_p - jQ_p}{(V_p^k)^*} - \sum_{q=1}^{p-1} Y_{pq}V_q^{k+1} - \sum_{q=p+1}^{n} Y_{pq}V_q^k \right]$$

$$V_{2,temp}^1 = \frac{1}{Y_{22}} \left[\frac{P_2 - jQ_2}{\left(V_2^0\right)^*} - Y_{21}V_1^1 - Y_{23}V_3^0 - Y_{24}V_4^0 \right]$$

$$= \frac{1}{3 - j2} \left[\frac{0.5 - j0.3531}{1.07 - j0} - (-2 + j6) \right.$$

$$\left. (1.05 + j0) - (0 \times V_3^0) - (-1 + j3)(1 + j0) \right]$$

$$= \frac{1}{3 - j9} [0.4673 - j0.33 + 2.1 - j6.3 + 1 - j3]$$

$$= \frac{3.5676 - j9.63}{3 - j9} = \frac{10.2695\angle -69.67}{9.4868\angle -71.57^\circ} = 1.0825\angle 1.9^\circ$$

$$\therefore \quad \delta_2^1 = \angle V_{2,\text{temp}}^1 = 1.9^\circ$$

$$\therefore \quad V_2^1 = |V_2|_{Spec} \angle \delta_2^1$$

$$= 1.07\angle 1.9^\circ \ P.u.$$

$$= 1.0694 + j0.0355 \ P.u.$$

The voltage of bus-3 and bus-4 are calculated using the following equation.

$$V_p^{k+1} = \frac{1}{Y_{pp}}\left[\frac{P_p - jQ_p}{\left(V_p^k\right)^*} - \sum_{q=1}^{p-1} Y_{pq}V_q^{k+1} - \sum_{q=p+1}^{n} Y_{pq}V_q^k\right]$$

$$V_3^1 = \frac{1}{Y_{33}} = \left[\frac{P_3 - jQ_3}{\left(V_3^0\right)^*} - Y_{31}V_1^1 - Y_{32}V_2^1 - Y_{34}V_4^0\right]$$

$$= \frac{1}{3 - j9}\left[\frac{-1 - j0.5}{1 - j0} - (-1 + j3)(1.05 + j0)\right.$$

$$\left. - (0 \times V_2^1) - (-2 + j6)(1 + j0)\right]$$

$$= \frac{1}{3 - j9}[-1 - j0.5 + 1.05 - j3.15 + 2 - j6]$$

$$= \frac{2.05 - j9.65}{3 - j9} = \frac{9.8653\lfloor -78.01^\circ}{9.4868\lfloor -71.57^\circ}$$

$$= 1.0399 \ \angle -6.44^\circ = 1.0333 - j0.1166 \ P.u$$

$$V_4^1 = \frac{1}{Y_{44}} = \left[\frac{P_4 - jQ_4}{\left(V_4^0\right)^*} - Y_{41}V_1^1 - Y_{42}V_2^1 - Y_{43}V_3^1\right]$$

$$= \frac{1}{4 - j11}\left[\frac{0.3 + j0.1}{1 - j0} - (-1 + j2)(1.05 + j0) - (-1 + j3)\right.$$

$$\left.(1.0694 + j0.0355) - (-2 + j6)(1.0333 - j0.1166)\right]$$

$$= \frac{1}{4 - j11}[0.3 + j0.1 + 1.05 - j2.1$$

$$-(-1.1759 + j3.1727) - (-1.367 + j6.433)$$

$$= \frac{3.8929 - j11.6057}{4 - j11} = \frac{12.2412\lfloor -71.46^\circ}{11.7047\lfloor -70.02^\circ}$$

$$V_4^1 = 1.0458\angle -1.44^\circ = 1.0455 - j0.0263 \ P.u$$

Result: $V_1^1 = 1.05 + j0 = 1.05\angle 0^\circ \ P.u$

$$V_2^1 = 1.0694 + j0.0355 = 1.07\angle 1.9°\ P.u$$

$$V_3^1 = 1.0333 - j0.1166 = 1.0399\ \angle -6.44°\ P.u$$

$$V_4^1 = 1.0455 - j0.0263 = 1.0458\angle -1.44°\ P.u$$

Example. 3.9: For the network shown in Fig. (3.9.1), obtain the complex bus bar voltages at the end of first iteration, using *G-S* method. Bus-1 is a slack bus with $V_1 = 1.0\angle 0°$, Take $P_2 + jQ_2 = -5.96 + j1.46$, $P_3 = 6.02$ and $|V_3| = 1.02$ Assume, $V_3^0 = 1.02\angle 0°$ and $V_2^0 = 1\angle 0°$

➡ Line impedances are in *P.u*

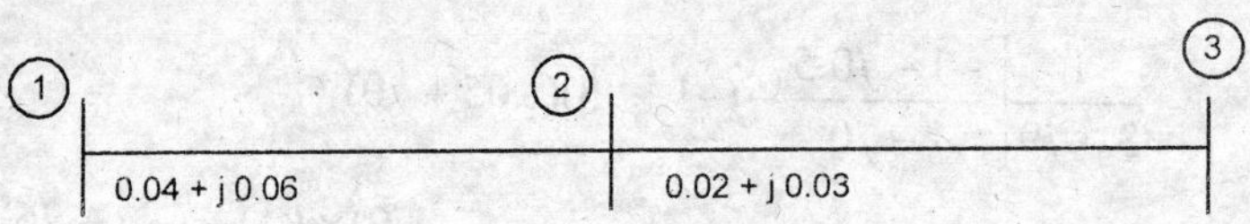

Figure 3.9.1

Solution: From Fig. 3.9.1, the line impedances are

$$Z_{12} = 0.04 + j0.06\ P.u$$
$$Z_{13} = 0.02 + j0.03\ P.u$$

The line admittances are given by the inverse of line impedances.

$$Y_{12} = \frac{1}{Z_{12}} = \frac{1}{0.04 + j0.06} = 7.692 - j11.538\ P.u$$

$$Y_{23} = \frac{1}{Z_{23}} = \frac{1}{0.02 + j0.03} = 15.385 - j23.077\ P.u$$

The elements of bus admittance matrix are formed from the line admittances as shown below

$$Y_{11} = Y_{12} = 7.692 - j11.538$$
$$Y_{22} = Y_{12} + Y_{23} = 7.692 - j11.538 + 15.385 - j23.077$$
$$= 23.077 - j34.615$$
$$Y_{33} = Y_{23} = 15.385 - j23.077$$
$$Y_{12} = Y_{21} = -Y_{12} = -(-7.692 - j11.538)$$
$$= -7.692 + j11.538$$
$$Y_{13} = Y_{31} = 0$$
$$Y_{23} = Y_{32} = -Y_{23} = -(15.385 - j23.077)$$
$$= -15.385 + j23.077$$

$$Y_{bus} = \begin{bmatrix} Y_{11} & Y_{12} & Y_{13} \\ Y_{21} & Y_{22} & Y_{23} \\ Y_{31} & Y_{32} & Y_{33} \end{bmatrix}$$

$$= \begin{bmatrix} 7.692 - j11.538 & -7.692 + j11.538 & 0 \\ -7.692 + j11.538 & 23.077 - j34.615 & -15.385 + j23.077 \\ 0 & -15.385 + 23.077 & 15.385 - j23.077 \end{bmatrix}$$

In the given system, bus-1 is slack bus, The bus-2 is load bus and bus 3 is generator bus. The initial value of the phase of V_3 is assunced as zero.

$$\therefore \quad V_1^0 = 1.0\angle 0°$$
$$V_2^0 = 1.0\angle 0°$$
$$V_3^0 = 1.02\angle 0°$$

The bus-1 is a slack bus and so it's voltage will not change

$$\therefore \quad V_1^1 = V_1^0 = V_1 = 1.0\angle 0° = 1 + j0 \ P.u$$

The bus-2 ia a load bus. The voltage of bus-2 is calculated using the following equation.

The first iteration, $K = 0$

$$V_p^{k+1} = \frac{1}{Y_{pp}}\left[\frac{P_p - jQ_p}{\left(V_p^k\right)*} - \sum_{q=1}^{p-1} Y_{pq}V_q^{k+1} - \sum_{q=p+1}^{n} Y_{pq}V_q^k\right]$$

$$V_2^1 = \frac{1}{Y_{22}}\left[\frac{P_2 - jQ_2}{\left(V_2^0\right)*} - Y_{21}V_1^1 - Y_{23}V_3^0\right]$$

$$= \frac{1}{23.077 - j34.615}\left[\begin{aligned}&\frac{-5.96 - j1.46}{1 - j0} - (-7.692 + j11.538)(1 + j0)\\ &-(-15.385 + j23.077)(1.02 + j0)\end{aligned}\right]$$

$$= \frac{1}{23.077 - j34.615}\left[-5.96 - j1.46 + 7.692 - j11.538 + 15.6927 - j23.5385\right]$$

$$= \frac{17.4247 - j36.5365}{23.077 - j34.615} = \frac{40.4788\angle -64.50°}{41.6022\angle -56.31°} = 0.973\angle - 8.19°$$

$$V_2^1 = 0.99607 - j0.13861 \ P.u$$

The bus-3 is a generator bsu.The magnitude of the voltage does not change in generator bus. The phase of the voltage can be calculateed from the following equation.

$$V_{p,temp}^{k+1} = \frac{1}{Y_{pp}}\left[\frac{P_p - jQ_p}{\left(V_p^k\right)*} - \sum_{q=1}^{p-1} Y_{pq}V_q^{k+1} - \sum_{q=p+1}^{n} Y_{pq}V_q^k\right]$$

The phase of V_3^1 is the phase of $V_{3\,,\,\text{temp}}^1$ and the magnitude of V_3^1 is the specified voltage.

i.e, $|V_3^1| = |V_3|_{spce} = 1.02\ P.u$

Let initial value of bus-2 voltage, $V_3^0 = 1.02 + j0$

In order to calculate the phase of V_3^1 we require the value of Q_3 which can be calculated using the following equation.

$$Q_{3,cal}^1 = -1 \times I_m\left\{\left(V_3^0\right)*\left[Y_{31}V_1^1 + Y_{32}V_2^1 + Y_{33}V_3^0\right]\right\}$$

$$= -1 \times I_m\left\{(1.02 - j0)\left[\begin{matrix}(0 \times V_1^1) + (-15.385 + j23.077) \\ (0.96307 - j0.13861) + (15.38 - j23.077)(1.02 + j0\end{matrix}\right]\right\}$$

$$= -1 \times I_m\ \{1.02\ [-11.6181 + j24.3573 + 15.6927 - j23.5385]\}$$

$$= -08351\ P.u$$

> ➡ For the generator bus the limits of reactive power are not specified and so the calculated reactive power can be used as such.

$$V_{3,temp}^1 = \frac{1}{Y_{33}}\left[\frac{P_3 - jQ_3}{\left(V_3^0\right)^*} - Y_{31}V_1^1 - Y_{32}V_2^1\right]$$

$$= \frac{1}{15.385 - j23.077}\left[\begin{matrix}\frac{6.02 + j0.8351}{1.02 - j0} - (0 \times V_1^1) - (-15.385 + j23.077) \\ (0.96307 - 13861)\end{matrix}\right]$$

$$= \frac{1}{15.385 - j23.077}[5.90196 + j0.81872 - (011.6181 + j24.3566)]$$

$$= \frac{17.52006 - j23.53788}{15.385 - j23.077} = \frac{29.3425\angle -53.34°}{27.73529\angle -56.31°} = 1.05795\angle 2.97°$$

$\therefore$ $\delta_3^1 = \angle V_{3\,,\,temp}^1 = 2.97°$

$\therefore$ $V_3^1 = |V_3|_{\text{spec}} \angle\delta_3^1 = 1.62\angle 2.97°\ P.u$

$V_3^1 = 1.08163 + j0.05285\ P.u$

Result: $V_1^1 = 1 + j0 = 1.0\angle 0°\ P.u$

$V_2^1 = 0.96307 - j0.13861 = 0.973\angle -8.19°\ P.u$

$V_3^1 = 1.01863 + j0.05285 = 1.02\angle 2.97°\ P.u$

Example 3.10: Fig. (3.10.1) shows a three bus power system.

Bus 1: Slack bus, $V = 1.05\angle 0°\ P.u$

Bus 2: PV bus, $|V| = 1.0$ P.u, $P_g = 3\ P.u$

Bus 3: PQ bus, $P_L = 4\ P.u,\ Q_L = 2\ P.u$

Carry out one iteration of Load flow solution by Gauss-seidel method. Neglect limits on reactive Power generation.

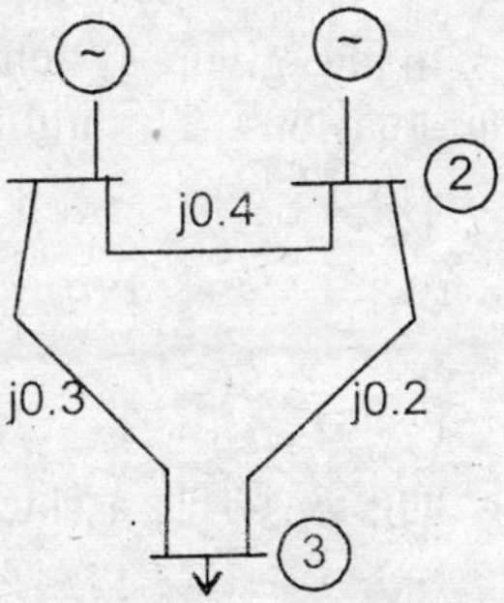

Figure 3.10.1

Solution: From Fig.(3.10.1),the line impedances are,

$$Z_{12} = j0.4\ P.u$$
$$Z_{13} = j0.3\ P.u$$
$$Z_{23} = +\ j0.2\ P.u$$

The line admittances are given by the inverse of line impedances.

$$Y_{12} = \frac{1}{Z_{12}} = \frac{1}{j0.4} = -\ j2.5\ P.u$$

$$Y_{13} = \frac{1}{Z_{13}} = \frac{1}{j0.3} = -\ j\ 3.333\ P.u$$

$$Y_{23} = \frac{1}{Z_{23}} = \frac{1}{j0.2} = -\ j\ 5\ P.u$$

The element of bus admittance matrix are formed from the line admittances as shown below.

$$Y_{11} = Y_{12} + Y_{13} = -\ j2.5 - j3.33 = -j5.8333$$
$$Y_{22} = Y_{12} + Y_{23} = -\ 2.5 - j5 = -\ j7.5$$
$$Y_{33} = Y_{13} + Y_{23} = -\ j3.33 - j5 = -\ j8.333$$
$$Y_{12} = Y_{21} = -\ Y_{12} = -(-\ j2.5) = j2.5$$
$$Y_{13} = Y_{31} = -Y_{13} = -(j3.333) = j3.333$$
$$Y_{23} = Y_{32} = -Y_{23} = -(-\ j5) = j5$$

$$Y_{bus} = \begin{bmatrix} Y_{11} & Y_{12} & Y_{13} \\ Y_{21} & Y_{22} & Y_{23} \\ Y_{31} & Y_{32} & Y_{33} \end{bmatrix}$$

$$= \begin{bmatrix} -j5.833 & j2.5 & j3.333 \\ j2.5 & -j7.5 & j5 \\ j3.333 & j5 & -j8.333 \end{bmatrix}$$

In the given system bus-1 is a slack bs, the bus-2 is generator bus and bus-3 is load bus. The initial values of bus voltages are as follows.

$$V_1^0 = 1.05\angle 0° = 1.05 + j0\ P.u \text{ (specified value)}$$
$$V_2^0 = 1.0\angle 0° = 1.0 + j0\ P.u \text{ (Initial phase is assumed zero)}$$
$$V_3^0 = 1 + j0 \text{ (Assumed value)}$$

The bus-1 is a slack bus and so its voltage will not change in any iteration.

$$\therefore \quad V_1^1 = V_1^0 = 1.05\angle 0° = 1.05 + j0\ P.u$$

For the bus-2, *ie.*, generator bus the magnitude of the voltage is the specified value. The phase of the bus-2 voltage in first iteration is given by phase of $V_{p,temp}^{k+1}$, when $P = 0$ and $K = 0$.

$$V_{p,temp}^{k+1} = \frac{1}{Y_{pp}}\left[\frac{P_p - jQ_p}{\left(V_p^k\right)^*} - \sum_{q=1}^{p-1} Y_{pq}V_q^{k+1} - \sum_{q=p+1}^{n} Y_{pq}V_q^k\right]$$

To calculate the phase using the above equation we have to estimate the reactive power Q_2. The reactive power of bus-2 is given by the following equation.

$$Q_{p,cal}^{k+1} = (-1)\times I_m\left\{\left(V_P^k\right)^*\left[\sum_{q=1}^{P-1} Y_{pq}V_q^{k+1} + \sum_{q=p}^{n} Y_{pq}V_q^k\right]\right\}$$

$$\therefore \quad Q_{2,cal}^1 = (-1)\times I_m\left\{\left(V_2^0\right)^*\left[Y_{21}V_1^1 + Y_{22}V_2^0 + Y_{23}V_3^0\right]\right\}$$

$$= -1\times I_m\left\{(1-j0)\left[j2.5(1.05+j0) + (-j7.5)(1+j0) + j5(1+j0)\right]\right\}$$

$$= -1 \times I_m\ \{j2.625 - j7.5 + j5\} = -0.125\ P.u$$

Now,
$$Q_2 = -0.125,$$
$$P_2 = 3$$
$$V_2^0 = 1 + j0$$
$$|V_2|_{spoc} = 1.0$$

$$\therefore \quad V_{2,\ temp}^1 = \frac{1}{Y_{22}}\left[\frac{P_2 - jQ_2}{\left(V_2^0\right)^*} - Y_{21}V_1^1 - Y_{23}V_3^0\right]$$

$$= \frac{1}{-j7.5}\left[\frac{3 + j0.125}{1 - j0} - (j2.5)(1.05 + j0) - (j5)(1 + j0)\right]$$

$$= \frac{1}{-j.5}[3 + j0.125 - j2.625 - j5]$$

$$= \frac{1}{-j7.5}[3 - j7.5]$$

$$V_{2,\ \text{temp}}^{1} = 1 + j0.4 = 1.077\angle 21.8°$$

$$\therefore \quad \delta_2^{\ 1} = \angle V_{2,\ temp}^{1} = 21.8°$$

$$\therefore \quad V_2^{\ 1} = |V_2|_{spec}\ \angle\delta_2^1 = 1.0\angle 21.8° = 0.92849 + j0.37137.$$

The bus-3 in load bus and its voltage in first iteration is given by the following equation when $p = 3$ and $k = 0$. The specified power are load powers, and so they are considered as negative powers.

$$\therefore \quad P_3 = -P_L = -4$$

and

$$Q_3 = -Q_L = -2$$

$$V_P^{k+1} = \frac{1}{Y_{pp}}\left[\frac{P_p - jQ_p}{\left(V_p^k\right)^*} - \sum_{q=1}^{p-1} Y_{pq}V_q^{k+1} - \sum_{q=p+1}^{n} Y_{pq}V_q^k\right]$$

$$V_3^{\ 1} = \frac{1}{Y_{33}}\left[\frac{P_3 - jQ_3}{\left(V_3^k\right)^*} - Y_{31}V_1^1 - Y_{32}V_2^1\right]$$

$$= \frac{1}{-j8.333}\left[\frac{-4+j2}{1-j0} - (j3.333)(1.05+j0) - (j5)(0.92849 + j0.371373\right]$$

$$= \frac{1}{-j8.333}\left[-4 + j2 - j3.49965 + 1.85685 - j4.64245\right]$$

$$= \frac{-2.14315 - j6.1421}{-j8.333} = \frac{6.50527\underline{|-109.24°}}{8.333\underline{|-90°}}$$

$$V_3^{\ 1} = 0.78064\angle -19.24° = 0.73704 - j0.25724\ P.u$$

Result:

$$V_1^{\ 1} = 1.05 + j0 = 1.05\angle 0°\ P.u$$

$$V_2^{\ 1} = 0.92849 + j.37137 = 1.0\angle 21.8°$$

$$V_3^{\ 1} = 0.73704 - j0.25724 = 0.78064\ \angle -19.24°\ P.u$$

3.6 LOAD FLOW SULUTION BY NEWTON RAPHSON METHOD

The newton-Raphson method of load flow analysis is an iterative method which approximates the set of non-linear simultaneous (load flow) equations to a set of linear simulataneous (load flow) equation using Taylor's series expansion and the terms are limited to first order approximation.

The load flow equations for newton-Rephson method are developed in equations (3.51), (3.52) and (3.53) are non Linear equations in terms of real and imaginary part of bus voltages.

$$P_p = \sum_{q=1}^{n} \left\{ e_p \left[e_p G_{pq} + f_q B_{pq} \right] + f_p \left[f_q G_{pq} - e_q B_{pq} \right] \right\} \quad ...(3.51)$$

$$Q_p = \sum_{q=1}^{n} \left\{ f_p \left[e_p G_{pq} + f_q B_{pq} \right] + e_p \left[f_q G_{pq} - e_q B_{pq} \right] \right\} \quad ...(3.52)$$

$$|V_p|^2 = e_p^{\,2} + f_p^{\,2} \quad ...(3.53)$$

Where, e_p = Reat part of V_p(Voltage of bus-P)

f_q = Imaginary Part of V_p(voltage of bus–P)

G_{pq}, B_{pq} = Conductance and Susceptance of admittance Y_{pq} respectively.

The voltage of slack bus will be a known quantity in a power system and so it need not be solved. For load bases P_p and Q_p will be specified and we have to solve V_p. For a generator bus P_p and $|V_p|$ will be specified and we have to solve Q_p and phase of V_p.

In order to calculate V_p of load bus and Q_p and phase of V_p for generator bus we have to first calculate the real and imaginary part of bus voltages. Therefore the unknown quantities are neal part (e_p) of bus boltages and imaginary prt(f_p) of bus voltages.

For a n-bus system, the bus-1 is considered as slack bus and the real and imaginary part of remaining (n – 1) bus voltages have to be solved. Hence we have 2(n – 1) Variables to be solved and they are e_2, e_3....e_n. f_2, f_3.....f_n. Thus to slolve the problem for 2(n – 1) variables wee need to solve 2(n – 1) set of equations.

Mathematical calculations for newton-raphson method

Let x_1, x_2..........x_n be a set of unknown variables and y_1, y_2.....y_0 be a set specified quantities. Now the specified quantities can be expressed as a non-linear function of unknown variables as shown below.

$$\left.\begin{aligned} Y_1 &= f_1(x_1, x_2 ... x_n) \\ Y_2 &= f_2(x_1, x_2 ... x_n) \\ \vdots \quad & \quad \vdots \\ Y_n &= f_n(x_1, x_2 ... x_n) \end{aligned}\right\} \quad ...(3.54)$$

Let as assume an approzimate initial solution $x_1^\circ, x_2^\circ, ...x_n^\circ$. For the above equations (3.54). The prefix zero refers to zeroth iteration in the process of solving the above non-linear equations.

Let $\Delta x_1^{\,0}$, $\Delta x_2^{\,0}$, $\Delta x_3^{\,0}$....$\Delta_n^{\,0}$ are the corrections required for x_1°, x_2°, ...x_n°.x_1°, x_2°, ...x_n°, respectively for the next better solution. Now the non-liners equations can be expressed as shown below, i.e,. they can be pressed as functions of modified variables,

$x_1^{\,0} + \Delta x_1^{\,0}$, $x_2^{\,0} + \Delta x_2^{\,0}$,.....$x_n^{\,0} + \Delta x_n^{\,0}$.

$$\left.\begin{aligned}Y_1 &= f_1(x_1^0 + \Delta x_1^0, x_2^0 + \Delta x_2^0 \ldots x_n^0 + \Delta x_n^0)\\ Y_2 &= f_2(x_1^0 + \Delta x_1^0, x_2^0 + \Delta x_2^0 \ldots x_n^0 + \Delta x_n^0)\\ &\vdots\\ Y_n &= f_n(x_1^0 + \Delta x_1^0, x_2^0 + \Delta x_2^0 \ldots x_n^0 + \Delta x_n^0)\end{aligned}\right\} \quad \ldots(3.55)$$

The above equations are lineraised about the initial guess using Taylor's series expansion. The Linearised equations with second order asn higher derivatives neglecter given below.

$$\left.\begin{aligned}y_1 &= f_1(x_1^0, x_2^0, \ldots. x_n^0) + \Delta x_1^0 \frac{\partial f_1}{\partial x_1} + \Delta x_2^0 \frac{\partial f_1}{\partial x_2} + \ldots. + \Delta x_n^0 \frac{\partial f_1}{\partial x_n}\\ y_2 &= f_2(x_1^0, x_2^0, \ldots. x_n^0) + \Delta x_1^0 \frac{\partial f_2}{\partial x_1} + \Delta x_2^0 \frac{\partial f_2}{\partial x_2} + \ldots. + \Delta x_n^0 \frac{\partial f_2}{\partial x_n}\\ &\vdots\\ y_n &= f_n(x_1^0, x_2^0, \ldots. x_n^0) + \Delta x_1^0 \frac{\partial f_n}{\partial x_1} + \Delta x_2^0 \frac{\partial f_n}{\partial x_2} + \ldots. + \Delta x_n^0 \frac{\partial f_n}{\partial x_n}\end{aligned}\right\} \quad \ldots(3.56)$$

The equation (3.56) can be rearranged as shown in equation (3.57)

$$\left.\begin{aligned}&y_1 - f_1(x_1^0, x_2^0, \ldots. x_n^0) + \Delta x_1^0 \frac{\partial f_1}{\partial x_1} + \Delta x_2^0 \frac{\partial f_1}{\partial x_2} + \ldots. + \Delta x_n^0 \frac{\partial f_1}{\partial x_n}\\ &y_2 - f_2(x_1^0, x_2^0, \ldots. x_n^0) + \Delta x_1^0 \frac{\partial f_2}{\partial x_1} + \Delta x_2^0 \frac{\partial f_2}{\partial x_2} + \ldots. + \Delta x_n^0 \frac{\partial f_2}{\partial x_n}\\ &\vdots\\ &y_n - f_n(x_1^0, x_2^0, \ldots. x_n^0) + \Delta x_1^0 \frac{\partial f_n}{\partial x_1} + \Delta x_2^0 \frac{\partial f_n}{\partial x_2} + \ldots. + \Delta x_n^0 \frac{\partial f_n}{\partial x_n}\end{aligned}\right\} \quad \ldots(3.57)$$

Let,

$$\begin{aligned}y_1 - f_1(x_1^0, x_2^0, \ldots., x_n^0) &= \Delta y_1\\ y_2 - f_2(x_1^0, x_2^0, \ldots., x_n^0) &= \Delta y_2\\ &\vdots\\ y_n - f_1(x_1^0, x_2^0, \ldots., x_n^0) &= \Delta y_n\end{aligned}$$

Now the equation (3.55) can be written as

$$\left.\begin{aligned}\Delta y_1 &= \Delta x_1^0 \frac{\partial f_1}{\partial x_1} + \Delta x_2^0 \frac{\partial f_1}{\partial x_2} + \ldots. + \Delta x_n^0 \frac{\partial f_1}{\partial x_n}\\ \Delta y_2 &= \Delta x_1^0 \frac{\partial f_2}{\partial x_1} + \Delta x_2^0 \frac{\partial f_2}{\partial x_2} + \ldots. + \Delta x_n^0 \frac{\partial f_2}{\partial x_n}\\ &\vdots\\ \Delta y_n &= \Delta x_1^0 \frac{\partial f_n}{\partial x_1} + \Delta x_2^0 \frac{\partial f_n}{\partial x_2} + \ldots. + \Delta x_n^0 \frac{\partial f_n}{\partial x_n}\end{aligned}\right\} \quad \ldots(3.58)$$

The equation (3.58) can be arranged in the matrix form as shown in equation (3.59)

$$\begin{bmatrix} \Delta y_1 \\ \Delta y_2 \\ \vdots \\ \Delta y_n \end{bmatrix} = \begin{bmatrix} \frac{\partial f_1}{\partial x_1} & \frac{\partial f_1}{\partial x_2} & \cdots & \frac{\partial f_1}{\partial x_n} \\ \frac{\partial f_2}{\partial x_1} & \frac{\partial f_2}{\partial x_2} & \cdots & \frac{\partial f_2}{\partial x_n} \\ \vdots & & & \\ \frac{\partial f_n}{\partial x_1} & \frac{\partial f_n}{\partial x_2} & \cdots & \frac{\partial f_n}{\partial x_n} \end{bmatrix} \begin{bmatrix} \Delta x_1^\circ \\ \Delta x_2^\circ \\ \vdots \\ \Delta x_n^\circ \end{bmatrix} \quad \text{...(3.59)}$$

In matrix (vector) notation, the equation (3.59) cam be expressed as

$$B = JC \quad \text{..(3.60)}$$

where $B = \begin{bmatrix} \Delta y_1 \\ \Delta y_2 \\ \vdots \\ \Delta y_n \end{bmatrix}$ $J = \begin{bmatrix} \frac{\partial f_1}{\partial x_1} & \frac{\partial f_1}{\partial x_2} & \cdots & \frac{\partial f_1}{\partial x_n} \\ \frac{\partial f_2}{\partial x_1} & \frac{\partial f_2}{\partial x_2} & \cdots & \frac{\partial f_2}{\partial x_n} \\ \vdots & & & \vdots \\ \frac{\partial f_n}{\partial x_1} & \frac{\partial f_n}{\partial x_2} & \cdots & \frac{\partial f_n}{\partial x_n} \end{bmatrix}$

$$C = \begin{bmatrix} \Delta x_1^0 \\ \Delta x_2^0 \\ \vdots \\ \Delta x_n^0 \end{bmatrix}$$

Here J is the first derivative matrix and it is called Jacobian matrix. The elements of tacobian matrix are obtained by evaluating the first derivatives at the assumed solution. the B matrix is called residual column vector. The elements of matrix B are the difference between the specified quantities and calculated quantities at the assumed solution. With the elements of J and B are known, the elements of matrix C are obtained by solving the matrix equation (3.58) the C matrix is called increment solution vector.

The solution of the matrix equation (3.60) gives $\Delta x_1^0, \Delta x_2^0, \ldots \Delta x_n^0$. The next better solution is obtained as follows.

$$\left.\begin{aligned} x_1^1 &= x_1^0 + \Delta x_1^0 \\ x_2^1 &= x_2^0 + \Delta x_2^0 \\ \vdots \quad & \quad \vdots \\ x_n^1 &= x_n^0 + \Delta x_n^0 \end{aligned}\right\} \quad \text{...(3.61)}$$

With the new solution (equation (3.61)) the process is repeated to find next new solution. the iterative process is terminated if any one of the following condition is saisfied.

(*i*) The largest (magnitude of the) element in the B matrix is less than a prespecified value.

(*ii*) The largest (mgnitude of the) element in the C matrix is less than a prespecified value.

In this method the initial values assumed should be close to actual solution. Otherwise, there are chances of the solution diveging rather than converging and it may not possible to get a solution. In Power System Problems, the choice of initial values may not pose a Problem because, a flat voltage profile, *i.e.*, $V_p = 1 + j\,0$ for $P = 1, 2, \ldots, n$ except the slack bus has been found to be satisfactory for almost all practical systems.

How to apply Newton-Raphson Technique to Load Flow Problem

Consider a Power System with n-buses. The bus-1 is usually selected as slack bus. The other buses (*i.e.*, bus-2 to bus-n) can be either generator bus (or) load bus. for load buses the specified quantifies are P_p and Q_p. For generator buses the specified quantities and P_p and $|V_p|$.

Formation of 2(n-1) Equations for a System with n Buses

Case (*i*) : When all the (n-1) buses are load buses

In this case, bus-1 is a slack bus and the bus-2 to bus-n are load buses.

Let $P_2, P_3, \ldots, P_n$ be the specified real powers and $Q_2, Q_3, \ldots, Q_n$ be the specified reactive power of $(n - 1)$ load buses. The unknown variables are real part of bus voltages $e_2, e_3, \ldots e_n$ and reactive part of bus voltages $f_2, f_3, \ldots f_n$. Now the matrix equation $B = JC$ for this power system problem will be in the form shown in equation (3.62).

$$\begin{bmatrix} \Delta P_2 \\ \Delta P_3 \\ \vdots \\ \Delta P_n \\ \Delta Q_2 \\ \Delta Q_3 \\ \vdots \\ \Delta Q_n \end{bmatrix} = \begin{bmatrix} \frac{\partial P_2}{\partial e_2} & \frac{\partial P_2}{\partial e_3} & \cdots & \frac{\partial P_2}{\partial e_n} & \frac{\partial P_2}{\partial f_2} & \frac{\partial P_2}{\partial f_3} & \cdots & \frac{\partial P_2}{\partial f_n} \\ \frac{\partial P_3}{\partial e_2} & \frac{\partial P_3}{\partial f_3} & \cdots & \frac{\partial P_3}{\partial e_n} & \frac{\partial P_3}{\partial f_2} & \frac{\partial P_3}{\partial f_3} & \cdots & \frac{\partial P_3}{\partial f_n} \\ \frac{\partial P_n}{\partial e_2} & \frac{\partial P_n}{\partial e_3} & \cdots & \frac{\partial P_n}{\partial e_n} & \frac{\partial P_n}{\partial f_2} & \frac{\partial P_n}{\partial f_3} & \cdots & \frac{\partial P_n}{\partial f_n} \\ \frac{\partial Q_2}{\partial e_2} & \frac{\partial Q_2}{\partial e_3} & \cdots & \frac{\partial Q_2}{\partial e_n} & \frac{\partial Q_2}{\partial f_2} & \frac{\partial Q_2}{\partial f_3} & \cdots & \frac{\partial Q_2}{\partial f_n} \\ \frac{\partial Q_3}{\partial e_2} & \frac{\partial Q_3}{\partial e_3} & \cdots & \frac{\partial Q_3}{\partial e_n} & \frac{\partial Q_3}{\partial f_2} & \frac{\partial Q_3}{\partial f_3} & \cdots & \frac{\partial Q_3}{\partial f_n}. \\ \frac{\partial Q_n}{\partial e_2} & \frac{\partial Q_n}{\partial e_3} & \cdots & \frac{\partial Q_n}{\partial e_n} & \frac{\partial Q_n}{\partial f_2} & \frac{\partial Q_n}{\partial f_3} & \cdots & \frac{\partial Q_n}{\partial f_n} \end{bmatrix} \begin{bmatrix} \Delta e_2 \\ \Delta e_3 \\ \vdots \\ \Delta e_n \\ \Delta f_2 \\ \Delta f_3 \\ \vdots \\ \Delta f_n \end{bmatrix}$$

Case (*ii*) : When the System has both Load and Generator Buses

In this also the bus-1 is a slack bus.

Let buses 2, 3, 4... m be load buses and the remaining buses be generator buses. Here buses are numbered such that the first bus is slack bus, bus-2 to bus-m are load buses and bus $(m + 1)$ to bus-n are generator buses.

Let $P_2, P_3, \ldots P_n$ be the specified real power of $(n - 1)$ buses. Let $Q_2, Q_3, \ldots Q_n$ be the specified reactive powers at load buses. Let $|V_{m+1}|, |V_{m+2}|, \ldots |V_n|$ be the specified magnitude of voltages of generator buses. The unknown variables are real part of bus voltages $e_2, e_3, \ldots e_n$ and imaginary part of bus voltages $f_2, f_3, \ldots f_n$. Now the matrix equation $B = JC$ for this case will be in the form shown in equation (3.61)

$$\underset{2(n-1)\times 1}{\begin{bmatrix} \Delta P_2 \\ \Delta P_3 \\ \vdots \\ \Delta P_n \\ \vdots \\ \Delta Q_2 \\ \vdots \\ \Delta Q_m \\ |\Delta V_{m+1}|^2 \\ \vdots \\ |\Delta V_n|^2 \end{bmatrix}} = \begin{bmatrix} \frac{\partial P_2}{\partial e_2} & \frac{\partial P_2}{\partial e_3} & \cdots & \frac{\partial P_2}{\partial e_n} & \frac{\partial P_2}{\partial f_2} & \frac{\partial P_2}{\partial f_3} & \cdots & \frac{\partial P_2}{\partial f_n} \\ \partial P_3/\partial e_2 & \partial P_3/\partial e_3 & \cdots & \partial P_3/\partial e_n & \partial P_3/\partial f_2 & \partial P_3/\partial f_3 & \cdots & \partial P_3/\partial f_n \\ \vdots & \vdots & \vdots & \vdots & \vdots & \vdots & \vdots & \vdots \\ \frac{\partial P_n}{\partial e_2} & \frac{\partial P_n}{\partial e_3} & \cdots & \frac{\partial P_n}{\partial e_n} & \frac{\partial P_n}{\partial f_2} & \frac{\partial P_n}{\partial f_3} & \cdots & \frac{\partial P_n}{\partial f_n} \\ \frac{\partial Q_2}{\partial e_2} & \frac{\partial Q_2}{\partial e_3} & \cdots & \frac{\partial Q_2}{\partial e_n} & \frac{\partial Q_2}{\partial f_2} & \frac{\partial Q_2}{\partial f_3} & \cdots & \frac{\partial Q_2}{\partial f_n} \\ \frac{\partial Q_m}{\partial e_2} & \frac{\partial Q_m}{\partial e_3} & \cdots & \frac{\partial Q_m}{\partial e_n} & \frac{\partial Q_m}{\partial f_2} & \frac{\partial Q_m}{\partial f_3} & \cdots & \frac{\partial Q_m}{\partial f_n} \\ \frac{\partial |V_{m+1}|^2}{\partial e_2} & \frac{\partial |V_{m+1}|^2}{\partial e_3} & \cdots & \frac{\partial |V_{m+1}|^2}{\partial e_n} & \frac{\partial |V_{m+1}|^2}{\partial f_2} & \frac{\partial |V_{m+1}|^2}{\partial f_3} & \cdots & \frac{\partial |V_{m+1}|^2}{\partial f_n} \\ \vdots & \vdots & \vdots & \vdots & \vdots & \vdots & \vdots & \vdots \\ \frac{\partial |V_n|^2}{\partial e_2} & \frac{\partial |V_n|^2}{\partial e_3} & \cdots & \frac{\partial |V_n|^2}{\partial e_n} & \frac{\partial |V_n|^2}{\partial f_2} & \frac{\partial |V_n|^2}{\partial f_3} & \cdots & \frac{\partial |V_n|^2}{\partial f_n} \end{bmatrix} \begin{bmatrix} \Delta e_2 \\ \Delta e_3 \\ \vdots \\ \Delta e_n \\ \Delta f_2 \\ \vdots \\ \Delta f_m \\ \Delta f_{m+1} \\ \vdots \\ \Delta f_n \end{bmatrix}$$

Computing the Elements of Matrices

The elements of Jacobian matrix (J) can be derived from the three load flow equation [equations (3.51) to (3.53)]. The elements of the matrices obtained by partially differentiating the load flow equations with respect to a unknown variable and then evaluating the first derivatives using the solution of previous iteration. For first iteration the initial assumed values are used as the solution of previous iteration. The initial values are e_p^0 and f_p^0 for $P = 2, 3, \ldots, n$.

The elements of the residual column (B) is the difference between the specified value of the quantity solution of previous iteration.

Let $P_{p,\,spec}$, $Q_{p,\,spec}$ and $|V_p|_{spec}$ be the specified quantities at the bus-p. For the initial solution the values of P_p^0, Q_p^0 and $|V_p^0|$ can be calculated using the load flow equations. Now for the first iteration

$$\Delta P_p = P_{p,\,spec} - P_p^0$$

$$\Delta P_p = Q_{p,\,spec} - Q_p^0$$

$$|\Delta V_p|^2 = |V_p|^2_{spec} - |V_p^0|^2$$

After calculating the elements of Jacobian matrix (J) and residual column vector (B), the elements of increment voltage vector (C) can be calculated by using any standard technique.

Now the next better solution will be

$$e_p^1 = e_p^0 + \Delta e_p^0$$

$$f_p^1 = f_p^0 + \Delta f_p^0$$

These values of voltages will be used in the next iteration. The process will be repeated and in general the new better estimates for bus voltages will be

$$e_p^{k+1} = e_p^k + \Delta e_p^k$$

$$f_p^{k+1} = f_p^k + \Delta f_p^k$$

The process is repeated till the magnitude of the largest element in the residual column vector is less than the prespecified value.

Procedure for Load flow Solution by Newton Raphson Method

Step 1 :Assume a flat voltage profile $1 + j0$ for all buses (nodes) except the slack bus. The voltage of the slack bus is the specified voltage end it is not modified in any iteration.

Step 2 :Assume a suitable value of $\in$ called convergence criterion. Hence $\in$ is a specified change in the residue that is used to compare the actual residues (ΔP and ΔQ or ΔV) at the end of each iteration.

Step 3 :Set iteration count $k = 0$, and assumed voltage profile of the buses are deneted as $V_1^0, V_2^0, V_3^0, ..., V_n^0$ except slack bus. The real part of the voltage are denoted as e_p^0 and the reactive part as f_p^0 for $P = 1, 2, 3, ..., n$.

Step 4 :Set bus count $P = 1$.

Step 5 :Check for slack bus. It is a slack bus than go to step-13, otherwise go to next step.

Step 6 :Calculate the real and reactive power of bus-*P* using the following equation.

$$P_p^k = \sum_{q=1}^{n}\left\{e_p^k(e_q^k G_{pq} + f_q^k B_{pq}) + f_p^k(f_q^k G_{pq} + e_q^k B_{pq})\right\}$$

$$Q_p^k = \sum_{q=1}^{n}\left\{f_p^k(e_q^k G_{pq} + f_q^k B_{pq}) + e_p^k(f_q^k G_{pq} - e_q^k B_{pq})\right\}$$

Step 7 :Calcualte the change in real power change in real power, $\Delta_p^k = P_{p,\,\text{spec}} - P_p^k$ where, $P_{p,\,\text{spec}}$ = specified real power for bus-*P*.

Step 8 :Check for generator bus. if it is a generator bus go to next step (step-9) otherwise (*i.e.*, if it is a load bus) go to step-12.

Step 9 :Check for reactive power limit violation of generator buses. For this compare the calculated reactive power Q_p^k with specified limits. If the limit is violated go to step-11, otherwise go to next step.

Step 10 :If the calculated reactive power is within the specified limits than consider this bus as generator bus. Now calculate the voltage residue (change in voltage) using the following equation.

$$|\Delta V_p^k|^2 = |V_p|_{\text{spec}}^2 - |V_p^k|^2$$

where $|V_p|_{\text{spec}}$ = specified voltage magnitude for generator bus then go to step-13.

Step 11 :If the reactive power limit is violated then treat this bus as a load bus. Now the specified reactive power for this bus will corresponds to the limit violted.

i.e., if $Q_p^k < Q_{p,\,\min}$

then Q_p, spec $= Q_{p,\,\min}$

or if $Q_p^k > Q_{p,\,\max}$

then $Q_{p,\,\text{spec}} = Q_{p,\,\max}$

Step 12 :Calculate the change in reactive power for load bus.

$$\Delta Q_p^k = |Q_{p,\text{ spec}}| - Q_p^k.$$

Step 13 :Repeat steps 5 to 12 until all residues (change in P and Q or V) are calculated. For this increment the bus count by 1 and go to step-5, unit the bus count is n.

Step 14 :Determine the largest of the absolute value of the residue (*i.e.*, find the largest among ΔP_p^k, ΔQ_p^k or $|\Delta V_p^k|^2$. Let this largest change be ΔE.

Step 15 :Compare ΔE and $\in$. It $\Delta E < \in$ then go to step-20. If $\Delta E > \in$ go to next step-16.

Step 16: Determine the elements of jacobian matrix (J) by partially differentiating the load flow by equations and evaluating the equations using the kth iteration values.

Step 17 :Calculate the icrements in real and reactive part of voltages, Δe_p^k and Δf_p^k by solving the matrix equation $B = JC$ (the element of B matrix are ΔP_p^k, ΔQ_p^k or $|\Delta V_p^k|^2$ calculated in the previous steps. The elements of C matrix are Δe_p^k and Δf_p^k which are voltage increments to be solved from the matrix equation ($B = JC$).

Step 18 :Calculate the new bus voltages as shown below

$$e_p^{k+1} = e_p^k + \Delta e_p^k; \qquad P, = 1, 2, 3, \ldots\ldots n \text{ except slack bus}$$

$$f_p^{k+1} = f_p^k + \Delta f_p^k; \qquad P = 1, 2, 3, \ldots n \text{ except slack bus}$$

$$|V_p^{k+1}| = \sqrt{(e_p^{k+1})^2 + (f_p^{k+1})^2}$$

and

$$\delta_p^{k+1} = \tan^{-1}\left(e_p^{k+1}/f_p^{k+1}\right)$$

$$\therefore \quad V_p^{k+1} = |V_p^{k+1}| \angle \delta_p^{K+1}$$

Step 19 :Advance the iteration count, *i.e.*, $K = K + 1$ and go to step-4.

Step 20 :Calculate the line flows. Now let we see the flow chart of N–R method.

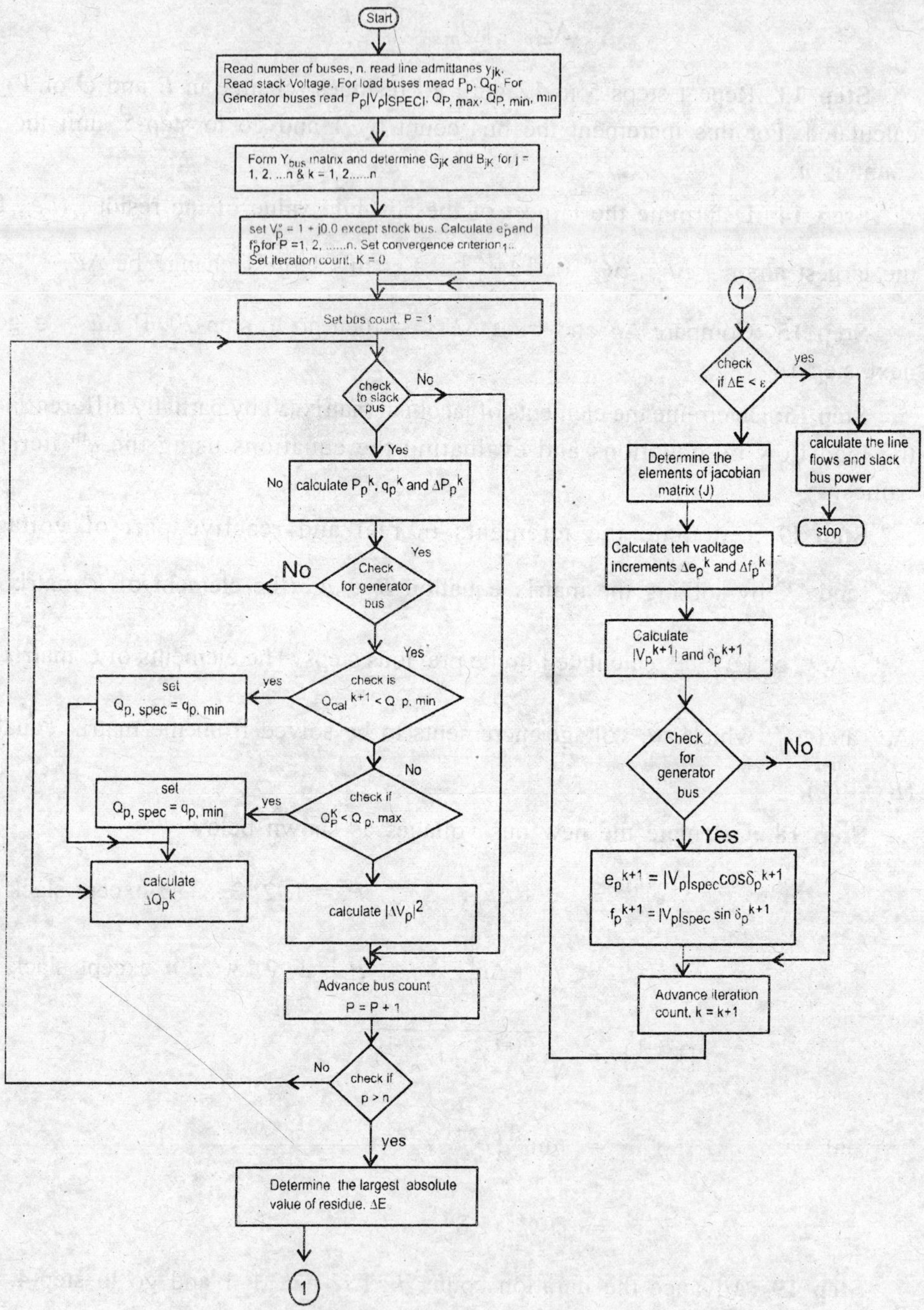

Figure 3.3 *Flow chart of Newton Raphson method.*

3.7 COMPARISON OF G-S AND N-R METHODS:

G–S method (Gauss seidel method	N–R method (Newton Raphson method)
1. Calculations are simple and so the Programming task is lesser.	1. Programming Logic is more complex than G–S method.
2. The memory requirements is less	2. Memory requirement is more
3. Number of calculations per iterations are lesser	3. Number of calculations per iterations are higher than G–S method.
4. Requires large number of iterations to reach convergence	4. Requires less number of iterations for convergence.
5. Convergence time increases with size of the system.	5. Convergene time independent of size of the system.
6. Useful for small size systems.	6. Useful for large size systems.
7. The variables are expressed in rectangular coordinates	7. The variables are expressed in in Polar Coordinates.
8. It has Linear convergence Characteristics	8. It has quadratic convergence Characteristics.
9. Convergence is affected by the choice of slack bus and the presence of series capacitors.	9. It is less sensitive to these factors.
10. It needs 30 (or) more iterations for accuracy	10. It needs 3 to 5 iterations for same Level of accuracy as G–S method.

EXERCISE

1. A two bus system is shown in Fig. (E. 3.1.1). Determine the voltage at the end of 3rd iteration by G–S method. The elements of bus impedance matrix are. $Y_{11} = Y_{22} = 1.5\angle{-86°}$ P.u

and $Y_{21} = Y_{12} = 1.8\angle 110°$ P.u

Figure E 3.1.1

2. A three bus system is shown in Fig. (E-3-2-1) Determine the bus-2 voltage at the and end of second iteration by G–S method.

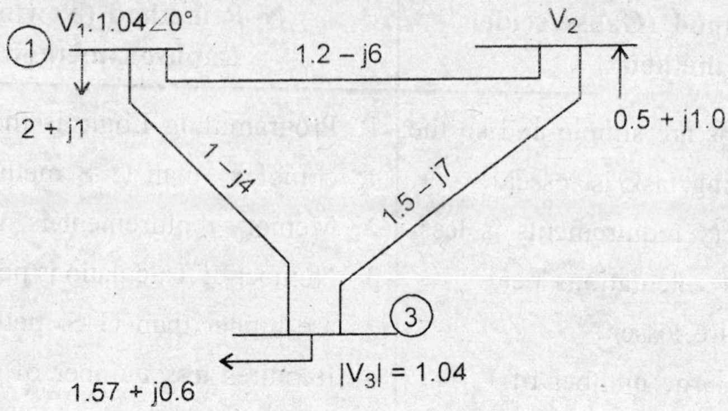

Figure E 3.2.1

➡ All elements are in P.u. admittances

3. The load flow data for a bus system are shown in table (T. 3.3.1) and (T. 3.3.2). The reactive power limit of bus-2 is $0.01 \le Q_2 \le 0.25$. Calculate the bus voltages at the and of second iteration by G–S method.

T. 3.3.1

Bus code	impedance
1 – 2	$0.07 + j0.2$
1 – 3	$0.01 + j0.05$
2 – 3	$0.02 + j0.15$

T. 3.3.2

Bus code	*P*	*Q*	*V*	Remarks
1	–	–	$1.06\angle 0°$	slack
2	0.2	–	1.05	*PV*
3	0.6	0.25	–	*PQ*

Short Questions and Answers

Q. 1. What is bus?

Ans. The meeting point of various components in a power system is called a bus, The bus is a conductor made of copper or aluminium having hegligible resistance. The buses are considered as point of constant voltage in a power system.

Q. 2. What is bus admittance matrix?

Ans. The matrix consisting of the self and mutual admittances of the network of a power system is called bus admittance matrix. It is given by the admittance matrix Y in the node basis matrix equation of a power system and it is denoted as Y_{bus}. The bus admittance matrix is symmetrical.

Q. 3. Name the diagonal and off-diagonal elements of bus admittance matrix.

Ans. Diagonal elements → self admittances.

of diagonal elements → mutual admittances

Q. 4. Write the equation to find the elements of new bus admittance matrix after eliminating n^{th} row and column in a $n \times n$ bus admittance matrix.

Ans. The element Y_{jk} of new bus admittance matrix is given by,

$$Y_{jk,\, new} = Y_{jk} - \frac{Y_{in} Y_{nk}}{Y_{nn}}$$

For $j = 1, 2, 3, \ldots..(n-1)$ and

$K = 1, 2, 3, \ldots..(n-1)$

where Y_{jk}, Y_{jn}, Y_{nk} and Y_{nn} are elements of original given bus admittance matrix of order $n \times n$.

Q. 5. Form the bus admittance matrix for the system whose reactance diagram is shown in Fig. (Q.5)

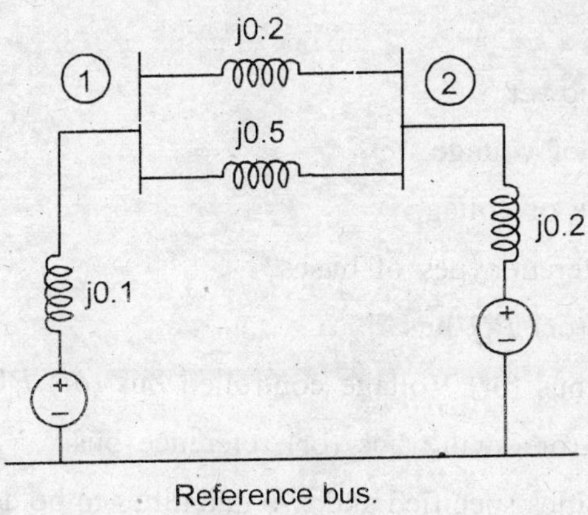

Figure Q. 3.5.1

Ans. $$Y_{bus} = \begin{bmatrix} \frac{1}{j0.1}+\frac{1}{j0.25}+\frac{1}{j0.5} & -\left(\frac{1}{j0.25}+\frac{1}{j0.5}\right) \\ -\left(\frac{1}{j0.25}+\frac{1}{j0.5}\right) & \frac{1}{j0.25}+\frac{1}{j0.5}+\frac{1}{j0.2} \end{bmatrix}$$

$$= \begin{bmatrix} -j16 & j6 \\ j6 & -j11 \end{bmatrix}$$

Q. 6. What is power flow study (or) load flow study?

Ans. The study of various methods of solution to power system network is referred to as load flow study. The solution provides the voltages at various buses, power flowing in various lines and line losses.

Q. 7. What are the informations which are obtained from power flow study?

Ans. Magnitude of phase voltage of bus

Real and reactive power flow in each bus

line losses

It also gives the initial

Conditions of the system when the transient behaviors of the system is to be studied.

Q. 8. What is the need for load flow study?

Ans. The Load flow study of a power system is essential to decide the best operation of existing system and for planning the feature expansion of the system. It is also essential for designing a new power system.

Q. 9. What are quantities that are associated with each bus in a system?

Ans. 1. real power

2. reactive power

3. magnitude of voltage

4. phase angle of voltage

Q. 10. What are different types of buses?

Ans. 1. Load bus (or) *PQ* bus

2. Generator bus (or) Voltage controlled bus (or) *PV* bus

3. Slack bus (or) swing bus (or) reference bus

Q. 11. List the quantities specified and two quantities to be determined from load flow study for various types of buses.

Ans.

Bus type	Quantities specified	Quantities to be obtained
load bus	P,Q	$\|V\|, \delta$
Generator bus	$P,\|V\|$	Q, δ
Shack bus	$\|V\|, \delta$	P,Q

Q. 12. Define voltage controlled bus.

Ans. A bus is called voltage controlled bus if the magnitude of voltage $|V|$ and real power (P) are specified for it. In a voltage controlled bus the magnitude of the voltage is not allowed to change. The other names for this bus is generator bus (or) PV bus.

Q. 13. What is PQ bus?

Ans. A bus is called P-Q bus (or) load bus when real and reactive components of power are specified for the bus. In a load bus the voltage is allowed to vary within permissible limits.

Q. 14. What is slack bus?

Ans. A bus is called slack bus when the magnitude and phase of bus voltage are specified for it. It is also called as reference bus. It is required for accounting line losses. Usually one of the generator bus is selected as slack bus.

Q. 15. What is the need of slack bus?

Ans. The slack bus is needed to account for transmission line losses. In a power system the total power generated will be equal to sum of power consumed by loads and losses. In a power system only the generated power and load power are specified for buses. The slack bus is assumed to generate the power required for losses. Since the losses are unknown the real and reactive power are not specified for slack bus they are estimated through the solution of load flow equations.

Q. 16. What are the operating constraints imposed in the load flow studies?

Ans. Reactive power limits for generator buses and allowable change in magnitude of voltage for load buses.

Q. 17. Write the expression for complex power injected to a bus.

Ans. The complex power injected to bus-p of a n-bus system is given by

$$S_P = P_P - jQ_P = V_P^* \sum_{q=1}^{n} Y_{P_q} V_q$$

The complex power S_p is the difference between power injected by generators connected to the bus and power drawn by the loads connected to the bus. In a pure generator bus, only loads will be connected to the bus and so P_p and Q_p corresponds to power drawn by loads (in this case P_p and Q_p and negative)

Q. 18. Discuss the effect of acceleration factor in the load how solution algorithm. How will you account for voltage controlled buses in this algorighm?

Ans. In load flow solution by iterative method the number of iterations can be reduced if the correction voltage at each bus is multiplied by some constant. The maltiplication of the constant will increase the amount of correction and bring the voltage closer to the value in which it is approaching. The multipliers which improves the convergence are called acceteration factors. An acceleration factor of 1.6 is normally used in load flow problems. The acceleration factor is a real quantity and it modify the magnitude of voltage alone. Since in voltage controlled bus (generator bus), the magnitude of bus voltage is not allowed to change, the acceleration factor is not used for voltage controlled bus. (i.e. acceleration factor is used only for load bus)

Q. 19. Why do we go for iterative methods to solve load flow problems?

Ans. The load flow equation are nonlinear algebraic equations so obtaining the solution is very difficult so it can be obtained only by iterative numerical techniques.

Q. 20. When the generator bus is treated as load bus?

Ans. If the reactive power of a generator bus violates the specified limits than the generator bus is treated as load bus.

Q. 21. What will be the reactive power and bus voltage when the generator bus is treated as load bus?

Ans. When the generator bus is treated as load bus, the reactive power of the bus is equated to the limit it has violated and the previous iteration value of bus voltage is used for calculating current iteration value.

Q. 22. How approximation is performed in newton-Raphson method?

The set of nonlinear simultaneous equations are approximated to a set of linear simultaneous equations using Taylor's series expansion and the terms are limited to first order approximation.

Q. 23. Write advantages of both *G-S* and *N-R* method.

(*i*) *G-S* method:

1. Calculations are simple and so the programming task is lesser
2. Memory requirement is less
3. Useful for small size system.

(*ii*) *N-R* method:

1. It is faster, more reliable and accurate.
2. Requires less number of iterations for convergence.
3. Suitable for large size system.

Q. 24. Write Disadvantages of both *G-S* and *N-R* method.

(*i*) *G-S* method:

1. Requires large number of iterations.
2. Not suitable for large size system.
3. Convergence time increases with size of the system.

(*ii*) *N-R* method:

1. Programming is complex.
2. Memory requirement is more.

Choose The Best Answer

1. In a load bus the magnitude of bus voltage can

(a) Vary within specified limits (b) Vary up to any value

(c) both a & b (d) none of the above

2. In a generator bus the Voltage is not allowed to vary in

(a) the begining

(b) each iteration of load flow solution

(c) the end step

(d) none of the above

3. The acceleration factor reduces

(a) the number of iterations (b) the number of buses

(c) both a & b (d) None of the above

4. The load flow equations are

(a) linear algebraic equations (b) Non linear algebraic equations

(c) Quadratic equations (d) None of the above

5. If rectangular coordinates are used in *N-R* method then

(a) It requires more memory (b) It requires less memory

(c) It does not need memory (d) None of the above

6. All equipments are designed to work

(a) at rated voltage (b) Rated bus

(c) Rated impedance (d) None of the above

7. In *N-R* method the number of iterations are

(a) independent of the size of the system

(b) Dependent of the size of the system

(c) No relation between number of iterations & size of the system.

(d) None of the above

8. When real and reactive powers are specified then

(a) Load bus (b) generator bus

(c) slack bus (d) None of the above

9. When the magnitude and phase of bus voltage are specified then

(a) slack bus (b) generator bus

(c) load bus (d) None of the above

10. Largest generating station bus is selected as

(a) slack bus (b) load bus

(c) generator bus (d) none of the above

ANSWERS

1. a	2. b	3. a	4. a	5. a
6. a	7. a	8. a	9. a	10. a

Fill in the Blanks

1. The bus is called ________ when real and reactive components of power are specified for the bus.

2. The bus is called ________ when real power and magnitude of bus voltage are specified for the bus.

3. In load flow solution it is assumed that the ________ generates the real and reactive power required for transmission line losses.

4. The normal value of acceleration factor is ________

5. The ________ bus is treated as ________ bus if the reactive power limits is violated.
6. In *N-R* method the nonlinear equations are approximated using ________
7. The elements of ________ are first derivatives of load flow equations.
8. In ________ method the convergence is not affected by the choice of slack bus.
9. The *G-S* method has ________ Convergence characteristic
10. The ________ variation in a bus is directly related to ________ power.

ANSWERS

1. load bus
2. generator bus
3. swing (or) slack bus
4. 1.6
5. generator, load
6. Tay'lors series expansion
7. Jacobian matrix
8. newton-Raphson
9. linear quadratic
10. voltage, reactive.

Chapter-4

POWER SYSTEM STABILITY

4.1 INTRODUCTION

The term stability refers to stable operation of the synchronous machines connected to a power system when they are subjected to sudden disturbances. Hence we can say that the stability is the ability of power system to return to stable operation when it is subjected to a disturbance. The instability refers to the loss of synchronism (or falling out of step) of sychronous machines in the power system. The stability studies are classified into 3-types

1. transient state study
2. Dynamic state study
3. steady state study

4.2 STEADY STATE STABILITY STUDY

The study of steady state stability is concerned with the determination of upper limit of loading synchronous machine. In these studies it is assumed that the machine is loaded gradually.

In a power system, small disturbances will continuously occur due to variations in load, changes in the speed of prime mover, fault in certain part of system, etc. These small disturbances may excite the system into a state of natural oscillations.

In dynamically stable systems the amplitude of the oscillations will be small and they die out quickly. In dynamically unstable system the amplitude of the oscillations is very large and they exist for a long time. This type of unstable behaviour may create serious problems in power system operation. (i.e. it may lead to unnecessary frequent load shedding).

Large disturbances may occur in a power system due to switching of heavy loads, switching of long transmission lines etc. The sudden large disturbances are characterized by large change in the speed of the rotor of synchronous machine, large changes in power angle and fast change in power transfer. These characteristic may lead to loss of synchronism. This type of instability is called transient instability. Hence the aim of transient stability is to determine whether the system remains in the synchronism (or) not.

4.3 STABILITY AND STABILITY LIMIT

Stability study means, analysing about the synchronous machines about their stable condition when they are subjected to sudden disturbances.

The concept of stability can be understood by considering a single synchronous machine connected to infinite bus. It is equivalent to two machine system. A practical power system may have a number of machines running in synchronism.

For theoretical analysis the multimachine system can be converted to an equivalent single machine system connected to infinite bus. Alternatively, the stability study of a multimachine system can be carried on a computer. The dynamic behaviour of a power system have the following basic features.

1. A synchronous machine has a maximum limit of power transfer when remains in synchronism. When the power transfer exceeds the limit it cannot stay in synchronism.
2. The synchronous machine connected to an infinite bus is basically a spring inertia oscillatory system. The inertia system is due to the mechanical part (rotor) and the spring action is due to the synchronous lie with infinite bus.
3. If δ is the anular displacement of the rotor in electrical radians then the power transfer is proportional to sin δ. Hence the mathematical equation governing the system dynamics is non-Linear. In order to simplify the computational task in stability studies, the following three assumptions are made in all stability studies.

   (*i*) The d.c off set currents and voltages are assumed to have fundamental component alone.

   (*ii*) The symmetrical components are used for the representation of unbalanced faults.

   (*iii*) It is assumed that the machine speed variations will not affect the generated voltage.

4.4 DERIVATION OF SWING EQUATION

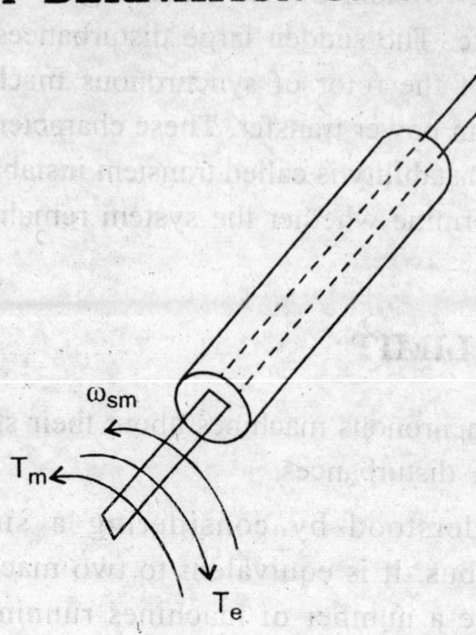

Figure 4.1 *Generator*

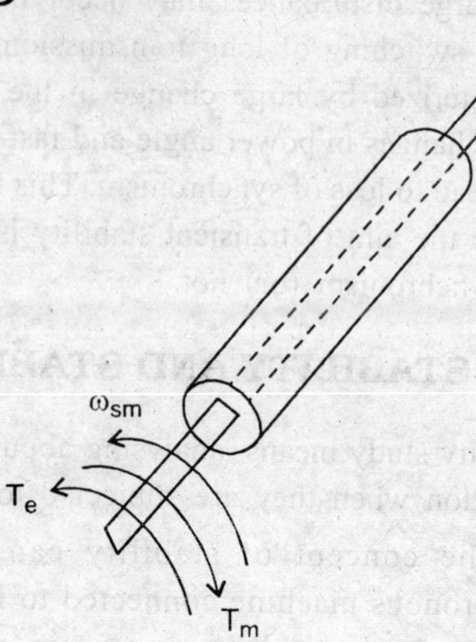

Figure 4.2 *Motor*

The rotor of a synchronous machine is subjected to two torques, T_e and T_m, which are acting in opposite directions as shown in Figure 4.1 and 4.2

where,

T_e = Net electrical or electromechanical torque in *N-M*

T_m = Mechanical (or) shaft torque supplied by the prime mover in *N-M*.

Under steady state operating condition the T_e and T_m are equal and the machine runs at constant speed, which is called synchronous speed. If there is a difference between the two torques then rotor will have an accelerating (or) deaccelerating torque, denoted as T_a.

$$\therefore \quad T_a = T_m - T_e \quad \text{...(4.1)}$$

Here T_m & T_e are positive for generators and T_m & T_e are negative for motors.

Let,

θ_m = Angular displacement of rotor with respect to stationary reference axis.

δ_m = Angular displacement of rotor with respect to synchronously rotating reference axis.

By newton's second law of motion, we can say that the accelerating torque, T_a is directly proportional to angular accderation and the constant of proportionality is the moment of inertia, *J*.

$$\therefore \quad T_a \propto \frac{d^2\theta_m}{dt^2} \text{ (or) } T_a = J\frac{d^2\theta_m}{dt^2} \quad \text{...(4.2)}$$

on substituting for T_a from equation (4.2) in (4.1) we get,

$$J\frac{d^2\theta_m}{dt^2} = T_m - T_e \quad ...(4.3)$$

The angular displacement q_m and d_m are related to synchronous speed by the following equation [equation 4.4]

$$\theta_m = \omega_{sm} + \delta_m \quad ...(4.4)$$

On differentiating the equation (4.4) with respect to time, t we get,

$$\frac{d\theta_m}{dt} = \omega_{sm} + \frac{d\delta_m}{dt} \quad ...(4.5)$$

On differentiating the equation (4.5) with respect to time, t we get,

$$\frac{d^2\theta_m}{dt^2} = \frac{d^2\delta_m}{dt^2} \quad ...(4.6)$$

From equation (4.5) we can say that the rotor angular velocity $d\theta_m/dt$ is constant and equal to ω_{sm} (synchronous speed) only if $\frac{d\delta_m}{dt}$ is zero. Hence $\frac{d\delta_m}{dt}$ represents the deviation of the rotor speed from synchronism.

From equation (4.3) and (4.6) we can write,

$$T\frac{d^2\delta_m}{dt^2} = T_m - T_e \quad ...(4.7)$$

Let,

$P_{m,act}$ = shaft power input to the machine. neglecting losses (in *MW*)

$P_{e,act}$ = Electrical power developed in rotor (in *MW*)

We know that, power, $P = \frac{2\pi NT}{60} = \omega T$

$$\therefore \quad P_{m,act} = \omega_{sm} T_m \text{ (or) } T_m = \frac{P_{m,act}}{\omega_{sm}} \quad ...(4.8)$$

$$P_{e,act} = \omega_{sm} T_e \text{ (or) } T_e = \frac{P_{e,act}}{\omega_{sm}} \quad ...(4.9)$$

On substituting for T_m and T_e from equation (4.8) & (4.9) in equation (4.7) we get,

$$T\frac{d^2\delta_m}{dt^2} = \frac{P_{m,act}}{\omega_{sm}} - \frac{P_{e,act}}{\omega_{sm}}$$

$$J\omega_{sm}\frac{d^2\delta_m}{dt^2} = P_{m,act} - P_{e,act} \quad ...(4.10)$$

The inertia constant, H is defined as the ratio of stored kinetic energy in MJ to the machine rating in MVA.

Let,

H = Inertia constant in MJ/MVA

S = Power rating of machine in MVA

Now, $$H = \frac{\frac{1}{2}J\omega_{sm}^2}{S} \quad ...(4.11)$$

$$\therefore \quad J\omega_{sm} = \frac{2HS}{\omega_{sm}} \quad ...(4.12)$$

➡ The kinetic energy $= \frac{1}{2}J\omega_{sm}^2$

From equations (4.10) and (4.12) we can write

$$\frac{2HS}{\omega_{sm}}\frac{d^2\delta_m}{dt^2} = P_{m,act} - P_{e,act} \quad ...(4.13)$$

We know that, $\omega_{sm} = \frac{2}{P}\omega_s$ and $\delta_m = \frac{2}{P}\delta$ ...(4.14)

where,

P = Number of poles in synchronous machine on substituting for ω_{sm} and δ_m from equation (4.14) in equation (4.13) we get,

$$\left(\frac{2HS}{\frac{2\omega_s}{P}}\right) \times \left(\frac{d^2(2\delta/p)}{dt^2}\right) = P_{m,act} - P_{a,act}$$

$$\frac{2HS}{\omega_s}\frac{d^2\delta}{dt^2} = P_{m,act} - P_{e,act} \quad ...(4.15)$$

On substituting, $\omega_s = 2\pi f$ in equation (4.15) we get,

$$\frac{2HS}{2\pi f}\frac{d^2\delta}{dt^2} = P_{m,act} - P_{e,act}$$

$$\therefore \quad \frac{Hs}{\pi f}\frac{d^2\delta}{dt^2} = P_{m,act} - P_{e,act} \quad ...(4.16)$$

In equation (4.16) the powers $P_{m,act}$ and $P_{e,act}$ are in MW. If the machine MVA rating S is chosen as base value then $P.u$ value of power is given by,

$P.u$ value of mechanical power, $P_m = \frac{P_{m,act}}{S}$

P.u. values of electrical power, $P_e = \frac{P_{e,act}}{S}$

From equation (4.16) we get

$$\frac{H}{\pi f} = \frac{d^2\delta}{dt^2} = \frac{P_{m,act}}{S} - \frac{P_{e,act}}{S}$$

$$\frac{H}{\pi f}\frac{d^2\delta}{dt^2} = P_m - P_e \quad ...(4.17)$$

The equation (4.17) is called swing equation. It is the fundamental equation which governs the dynamics of the synchronous machine rotor. The swing equation is second order differential equation. Swing equation is a nonlinear second order differential equation.

4.4.1 Steady state stability

In steady state every synchronous machine has a limit for power transfer to a receiving system. The steady state limit of a machine or transmitting system is defined as the maximum power that can be transmitted to the receiving system without loss of synchronism.

Let us consider single synchronous machine delivering power to a large system through a transmitting system. The power angle equation of such a system have been developed in section 4.5, and it can be used for the analysis of steady state stability if the transient emfs are replaced by steady state emfs.

Let $|E|$ = magnitude of steady state internal emf of synchronous machine

$|V|$ = Magnitude of voltage of receiving system.

X = Transfer reactance between the synchronous machine and receiving system.

Now,
$$P_{max} = \frac{|E||V|}{X} \quad ...(4.18)$$

∴ Real power injected by machine to system,

$$P_e = P_{max} \sin\delta \quad ...(4.19)$$

Let the system be operating with steady power transfer with a torque angle δ_0. In this operating condition, Let the electrical power output be p_{eo}. Now, the mechanical power input p_m is equal to p_{eo} under ideal conditions.

With the power input (p_m) remaining same let us assume that the electrical power output increases by a small amount Δp. Now the torque angle change by a small amount $\Delta\delta$. Therefore the new value of torque angle is ($\delta_0 + \Delta\delta$). The electrical power output for this new torque angle can be obtained from equation (4.19).

The electrical power output for a torque angle of $(\delta_0+\Delta\delta)$ $\Big\} = P_{e0} + \Delta P = P_{\max} \sin(\delta_0 + \Delta\delta)$

$$= P_{\max} [\sin \delta_0 \cos \Delta\delta + \cos \delta_0 \sin \Delta\delta] \quad ...(4.20)$$

since $\Delta\delta$ is small incremental displacement from δ_0,

$\sin \Delta\delta \approx \Delta\delta$ and $\cos \Delta\delta \approx 1$ ...(4.21)

$$\therefore \quad P_{e0} + \Delta P = P_{\max} \sin \delta_0 + (P_{\max} \cos \delta_0)\, \Delta\delta \quad ...(4.22)$$

From equation (4.19) $\delta = \delta_0$ we get,

$$P_e = P_{e0} = P_m \sin \delta_0 \quad ...(4.23)$$

From equations (4.22) and (4.23) we can write,

$$\Delta P = (P_{\max} \cos \delta_0)\, \Delta\delta \quad ...(4.24)$$

The swing equation for the system discussed here, have been developed in section (4.4) [The swing equation is given by equation (4.17)]. The nonlinear swing equation can be linearized about the operating point for steady state stability analysis. The nonlinear swing equation is

$$\frac{H}{\pi f} \frac{d^2\delta}{dt^2} = P_m - P_e \quad ...(4.25)$$

We know that the p.u value of $M = \dfrac{H}{\pi f}$. Hence the equation (4.25) can be written as

$$\frac{Md^2\delta}{dt^2} = P_m - P_e \quad ...(4.26)$$

For a torque of $\delta = (\delta_0 + \Delta\delta)$, The electrical power, P_e in the equation (4.26) should be replaced by $(P_{e0} + \Delta P)$

$$\therefore \quad \frac{Md^2(\delta_0 + \Delta\delta)}{dt^2} = P_m - (P_{e0} + \Delta P) \quad ...(4.27)$$

Since δ_0 is constant and $P_m = P_{e0}$, the equation (4.27) can be written as,

$$\frac{Md^2\Delta\delta}{dt^2} = -\Delta P \quad ...(4.28)$$

On substituting for ΔP from equation (4.24) in equation (4.27) we get,

$$\frac{Md^2\Delta\delta}{dt^2} = -(P_{\max} \cos \delta_0)\, \Delta\delta$$

$$\frac{Md^2}{dt^2}\Delta\delta + P_{\max} \cos \delta_0\, \Delta\delta = 0 \quad ...(4.29)$$

Let $$\frac{d^2}{dt^2} = X^2 \text{ and } P_{max} \cos \delta_0 = C$$

$$\therefore \quad Mx^2 \, \Delta\delta + C\Delta\delta = 0 \quad \text{...(4.30)}$$

$$(Mx^2 + C) \, \Delta\delta = 0$$

In equation (4.30) $\sin \Delta\delta \neq 0$, $MX^2 + C = 0$ ...(4.31)

The equation (4.31) is the characteristic equation of the system for small changes. The stability is determined by the roots of the characteristic equation.

$$\therefore \quad X^2 = \frac{-C}{M} \text{ (or) } X = \pm\sqrt{\frac{-C}{M}} \quad \text{...(4.32)}$$

Case (*i*): When *C* is positive, (ie, $P_{max} \cos \delta_0 > 0$)

In this case the roots are purely imaginary and conjugate. Hence the system behaviour is oscillatory about δ_0. In this analysis the resistances in the system have been neglected. If we include the resistances in the analysis then the roots will be complex conjugate and the response will be damped oscillatory. Therefore in a practical system the system is stable for small increment in power provided, $P_{max} \cos \delta_0 > 0$.

Case (*ii*): When *C* is negative, (ie, $P_{max} \cos \delta_0 < 0$)

In this case the roots are real and equal in magnitude. One of the root is positive and other one is negative. Due to the positive root the torque angle increases without bound when there is a small increment in power and the machine will loose synchronism. Hence the machine becomes unstable for small changes in power provided $P_{max} \cos \delta_0 < 0$.

Steady state limit:

The term $P_{max} \cos \delta_0$ decides the steady state stability of the system and so it is called synchronizing coefficient or stiffness of synchronous machine. From the power angle curve for generator action shown in Figure 4.3, we get the range of δ_0 as 0 to π. [The Figure (4.3) is drawn by using the equation $P_e = P_{max} \sin \delta$]

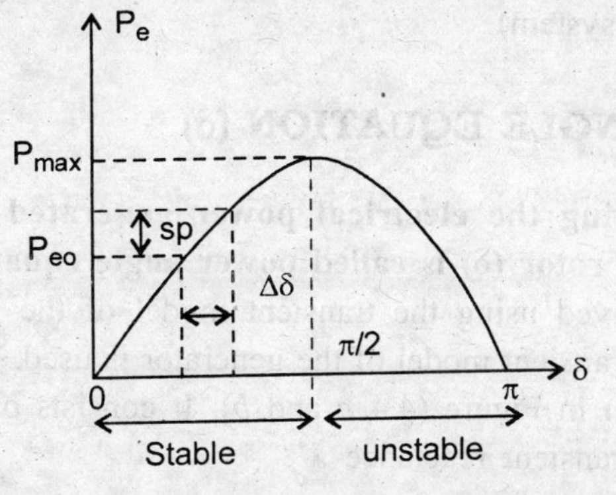

Figure 4.3 *Power angle curve for steady state operation of generator*

when $0 \le \delta_0 \le \frac{\pi}{2}$; $P_{max} \cos \delta_0$ and P_e are +ve

when $\delta_0 = \frac{\pi}{2}$; $P_{max} \cos \delta_0 = 0$ and $P_e = P_{max}$

when; $\frac{\pi}{2} \le \delta_0 \le \pi$ $P_{max} \cos \delta_0$ is negative and P_e is +ve

From the above discussion we say that the synchronizing coefficient ($P_{max} \cos \delta_0$) and real power injected to the system (P_e) are positive when δ is in the range of 0 to $\frac{\pi}{2}$. Therefore the maximum power that can be transmitted without loss of stability occurs $\delta = \delta_0 = \frac{\pi}{2} = 90°$. The maximum power transmitted is P_{max}.

where,

$$P_{max} = \frac{|E||V|}{X}$$

The Figure 4.3 shows the stable and unstable steady state operating region of generator. The concepts developed in this section is also applicable for power transfer from one system to another system if the transmitting system is represented by single equivalent generator.

In stable operating region of the system the damping should be sufficient to reduce the oscillations developed due to small changes in loads. If the oscillations exist for a long time then it may cause a problem to system security.

Practically, the system has to be operated below the steady state stability limit. This limit can be improved by reducing the reactance X or by increasing the voltages at sending end and/or at receiving end. The reactance can be reduced by introducing series capacitors in the transmission line. Alternatively the line reactance can be reduced by using two parallel transmission lines. (The two parallel lines improves the reliability of the system)

4.5 POWER ANGLE EQUATION (δ)

The equation relating the electrical power generated (P_e) to the angular displacement of the rotor (δ) is called power angle equation. The power angle equation can be derived using the transient model of the generator, because for stability studies the transient model of the generator is used. The transient model of a generator, is shown in Figure (4.4 *a* and *b*). It consists of the transient internal emf in series with transient reactance X'_d.

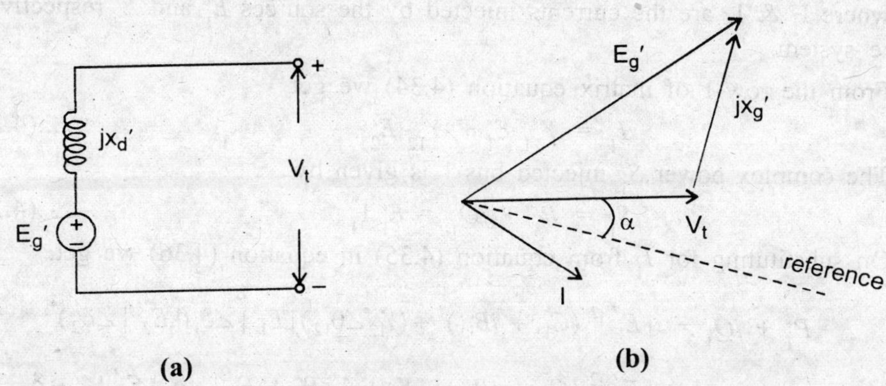

Figure 4.4 *Transient circuit model and vector diagram of generator used for stability studies*

Consider a single generator supplying power through a transmission system to a load or to a large system at other end. Such system can be represented by a 2 bus network as shown in Figure 4.5. The rectangular box represents the linear passive components (reactances) of the system including the transient reactance of the generator.

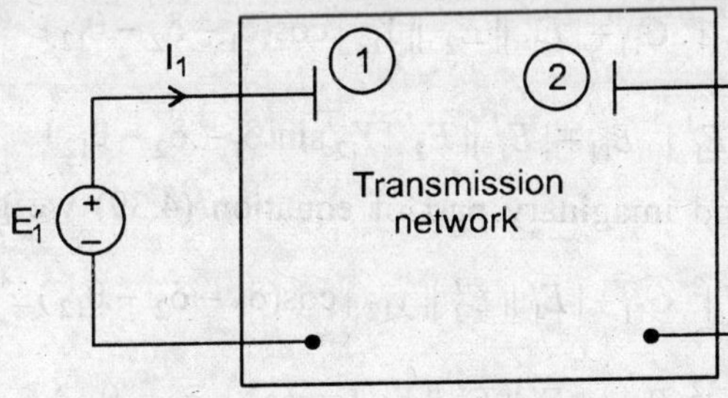

Figure 4.5

Here E_1' = Transient internal voltage of the generator at bus-1

E_2' = Voltage at the receiving end (This may be the voltage at infinite bus (or) transient internal voltage of synchronous motor at bus-2)

The node basis matrix equation of 2-bus system of Figure 4.5.Can be written as shown below.

$$I = Y_{bus} V \qquad ...(4.33)$$

In the expanded from the equation (4.33) can be written as,

$$\begin{bmatrix} I_1 \\ I_2 \end{bmatrix} = \begin{bmatrix} Y_{11} & Y_{12} \\ Y_{21} & Y_{22} \end{bmatrix} \begin{bmatrix} E_1' \\ E_2' \end{bmatrix} \qquad ...(4.34)$$

where I_1 & I_2 are the currents injected by the sources E_1' and E_2' respectively to the system.

From the row-1 of matrix equation (4.34) we get,

$$I_1 = Y_{11} E_1' + Y_{12} E_2' \qquad ...(4.35)$$

The complex power S_1 injected bus-1 is given by

$$S_1 = P_1 + jQ_1 = E_1' I_1^* \qquad ...(4.36)$$

On substituting for I_1 from equation (4.35) in equation (4.36) we get,

$$P_1 + jQ_1 = |E_1'|^2 (G_{11} + jB_{11})^* + (Y_{12}\angle\theta_{12}^*)|E_1'|\angle\delta_1(|E_2'|\angle\delta_2)^*$$

$$= |E_1'|^2 (G_{11} - jB_{11}) + |Y_{12}|\angle-\theta_{12})|E_1'|\angle\delta_1|E_2'|\angle-\delta_2$$

$$= |E_1'|^2 (G_{11} - jB_{11}) + |E_1'||E_2'|Y_{12}|\angle(\delta_1 - \delta_2 - \theta_{12})$$

$$= |E_1'|^2 (G_{11} - jB_{11}) + |E_1'||E_2'||Y_{12}|\cos(\delta_1 - \delta_2 - \theta_{12})$$

$$+ j|E_1'||E_2'|Y_{12}|\sin(\delta_1 - \delta_2 - \theta_{12}) \qquad ...(4.37)$$

On equating the real and imaginary part of equation (4.37) we get,

$$P_1 = |E_1'|^2 G_{11} + |E_1'||E_2'||Y_{12}|\cos(\delta_1 - \delta_2 - \theta_{12}) \qquad ...(4.38)$$

$$Q_1 = -|E_1'|^2 B_{11} + |E_1'||E_2'|Y_{12}\sin(\delta_1 - \delta_2 - \theta_{12}) \qquad ...(4.39)$$

On equating the real and imaginary part of equation (4.39) we get,

$$P_1 = |E_1'|^2 G_{11} + |E_1'||E_2'||y_{12}|\cos(\delta_1 - \delta_2 - \theta_{12}) \qquad ...(4.40)$$

$$Q_1 = |E_1'|^2 B_{11} + |E_1'||E_2'||Y_{12}|\sin(\delta_1 - \delta_2 - \theta_{12}) \qquad ...(4.41)$$

The equation (4.40) and (4.41) are equations of real and reactive power injected by a generator at bus-1.

Let, $\delta = \delta_1 - \delta_2$

$$\gamma = \theta_{12} - \frac{\pi}{2} \qquad \left(\text{and } \theta_{12} = \gamma + \frac{\pi}{2}\right)$$

$$P_c = |E_1'|^2 G_{11}$$

$$P_{max} = |E_1'||E_2'||Y_{12}|$$

$$P_1 = P_e$$

By using the terms difined above, the real power P_1 (*i.e*, equation (4.39) can be written as shown below.

$$P_e = P_c + P_{max} \cos\left(\delta - \gamma - \frac{\pi}{2}\right)$$

$$\therefore \quad P_e = P_c + P_{max} \sin(\delta - \gamma) \qquad \text{...(4.42)}$$

The equation (4.42) is called power angle equation. In this equation the term P_c represents power loss in the system and P_{max} represents the maximum real power that can be delivered by the generator to an infinite bus.

> ➡ Here the bus-2 is considered as infinite bus and so $\delta_2 = 0$ therefore $\delta = \delta_1$.

The power angle equation can be further simplified by considering the network as purely reactive network (by neglecting the resistances).

When resistances are neglected $G_{11} = 0$, $\theta_{12} = \frac{\pi}{2}$ and so $\gamma = 0$.

On substituting these conditions in equation (4.42) we get,

$$P_e = P_{max} \sin\delta \qquad \text{...(4.43)}$$

where, $P_{max} = \dfrac{|E_1'||E_2'|}{X_{12}}$

and X_{12} = Transfer reactance between bus-1 and 2

The equation (4.43) is called simplified power angle equation. The graph of P_e as a function δ is called the power angle curve. A typical power angle curve is shown in Figure 4.6.

On substituting the expression for p_e from equation (4.43) in the swing equation (i.e, equation (4.17)), we get,

$$\frac{H}{\pi f}\frac{d^2\delta}{dt^2} = P_m - P_{max} \sin\delta \qquad \text{...(4.44)}$$

The equation (4.44) is the swing equation in which the electrical power is expressed as a function of δ.

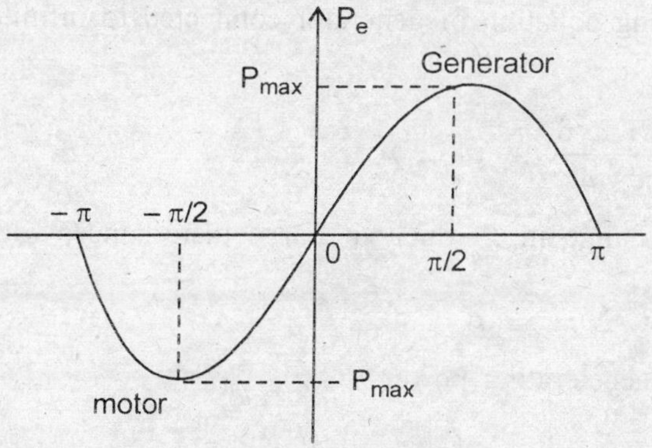

Figure 4.6 *Power angle curve of synchronous machine.*

4.6 TRANSIENT STABILITY BY EQUAL AREA CRITERION

The transient stability of a system is concerned with the study of system behaviour for large disturbances. The short circuits (fault conditions) and switching heavy loads can be treated.

The dynamics of the system under transient state are governed by the non-linear swing equation. Since the changes in δ is very large in the transient state, the swing equation cannot be linearized for a general solution and so solution has to be obtained by using any one of the numerical techniques. The most popular methods available for the solution of swing equation are step-by-step method and Runga-kutta method.

The transient stability of a single machine connected to infinite bus bar can be easily determined by a simple criterion called equal area criterion.

The transient stability analysis of simple system can be performed by using a simple criterion called equal area criterion. During the transient state of a power system we may come across the following two situations for changes in δ (torque angle) with respect to time.

(*i*) The δ may increase to a maximum value and then decrease to a stable value. The system is considered as stable.

(*ii*) The δ may keep on increasing indefinitely. In this case the system is unstable.

The above facts can be stated as stability criterion as given below.

(*i*) The system is stable if, $\frac{d\delta}{dt} = 0$ at some time instant

(*ii*) The system is unstable if, $\frac{d\delta}{dt} > 0$ for a sufficiently long time (typically 1 second or more)

For a single machine-infinite bus bar system, the stability criterion stated above can be converted to a simple condition as shown below.

Consider the swing equation of generator connected to infinite bus. (equation (4.17))

$$\frac{H}{\pi f}\frac{d^2\delta}{dt^2} = P_m - P_e \quad \text{...(4.45)}$$

Let there be a change in P_e due to Large disturbance, with P_m remaining constant.

Now, $$P_m - P_e = P_a \quad \text{...(4.46)}$$

where P_a is the accelerating power.

Also we know that P.u value of $M = \frac{H}{\pi f}$.

Hence the equation (4.45) can be written as,

$$\frac{Md^2\delta}{dt^2} = P_a$$

$$\therefore \quad \frac{d^2\delta}{dt^2} = \frac{P_a}{M} \quad \text{...(4.47)}$$

On multiplying the equation (4.47) by $\frac{2d\delta}{dt}$ we get,

$$2\frac{d\delta}{dt}\frac{d^2\delta}{dt^2} = 2\frac{d\delta}{dt}\frac{P_a}{M}$$

$$2\frac{d\delta}{dt}\frac{d}{dt}\frac{d\delta}{dt} = \frac{2}{M}P_a\frac{d\delta}{dt}$$

$$2\frac{d}{dt}\left(\frac{d\delta}{dt}\right)^2 = \frac{2}{M}P_a\frac{d\delta}{dt}$$

$$\therefore \quad 2d\left(\frac{d\delta}{dt}\right)^2 = \frac{2}{M}P_a d\delta \quad \text{...(4.48)}$$

On integrating the equation (4.48) we get,

$$\left(\frac{d\delta}{dt}\right)^2 = \frac{2}{M}\int_{\delta_0}^{\delta} P_a d\delta \quad \text{...(4.49)}$$

Where δ_0 is the initial value of torque angle or torque angle or rotor angle on taking square not of equation (4.49) we get,

$$\frac{d\delta}{dt} = \sqrt{\frac{2}{M}\int_{\delta_0}^{\delta} P_a d\delta} \quad \text{...(4.50)}$$

The equation (4.50) is zero if the integral of P_a is zero.

Hence we can say that,

$$\text{for} \quad \frac{d\delta}{dt} = 0, \text{ the term } \int_{\delta_0}^{\delta} P_a d\delta = 0 \quad \text{...(4.51)}$$

The physical meaning of integrating is the estimation of the area under the curve. Hence the integral of P_a equal to zero refers to zero area. Hence the condition of stability can be δ.

This is possible only if the positive (accelerating) area under P_a - δ curve is equal to the negative (decelerating) area under P_a - δ curve for a finite change in δ. Hence this stability criterion is called equal area criterion of stability.

The equal area criterion of stability can be applied to any type of disturbances that may occur in a single machine-infinite bus bar system. Let we see the transient stability study for a sudden change in mechanical input using equal are criterion.

4.6.1 Transient Stability Analysis for a Sudden Change in Mechanical Input

Consider a single generator feeding energes to infinite bus as shown in Figure 4.7. The electrical power transmitted by the generator is given by

$$P_e = \frac{|E'||V|}{X} \sin\delta = P_{max} \sin\delta \qquad \text{...(4.52)}$$

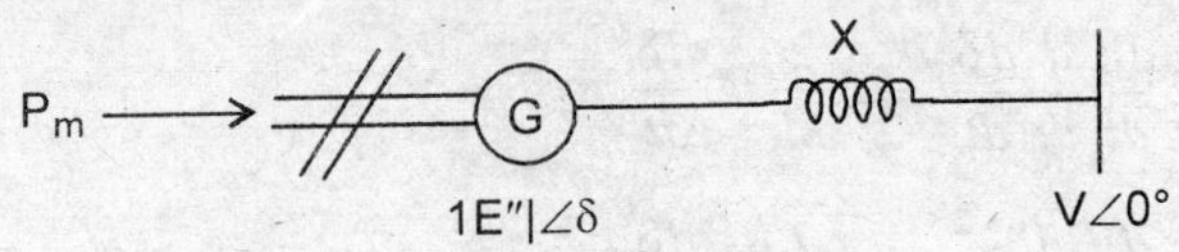

Figure 4.7

where, $$P_{max} = \frac{|E'|V|}{X}$$

Let the generator be operating in steady state with a torque angle δ_0. At this condition the mechanical power input is P_{m0} and electrical power output is p_{e0}. Under ideal conditions $P_{m0} = P_{e0}$. Hence from equation (4.52) for this operating condition we can write,

$$P_{m0} = P_{e0} = P_{max} \sin \delta_0 \qquad \text{...(4.53)}$$

The power angle curve of the generator is shown in Figure 4.8. In this the steady state operating point as described by equation 4.53 is the point *a*

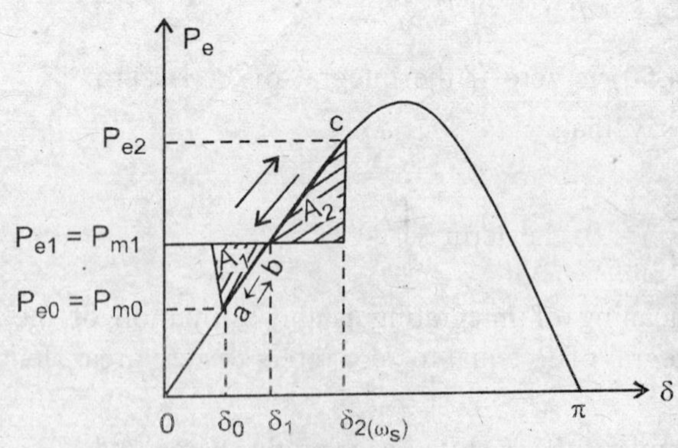

Figure 4.8

Let the mechanical input to the generator rotor be suddenly increased to P_{m1} (by adjustig prime mover). Since the mechanical power is more than electrical power, the generator will have an accelerating power, P_a is given by

$$P_a = P_{m1} - P_e \qquad ...(4.54)$$

where, $P_e = P_{max} \sin \delta$

Due to accelerating power the rotor speed increases and so the rotor angle also increases. This results in increased electrical power generation. Therefore the operating point will move upwards along the power angle curve. At point-*b* again the mechanical power P_{m1} equals the electrical power P_{e1}, where P_{e1} is the electrical power output corresponding to torque angle δ_1.

Now the rotor angle cannot stay at this point, because the inertia of the rotor will make the rotor to oscillate with respect to point-*b*. Hence the torque angle will continue to increase till point-*c*, when the operating point moves from *b* to *c*, the electrical power is more than mechanical power. Therefore the power, P_a given by equation (4.54) is negative and it is called decelerating power.

In this region (i.e., from point-*b* to *c*) the rotor angle δ increases but the rotor speed decreases due to decelerating power. The point-*c* is decided by the damping of the system (because the oscillations in the rotor are reduced by damping). At point-c the speed of rotor will be equal to synchronous speed.

➥ At point-*a* the speed is synchronous speed (w_s). From point *a* to *b* the speed increases and then from point *b* to *c* the speed decreases. Once again at point -*c* the speed is equal to synchronous speed (w_s). Thus the rotor oscillates between point-a and point-*c* before setting to point-*b*.

In Figure 4.8, the area A_1 is the acceleration area and area A_2 is the decceleration area. The equal area criterion says that, the system is stable if,

$$\int_{\delta_0}^{\delta_2} P_a d\delta = 0 \qquad ...(4.55)$$

To satisfy equation (4.55) the acceleration area A_1, should be equal to deceleration area A_2. when the oscillations die out the system will settle to a new state. In this new steady state, $P_{m1} = P_{e1}$.

$$\therefore \quad P_{m1} = P_{e1} = P_{max} \sin \delta_1 \qquad ...(4.56)$$

The areas A_1 and A_2 can be eveluated as shown below.

$$A_1 = \int_{\delta_0}^{\delta_1} (P_{m1} - P_e) d\delta \qquad ...(4.57)$$

$$A_2 = \int_{\delta_1}^{\delta_2} (P_e - P_{m1}) d\delta \qquad ...(4.58)$$

where, $P_e = P_{max} \sin \delta$

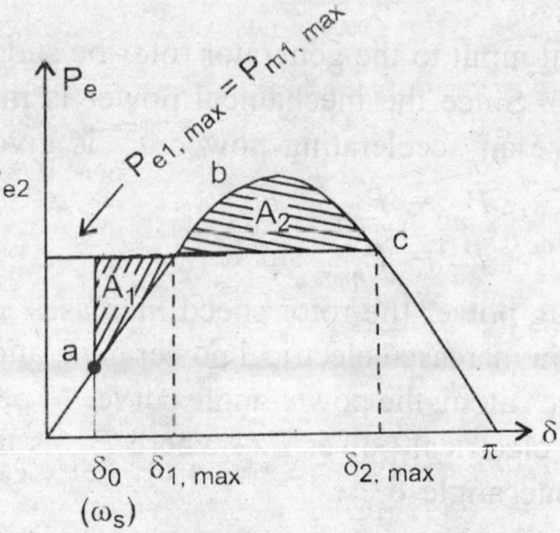

Figure 4.9 *Transient stability limit for sudden change in mechanical input.*

From above discussions we can say that there is a upper limit for increase in mechanical power input P_m. As the mechanical power is increased, a limiting condition is finally reached at a point where the area A_1 equals the area above the $P_{m1,max}$ line as show in Figure 4.9.

The corresponding δ can be $\delta_{1,max}$.

Under this condition δ_2 takes a maximum value of $\delta_{2,max}$.

Here, $$\delta_{2,max} = \pi - \delta_{1,max} \quad ...(4.59)$$

From equation (4.52) we can write,

$$P_{e1,max} = P_{m1,max} = P_{max} \sin \delta_{1,max}$$

$$\therefore \quad \sin \delta_{1,max} = \frac{P_{m,max}}{P_{max}} \text{ (or) } \delta_{1,max} = \sin^{-1}\left[\frac{P_{m1,max}}{P_{max}}\right] \quad ...(4.60)$$

From equations (4.59) and (4.60) we can write,

$$\delta_{2,max} = \pi - \sin^{-1}\left[\frac{P_{m1,max}}{P_{max}}\right]$$

From Figure 4.9 we can say that, any further increase in $P_{m1,max}$ will make the area A_2 less than area A_1. This means that the acceleration power is more than the deceleration power. Hence the system will have an excess kinetic energy which causes δ to increase beyond point-*c*. If the δ increases beyond point-*c* the deceleration power changes to acceleration power and so the system become unstable.

➡ It is important to note that system will remain stable even though the rotor may oscillate beyond δ = 90° as long as the equal area criterian is met. Hence the condition of δ = 90° for stability is applicable only for steady state stability and not for transient stability.

4.6.2 Critical Clearing Time and Critial Clearing Angle

Consider a single machine system shown in Figure 4.10.

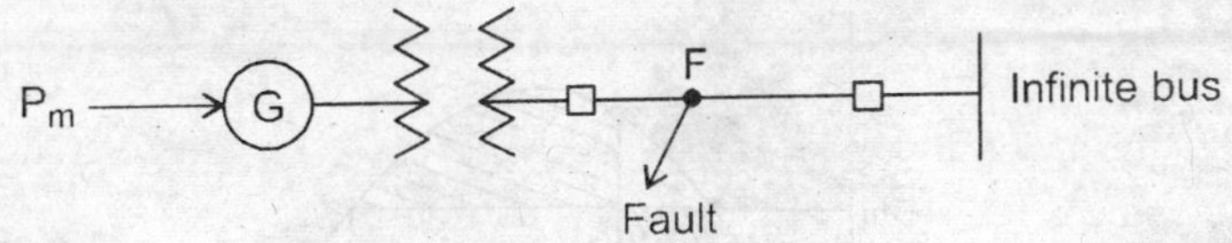

Figure 4.10

Let the mechanical input be P_m and Let machine is operating in steady state. with torque angle, δ_0. The power angle curve for this system is shown in Figure 4.11. The operating point is shown as point-*a*.

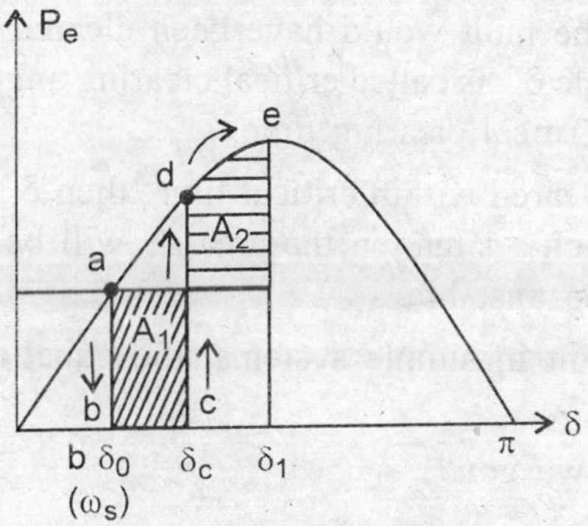

Figure 4.11

Let a three phase fault occur at point *F* in the system (refer fig (4.10)). Now $P_e = 0$ and the operating point drops to *b*. This means that the power transferred to infinite bus is zero and the entire power generated is flowing through fault. Now the operating point moves along bc.

Let the fault be transient in nature and so the fault be cleared by opening of the circuit breaker at Point-*c*, where $\delta = \delta_c$ and the corresponding time be t_c. Here t_c is called clearing time and δ_c is called clearing angle.

It is assumed that the circuit breaker closes its contact immediately after opening. Hence the normal operation is restored. Now the operating point shifts to point-*d*. The rotor now deaccelerates and the operating point moves along *de*.

For this transient state, if an angle δ_1, can be found such that, $A_2 = A_1$, then the system is found to be stable. The stable system may finally settles down to the steady operating point-*a* in an oscillatory manner due to damping in the system.

In the above discussion it is assumed that the fault is cleared at δ_c, but if the fault clearing is delayed then the angle δ_1 continue to increase to an upper limit, δ_{max}.

This corresponds to a point where equal areas for A_1 and A_2 can be found for a given P_m as shown in Figure 4.12

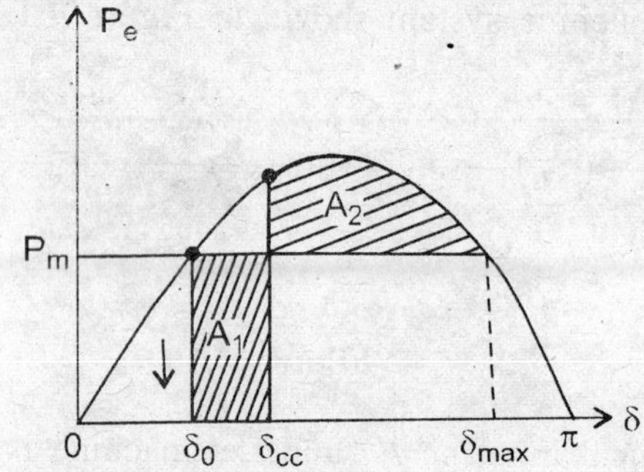

Note: $A_2 < A_1$

Figure 4.12

For this situation the fault would have been cleared at an angle δ_{cc}, as shown in Figure 4.12. This angle δ_{cc} is called critical clearing angle. The time corresponding to this angle is called critical clearing time, t_{cc}.

If the fault is not cleared within critical time, then δ_1 would increase to a value greater then δ_{max}. In such a situation the area A_2 will be less than the area A_1 and so the system would be unstable.

For three phase fault in simple systems, the equations for δ_{cc} and t_{cc} can be obtained as shown below.

From Figure 4.12 we get,

$$\delta_{max} = \pi - \delta_0 \qquad \text{...(4.61)}$$

Under steady state, for a given δ_0, $P_m = P_e$ and it is constant

$$\therefore \quad P_m = P_e = P_{max} \sin \delta_0 \qquad \text{...(4.62)}$$

The acceleration power,

$$P_a = P_m - P_e \qquad \text{...(4.63)}$$

When a three phase fault occurs,

$$P_e = 0$$

$$\therefore \quad P_a = P_m = \text{constant} \qquad \text{...(4.64)}$$

The accderation area A_1, can be evaluated by integration area A_1 can be evaluated by integrating P_a of equation (4.64) from $\delta = \delta_0$ to $\delta = \delta_{cc}$.

$$\therefore \quad A_1 = \int_{\delta_0}^{\delta_{cc}} P_m d\delta = P_m[\delta]_{\delta_0}^{\delta_{cc}} = P_m(\delta_{cc} - \delta_0) \qquad \text{...(4.63)}$$

When the power feeding is resumed after the fault,

$$P_e = P_{max} \sin \delta$$

Now, $P_a = P_e - P_m = P_{max} \sin \delta - P_m$ ...(4.64)

The deacceleration area A_2 can be evaluated by integrating P_a of equation (4.64) from $\delta = \delta_{cc}$ to $\delta = \delta_{max}$.

$$\therefore \quad A_2 = \int_{\delta_{cc}}^{\delta_{max}} (P_{max} \sin \delta - P_m) d\delta = \left[-P_{max} \cos \delta - P_m \delta\right]_{\delta_{cc}}^{\delta_{max}}$$

$$= \left[-P_{max} \cos \delta_{max} - P_m \delta_{max} + P_{max} \cos \delta_{cc} + P_m \delta_{cc}\right]$$

$$= P_{max} \left(\cos \delta_{cc} - \cos \delta_{max}\right) - P_m \left(\delta_{max} - \delta_{cc}\right) \quad ...(4.65)$$

For a stable system $A_1 = A_2$. Hence the equations (4.63) and (4.65) can be equated to solve δ_{cc}.

On equating (4.63) and (4.65) we get

$$P_{max} \left(\cos \delta_{cc} - \cos \delta_{max}\right) - P_m \left(\delta_{max} - \delta_{cc}\right) = P_m \left(\delta_{cc} - \delta_0\right)$$

$$\therefore \quad P_{max} \cos \delta_{cc} = P_m \delta_{cc} - P_m \delta_0 + P_{max} \cos \delta_{max} + P_m \delta_{max} - P_m \delta_{cc}$$

$$P_{max} \cos \delta_{cc} = P_m \left(\delta_{max} - \delta_0\right) + P_{max} \cos \delta_{max}$$

or

$$\therefore \quad \delta_{cc} = \cos^{-1} \left[\frac{P_m}{P_{max}} (\delta_{max} - \delta_0) + \cos \delta_{max}\right] \quad ...(4.66)$$

The equation (4.66) can be used to estimate the value of critical clearing angle, δ_{cc}. Consider the swing equation of single machine system,

$$\frac{H}{\pi f} \frac{d^2 f}{dt^2} = P_m - P_e \quad ...(4.67)$$

During three phase fault, $P_e = 0$

$$\therefore \quad \frac{H}{\pi f} \frac{d^2 \delta}{dt^2} = P_m$$

$$\frac{d^2 \delta}{dt^2} = \frac{\pi f \, P_m}{H} \quad ...(4.68)$$

On integrating equation (4.68) twice we get,

$$\delta = \frac{\pi f}{2H} P_m t^2 + \delta_0 \quad ...(4.69)$$

Where δ_0 is the integral constant

In equation (4.69) when $\delta = \delta_{cc}$, $t = t_{cc}$.

$$\therefore \text{ at } \quad t = t_{cc},\ \delta_{cc} = \frac{\pi f}{2H} P_m t_{cc}^2 + \delta_0$$

$$\frac{\pi f}{2H} P_m t_{cc}^2 = \delta_{cc} - \delta_0$$

$$\therefore \quad t_{cc} = \sqrt{\frac{2H(\delta_{cc} - \delta_0)}{\pi f P_m}} \quad ...(4.70)$$

The equation (4.70) can be used to estimate the value of critical clearing time, t_{cc}.

> ➡ Since $P_e = 0$, it is possible *to* find an solution for δ by integrating the swing equation.

4.7 STEP-BY-STEP METHOD

Consider the swing equation of a power system.

$$\frac{H}{\pi f} \frac{d^2\delta}{dt^2} = P_m - P_e \quad ...(4.71)$$

We know that,
$$M = \frac{H}{\pi f} \quad ...(4.72)$$

$$P_e = P_{max} \sin \delta \quad ...(4.73)$$

$$P_a = P_m - P_e = P_m - P_{max} \sin \delta \quad ...(4.74)$$

$$\therefore \quad M\frac{d^2\delta}{dt^2} = P_a$$

$$\frac{d^2\delta}{dt^2} = \frac{P_a}{M} \quad ...(4.75)$$

The equation (4.75) is a nonlinear equation. During transient state that the δ is a function of time, *t* and so it can be denoted as δ(*t*). In step-by-step method, the solution of δ(*t*) is obtained by dividing the time into small equal values of Δ*t*. ie, the entire time (range) of interest is divided into number of small equal interval Δ*t*.

The accelerating power and the change in speed are also continuous function of time. They are discretized as follows

1. The accelerating power P_a computed at the beginning of an interval is assumed to remain constant from the middle of the preceding interval to the middle of the interval being considered.

2. The angular velocity is assumed constant throughout any interval. This constant value is the value corresponding to the midpoint of concerned interval.

The discretization of P_a and ω are shown in Figure 4.13

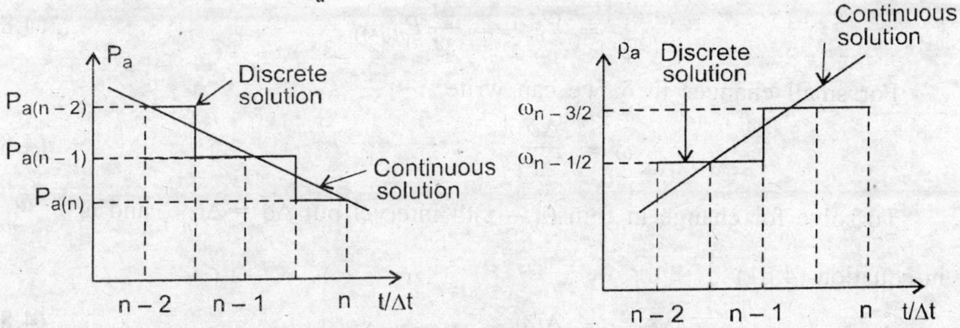

Figure 4.13 *Discretization of P_a and ω*

For each discrete interval the value of P_a, ω and δ are calculated as shown below. The solution starts from the initial condition values. The solution starts from the initial condition values, which corresponds to a stable operating point. Let δ_0 be the angle corresponding to initial operating point.

Let us assume that, the values ω and δ for (*n* – 1)th interval are known

δ_{n-1} = The value of δ at the end of (*n* – 1)th interval

$\omega_{n-\frac{1}{2}}$ = The value of ω at the end of (*n* – 1)th interval

$P_{a(n-1)}$ = The value of P_a at the end of (*n* – 1)th interval

Now from equation (4.75) we get,

$$P_{a(n-1)} = P_m - P_{max} \sin \delta_{n-1} \quad ...(4.76)$$

The equation (4.75) can be written in the modified form as,

$$\frac{d\omega}{dt} = \frac{P_a}{M} \quad ...(4.77)$$

where, $\omega = \frac{d\delta}{dt}$ and $\frac{d\omega}{dt} = \frac{d^2\delta}{dt^2}$

For small changes in δ, the equation (4.77) can be Linearized as shown below.

$$\frac{\Delta\omega}{\Delta t} = \frac{P_a}{M} \quad ...(4.78)$$

$$\therefore \quad \Delta\omega = \frac{\Delta t}{M} P_a \quad ...(4.79)$$

Let $W_{n-3/2}$ = The value of ω at the end of nth internal.

For calculating n^{th} internal value of ω, put,

$\Delta\omega = \omega_{n-\frac{1}{2}} - \omega_{n-\frac{3}{2}}$ and $P_a = P_{a(n-1)}$ in equation (4.79)

$$\therefore \quad \omega_{n-\frac{1}{2}} - \omega_{n-\frac{3}{2}} = \frac{\Delta t}{M} P_{a(n-1)}$$

$$\therefore \quad \omega_{n-\frac{3}{2}} = \omega_{n-\frac{1}{2}} - \frac{\Delta t}{M} P_{a(n-1)} \qquad ...(4.80)$$

For small changes in δ, we can write

$$\omega = \frac{\Delta\delta}{\Delta t} \qquad ...(4.81)$$

To solve for change in δ in $(n - 1)$th intervel put $\Delta\delta = \Delta\delta_{n-1}$ and $\omega = \omega_{n-\frac{3}{2}}$ in equation (4.81)

$$\therefore \quad \Delta\delta_{n-1} = \Delta t \omega_{n-\frac{3}{2}} \qquad ...(4.82)$$

Similarly for change in δ in n^{th} interval put

$\Delta\delta = \Delta\delta_n$ and $\omega = \omega_{n-\frac{1}{2}}$ in equation (4.81)

$$\therefore \quad \Delta\delta_n = \Delta t \omega_{n-\frac{1}{2}}$$

Let δ_n = The value of δ at the end of nth interval.

Now, $$\delta_n = \delta_{n+1} + \Delta\delta_n \qquad ...(4.83)$$

The above process of computation is repeated to obtain $P_{a(n)}$, $\Delta\delta_{(n+1)}$ and $\delta_{(n+1)}$. The solution of $\delta(t)$ is thus obtained in discrete form over the desired length of time. The normal desired length of time is 0.5sec.

The continuous form of solution is obtained by drawing smooth curve through discrete values as shown in Figure 4.14

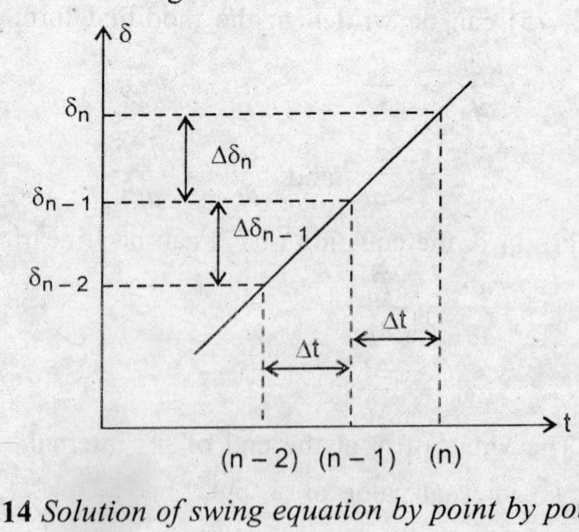

Figure 4.14 *Solution of swing equation by point by point method.*

4.8 METHODS OF IMPROVING TRANSIENT STABILITY

The transient stability depends on the type and location of a fault. Let us consider the most severe type of fault. (i.e., three phase fault)

Note: The following discussions are also applicable for transient conditions due to three phase fault.

An increase in the inertia constant *M* of a single machine connected to infinite bus, reduces the angle through which the rotor swings in a given time interval. Hence stability can be improved by increasing *M*.

However this connot be employed in practice because of economic reasons. Also, increasing *M* will have an undesirable effect of slowing down the response of the speed governor loop.

It can be proved that for given clearing angle the accderation area decreases but the deacceleration area increases the maximum power limit of power angle curve is raised which improves the transient stability limit of the system. The above conclusion suggest the following methods of improving the transient stability of a power system.

1. Increase of system voltages, use of *AVR* (automatic voltage regulator).
2. Use of high speed excitation systems.
3. Reduction in system transfer reactance.
4. Use of high speed reclosing breakers.

During fault, the voltage reduction of all buses at the generator terminals are sensed by the automatic voltage regulators which help to restore the generator voltage. Modern exciter systems having solid state controls which quickly respond to bus voltage reduction.

The transfer reactance can be reduced to improve the stability limit. Incidentally this also raised system voltage profile. The reactance of a transmission line can be decreased (i) by reducing the conductor spacing and (ii) by increasing conductor diameter.

Compensation for line reactance by series capacitors is an effective and economical method of increasing stability limit specially for transmission distances of more than 350 Km. Switched series capacitors simultaneously decrease fluctuation of load voltages and raise the transient stability limit to a value almost to the steady state limit.

The transfer reactance can also to reduce by increasing the number of parallel lines between transmission points. Rapid switching and isolation of unhealty lines followed by reclosing also improves stability margins.

Some of the important methods to improve stability are,

1. **HVDC links:** Increased use of *HVDC* lines employing thyristors can improve the stability problem. A dc link is asynchronous.
2. **Breaking resistors:** For improving stability where clearing is delayed or large load is suddenly lost, a resistive load called a breaking resistor is connected at or near the generator bus. This load compensates for some of the reduction of load on the generators.
3. **By pass valving:** The stability of a unit is improved by decreasing the mechanical input power to the turbine. i.e., some of the mechanical power is by passed which is given to turbine.
4. **Full load rejection technique:** Fast valving combined with high-speed clearing time will be sufficient to matain stability in most cases. However, there are still situations where stability is difficult to maintain. To remedy these situations, full load rejection scheme could be ultilized after the unit is separated from the system. To do this, unit has to be equiped with a large steam by pass system, After system has recovered from the shock caused by the fault the unit could be resynchronized and reloaded. The main disadvantage is the extra cost of the large by pass system.

Example 4.1: The generator shown in Figure 4.1.1 is delivering power to infinite bus. Take $|V_t|$ = 1.1 p.u find the maximum power that can be transferred when (i) The system is healthy (ii) on line is open

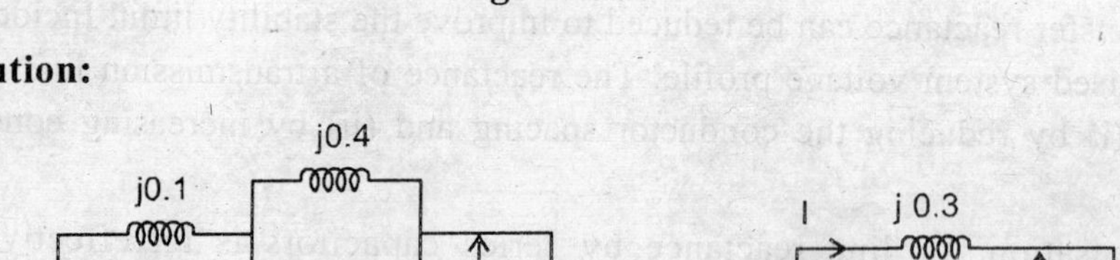

Figure 4.1.1

Solution:

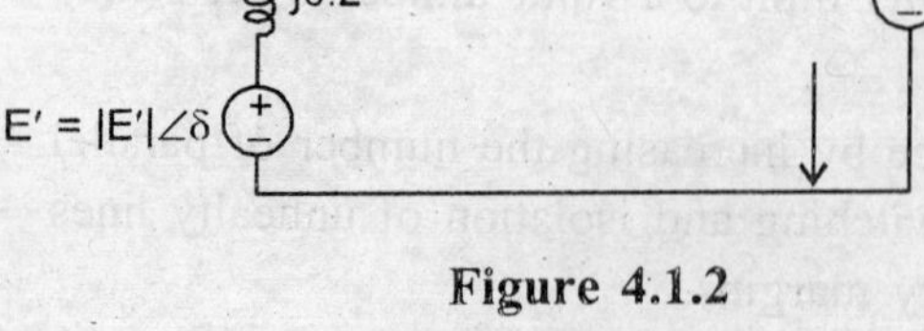

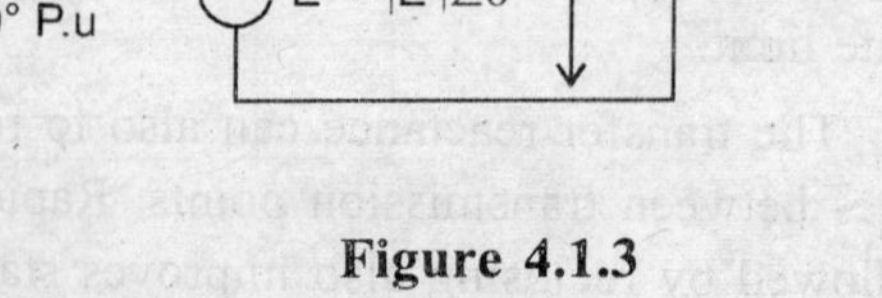

Figure 4.1.2 **Figure 4.1.3**

The reactance diagram of the system is shown in Figure 4.1.2 & 4.1.3

Here, $V_t = |V_t| \angle\theta$ and $V = 1\angle 0°$ P.u

Let electrical power transferred at steady state be equal to P_m.

$$\therefore \quad P_e = P_m = 1 \text{ P.u}$$

we know that, $$P_e = \frac{|V_t||V|}{X} \sin\theta$$

$$\theta = \sin^{-1}\left(\frac{P_e X}{|V_t||V|}\right) = \sin^{-1}\left(\frac{1\times 0.3}{1.1\times 1}\right) = 15.8°$$

$$\therefore \quad V_t = |V_t|\angle\theta = 1.1\angle 15.18° = 1.058 + j\,0.3$$

Current, $$I = \frac{V_t - V}{jX} = \frac{1.058 + j0.3 - 1}{j0.3} = \frac{0.058 + j0.3}{0.3\lfloor 90°}$$

$$= \frac{.3056\lfloor 79°}{0.3\lfloor 90°} = 1.0187\lfloor -11° \text{ p.u}$$

In Figure 4.1.3 using KVL,

$$E' = j\,0.2\,I + V_t$$

$$= j\,0.2 \times 1.0187\ \angle -11° + 1.1\lfloor 15.8°$$

$$= 0.2 \times 1.0187\angle -11° + 1.11\angle 15.8$$

$$= 0.2\lfloor 90° \times 1.0187\lfloor -11° + 1.058 + j0.3$$

$$= 0.2037\lfloor 79° + 1.058 + j0.3$$

$$= 0.039 + j0.2 + 1.058 + j0.3$$

$$= 1.097 + j0.5 = 1.2056\lfloor 24.5° p.u$$

$$\therefore \quad |E'| = 1.2056\ p.u \text{ and } \delta = \delta_0 = 24.5°$$

Case (*i*): Healthy system

Let X_{12} be the reactance between the transient internal emf and the source representing infinite bus.

From Figure 4.13, $X_{12} = 0.5\ p.u$

$$\therefore \quad P_{max} = \frac{|E'||V|}{X_{12}} = \frac{1.2056\times 1}{0.5} = 2.4112\ p.u$$

$$\therefore \quad P_e = P_{max} \sin\delta = 2.4112 \sin\delta.$$

Case (*ii*): When one line is open

The reactance diagram when one line is open is shown in Figure 4.14

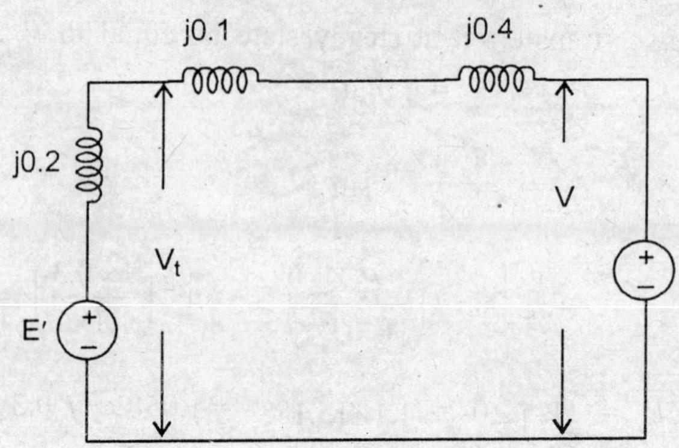

Figure 4.14

Let, $X_{12} = 0.2 + 0.1 + 0.4 = 0.7 \; p.u$

$\therefore$ $$P_{max} = \frac{|E'||V|}{X_{12}} = \frac{1.2056 \times 1}{0.7}$$

$$P_{max} = 1.722 \; p.u$$

$\therefore$ $$P_e = P_{max} \sin\delta = 1.722 \sin\delta$$

Result: (*i*) $P_e = 2.4112 \sin\delta$

(*ii*) $P_e = 1.722 \sin\delta$

Example 4.2: Find the steady state power limit of a system consisting of a generator with reactance of 0.8 *p.u.* The terminal of the generator is 1.15 p.u and the voltage of infinite bus is $1\underline{|0°}$ *p.u*

Solution: The reactance diagram of the system is shown in Figure 4.2.1

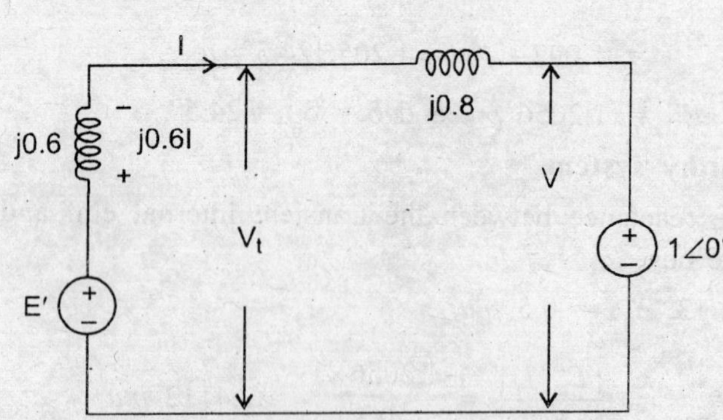

Figure 4.2.1

Given that, $V = 1\angle 0° \;\&\; |V_t| = 1.15$

Let, $V_t = 1.15\angle\theta$

From Figure 4.2.1 we get,

$$I = \frac{V_t - V}{j0.8}$$

$$E' = j\,0.6\,I + V_t$$

$$= j0.6\left[\frac{V_t - V}{j0.8}\right] + V_t = 0.75\,(V_t - V) + V_t$$

$$= 1.75V_t - 0.75V$$

$$= 1.75 \times 1.15\angle\theta - 0.75 \times 1\angle 0°$$

$$= 2.0125\angle\theta - 0.75 = 2.0125\cos\theta + j2.0125\sin\theta - 0.75$$

$$= (2.0125\cos\theta - .75) + j(2.0125\sin\theta) \qquad \text{...(4.2.1)}$$

Consider the transient internal *emf* in polar form.

i.e, $$E' = |E'|\angle\delta = |E'|\cos\delta + j|E'|\sin\delta \qquad \text{...(4.2.2)}$$

When the steady state power limit is reached, the value of δ is 90°, when δ = 90°, $\cos\delta = 0$ and so the real part of equation (4.2.2) is zero. Therefore at steady state limit, the real part of equation (4.2.1) is also zero.

$$\therefore\; 2.0125\cos\theta - 0.75 = 0$$

$$2.0125\cos\theta = 0.75$$

$$\theta = \cos^{-1}\left[\frac{0.75}{2.0125}\right] = 68.1°$$

$$V_t = |V_t|\angle\theta = 1.15\angle 68.1°\,p.u$$

Now, $$I = \frac{V_t - V}{j0.8} = \frac{1.15\angle 68.1° - 1\angle 0°}{0.8\angle 10°} = 1.51125\angle 28.2°$$

$$= 1.333 + j\,0.715\;p.u$$

$$E' = j\,0.6I + V_t = 0.6\angle 90° \times 1.5125\angle 28.2° + 1.15\angle 68.1°$$

$$= 0.9075\angle 118.2° + 1.15\angle 68.1° = -0.4288 + j0.8$$

$$+ .4289 + j1.067$$

$$= 0.0001 + j1.867 = 1.867\angle 90°\,p.u$$

steady state power limit,

$$P_{max} = \frac{|E'||V|}{0.6+0.8} = \frac{1.867\times 1}{1.4}$$

$$P_{max} = 1.336\ P.u$$

Result: $P_{max} = 1.336\ P.u$

Example 4.3: A synchronous generator having a reactance of 1 $P.u$ is connected to an infinite bus through a transmission system with a reactance of 0.7 $P.u.$ The generator is running at no-load with a voltage of 1.1 $P.u.$ Take $H = 4.5\ MW\text{–}s/MVA$. The voltage of infinite bus is $1\angle 0° P.u.$ and its frequency is 50Hz. Calculate the frequency of natural oscillations if the machine is suddenly loaded to 60% of its maximum power limit. Neglect the resistance and machine damping.

Solution: Maximum power limit,

$$P_{max} = \frac{|E||V|}{X}$$

Given that, $|E| = 1.1\ P.u$; $|V| = 1\ P.u$; $X = (1 + 0.7)\ P.u.$

$$\therefore \quad P_{max} = \frac{1.1\times 1}{1.7} = 0.6471\ P.u.$$

The incrtia constant, $M = \dfrac{H}{\pi f}$ in $P.u.$

Given that, $H = 4.5\ MW\text{–}s/MVA$ and $f = 50$ Hz

$$\therefore \quad M = \frac{4.5}{\pi\times 50} = 0.0287\ P.u.$$

at 60% Loading

The electrical power, $P_e = P_{max} \sin\delta$

For 60% loading, Let $\delta = \delta_1$ and $P_e = P_{el}$

It is given that, $\dfrac{P_{el}}{P_{max}} = 60\% = 0.6$

$$\therefore \quad P_{el} = P_{max} \sin\delta_1 \text{ (or) } \sin\delta_1 = \frac{P_{el}}{P_{max}} = 0.6$$

$$\therefore \quad \delta_1 = \sin^{-1} 0.6 = 36.9° \text{ (electrical degree)}$$

The characteristic equation of the system during small disturbances is given by

$$MX^2 + C = 0$$

where, $C = P_{max,} \cos\delta$

The natural frequency of oscillation of the system for small disturbances is given by the roots of characteristic equation.

The root of characteristic equation,

$$X = \pm\sqrt{\frac{-C}{M}} = \pm j\sqrt{\frac{P_{max}\cos\delta_1}{M}}$$

$$= \pm j\sqrt{\frac{0.6471\times\cos 36.9°}{0.0287}} = \pm j4.25$$

∴ Natural frequency of oscillation, ω = 4.25 rad/sec.

$$\therefore \qquad f = \frac{\omega}{2\pi} \Rightarrow = \frac{4.25}{2\pi} = 0.6764Hz$$

Result: Frequency of natural oscillation is 0.6764 *Hz.*

Example 4.4: Two power stations *A* and *B* are located close together. station A has four identical generator sets each rated 100 *MVA* and having an inertia constant of 9*MJ/MVA* where as the station *B* has 3 sets each rated 200 *MVA*, 4 *MJ/MVA*. Calculate the inertia content of a single equivalent machine on a base of 100 *MVA*.

Solution: Let us assume that the machines are swinging coherently. For two machines swinging coherently the equivalent inertia constant, H_{eq} is given by.

$$H_{eq} = \frac{H_{1,mach}S_{1,mach}}{S_{sys}} + \frac{H_{2,mach}S_{2,mach}}{S_{sys}} \qquad ...(4.4.1)$$

where,

S_{sys} = *MVA* rating of system (or specified base)

$S_{1,mach}$ & $H_{1,mach}$ = *MVA* rating and inertia constant of machine-1

$S_{2,mach}$ & $H_{2,mach}$ = *MVA* rating and inertia constant of machine-2

The equation (4.4.1) can be extended to any number of machines swinging cohenently

Section-A

The station-*A* has four machines of identical rating

∴ Equivalent inertia constant of station $A = H_A = 4\left[\frac{H_{mach}S_{mach}}{S_{sys}}\right]$

Given that,

$$H_{mach} = 9\ MJ/MVA;$$

$$S_{mach} = 100\ MVA \text{ and } S_{sys} = 100\ MVA$$

$$H_A = 4\left[\frac{9\times 100}{100}\right] = 36\ MJ/MVA$$

Section-B

The station-B has three machines of identical rating

$$\therefore \text{ Equivalent inertia constant of station } B = H_B = 3\left[\frac{H_{mach}S_{mach}}{S_{sys}}\right]$$

Given that, $H_{mach} = 4MJ/MVA;\ S_{mach} = 200\ MVA$

and $S_{sys} = 100MVA$

$$\therefore \quad H_B = 3\left[\frac{4\times 200}{100}\right] = 24\ MJ/MVA$$

H_{eq} of system

The inertia constant of station-A and station-B are calculated on a common system base MVA (of 100MVA). Hence the equivalent inertia constant of the system is given by the sum of inertia constents of the two stations.

$\therefore$ Equivalent inertia constant of the system $= H_{eq} = H_A + H_B$

$$= 36 + 24 = 60\ MJ/MVA$$

Example 4.5: A 2 pole, 50 Hz, 11KV turbo alternator has a ratio of 100 MW, power factor 0.85 lagging. The rotor has a moment of inertia of 10,000 kgm^2. Calculate H & N.

Given data:

Pole, $P = 2$

$f = 50\text{Hz}$

$P.f = 0.85$ lag

$V = 11KV$

$P = 100MW$

$J = 10{,}000\ \text{Kgm}^2$

Solution: Synchronous speed in *rps*,

$$n_s = \frac{2f}{P} = \frac{2\times 50}{2} = 50rps$$

Synchronous speed in rad/sec,

$$\omega_s = 2\pi ns = 2\pi \times 50 = 314.16 \text{ elect. rad/sec}$$

➡ One revolution is equal to 2π radians

Inertia constant, $$M = J\left(\frac{2}{P}\right)^2 \omega_s \times 10^{-6} \text{ in } MJ - \text{s/elect.rad}$$

MVA rating of machine,

$$S = \frac{P}{\text{Power factor}} = \frac{100}{0.85} = 117.675\ MVA$$

Let us choose the rated Voltage and *MVA* as base values

$$\therefore \quad KV_b = 11KV \text{ and } MVA_b = 117.65\ MVA$$

Inertia constant, *Min pu*,

$$M_{pu} = \frac{MinMJ - s/elect.rad}{MVA_b}$$

$$= \frac{3.1416}{117.65} = 0.0267\ p.u$$

The inertia constant, $H = \pi f\, M_{pu}$

$$= \pi \times 50 \times 0.0267 = 4.194\ MW\text{–}s/MVA$$

Result: $M = 3.1416\ MJ\text{-}s/\text{elec.rad (or) } 0.0267\ P.u$

Inertia constant, $H = 4.194\ MW\text{–}s\ /\ MVA$

EXERCISE

1. A 50 Hz four-pole turbo generator rated 30*MVA*, 13.2 *KV* has an inertia constant of *H* = 6.0 KW sec/*KVA*. Determine the kinetic energy stored in the rotor at synchronous speed. Determine the accderation if the rotational losses is 28000 *HP* and the electric power developed is 20000 *KW*. If the accderation computed for the generator is constant for a period of 15 cycles, determine the change in torque angle in that period and the rpm at the end of 25 cycles. Assume that the generator is synchronized with a large system and has no accelerating torque before the 25 cycle period begins.
2. Determine the critical fault clearing angle for the network in Figure 4.2.1 when a three phase fault takes place at *B* and the breakers at *A* and *B* operate simultaneously. The generator is delivering 1 *p.u* power before the fault takes place. Assuming the inertia constant *H* = 4, (*a*) determine the critical clearing time (*b*) plot swing curves both under sustained fault and when the breakers have operated. Used step-by-step method taking time internal of 0.05 sec.

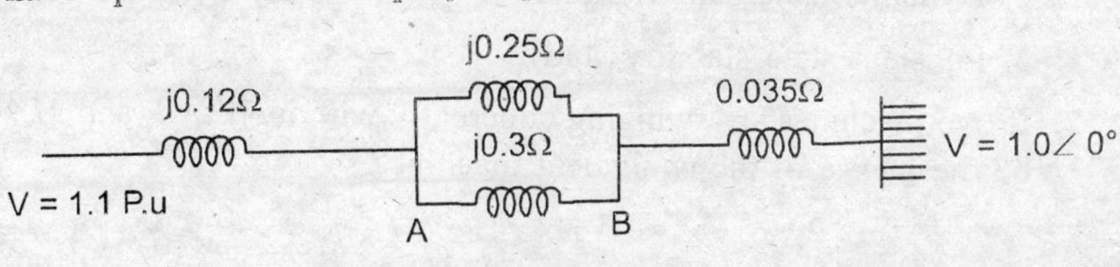

Figure E 4.2.1

3. Calculate the steady state power limit consisting of a generator with equivalent reactance 0.8 *P.u* connected to an infinite bus through a series reactance of 1.0 *P.u.* The generator *emf* is held fixed at 1.2 *P.u.* and the voltage of the infinite bus is 1.0 *P.u.*

4 A 50 Hz, 8 pole generator rated 80 *MVA*, 11*KV* has an inertia constant of 7*MJ*/*MVA* find the moment of inertia *M*.

Short Questions and Answers

Q. 1. Define stability.

Ans. The stability of a system is defined as the ability of power system to return to stable (synchronous) operation when it is subjected of a disturbance.

Q. 2. Define steady state stability.

Ans. The steady state stability is defined as the ability of a power system to remain stable i.e. without loosing syschronism for small disturbances.

Q. 3. Define transient stability.

Ans. The transient stability is defined as the ability of power system to remain stable (i.e. without loosing synchronism) for large disturbances.

Q. 4. What is steady state stability limit?

Ans. The amount of power maximum power that can be transmitted by a machine (or transmitting system) to a receiving system without loss of synchronism. In steady state the power transferred by synchronous machine (or power system) is always less that the steady state stability limit.

Q. 5. What is transient stability limit?

Ans. The transient stability limit is the maximum power that can be transmitted by a machine (or transmitting system) to a fault or a receiving system during a transient state without loss of synchronism. The transient stability limit is always less than the steady state stability limit.

Q. 6. How stability studies are classified, what they are?

Ans. Stability studies are classified depending on the nature of the disturbance.

1. Steady state stability study
2. Dynamic state stability study
3. transient state stability study

Q. 7. If two machines are swinging coherently with inertia M_1 and M_2, what will be the inertia of the equivalent machine?

Ans. The equivalent moment of inertia, $M_{eq} = \frac{M_1 S_1}{S_b} + \frac{M_2 S_2}{S_b}$

where S_1 & S_2 = MVA rating of machine 1 & 2

S_b = Base MVA

Q. 8. Define synchronizing coefficient. For what value of synchronizing coefficient the system remain stable.

Ans. The term $P_m \cos \delta_0$ called synchronizing (or) stiffness at synchronous machine. The system is stable if the synchronizing coefficient positive.

It is positive when $0 \leq \delta_0 \leq \frac{\pi}{2}$.

Q. 9. Define swing curve. what is the use of swing curve?

Ans. The swing curve is the plot (or) graph between the power angle δ and time *t*. It is usually plotted for a transient state to study the nature of variation in δ for a sudden large disturbance. From the nature of variations of δ the stability of a system for any disturbance can be determined.

Q. 10. Define power angle.

Ans. The power angle (or) torque angle is defined as the angular displacement of the rotor from synchronously rotating reference frame.

Q. 11. For the swing equation $\frac{Md^2\delta}{dt^2} = P_a$ what will be the value of P_a decceleration during steady state operation.

Ans. During steady state operation, there is no acceleration (or) decceleration on the rotor. Therefore, P_a = 0. (under this condition δ is a constant).

Q. 12. Name the two ways by which transient stability study can be made in a system where one machine is swinging with respect to an infinite bus.

Ans. For single machine-infinite bus system, the following two methods can be employed for transient stability studies.

1. Equal area criterion
2. Step-by-step method

Q. 13. Define critical clearing time and critical clearing angle.

Ans. The critical clearing angle, δ_{cc} is the maximum allowable change in the power angle δ before clearing the fault without loss of synchronism. The time corresponding to this angle is called critical clearing time, t_{cc}.

The critical clearing time, t_{cc} can be defined as the maximum time delay that can be allowed to clear a fault without loss of synchronism.

Q. 14. List the methods of improving the transient stability limit of a power system.

Ans. (*i*) Increase of system voltage and use of *AVR*

(*ii*) Use of high speed excitation systems.

(*iii*) Reduction in system transfer reactance

(*iv*) Use of high speed reclosing breakers.

Q 15. State equal area criterion.

Ans. The system is stable if the area under P_a - δ curve reduces to zero at some value of δ. This is possiblc only if the +ve area under (P_a – δ) curve is equal to the –ve area under P_a – δ curve for a finite change in δ. Hence the stability criterion is called equal area criterion.

Choose The Best Answer

1. A system is said to be stable if

(a) $\frac{dP}{d\delta}$ is positive (b) $\frac{dP}{d\delta}$ is negative

(c) both *a* & *b* (d) none of the above

2. If the loading of machine is large, the torque angle will not be

(a) less (b) more

(c) more (or) less (d) none of the above

3. Equal area criterion method is not applicable to a

(a) multi machine system (b) one machine

(c) both *a* & *b* (d) none of the above

4. In order to increase the transient stability limit of the system

(a) series reactance of transmission lines is reduced

(b) series reactance of transmission lines is increased

(c) series reactance of transmission lines is Kept constant.

(d) None of the above.

5. In step-by-step method, the computed swing curve approaches the true curve, only when the time interval is

(a) increased (b) decreased

(c) constent (d) none of the above

ANSWERS

1. a 2. a 3. c. 4. b 5. b

Fill in the Blanks

1. The angle between relative position of rotor axis and the stator magnetic field axis is ________

2. If transmitted power is increased beyond ________, then the synchronism will be lost.
3. The ratio of Ke. at rated speed to the rated apparent power of the machine is known as ________
4. The swing equation can be solved by ________ and ________ methods.
5. In a two machine system, __________ grounding at the sending end and ________ grounding at the receiving end is used.

ANSWERS

1. Load angle (or) torque angle
2. Steady state stability limit.
3. Inertia constant
4. Runge-Kutta, modified euler's
5. Resistance, Reactance.

Chapter-5

TRAVELLING WAVES

5.1 INTRODUCTION

As we know that the transmission lines are subjected to many disturbances. Due to all kinds of disturbances there will be sudden rise in voltage. For example lightning is the most dangerous disturbance to the transmission Lines. When Lighting strikes the transmission line then there will be a great rise in voltage magnitude with different wave-shapes. These waves will damage the equipments because they travel along the lines. The study of travelling waves is very important to protect the equipments.

5.2 WAVE EQUATION FOR UNIFORM TRANSMISSION LINES

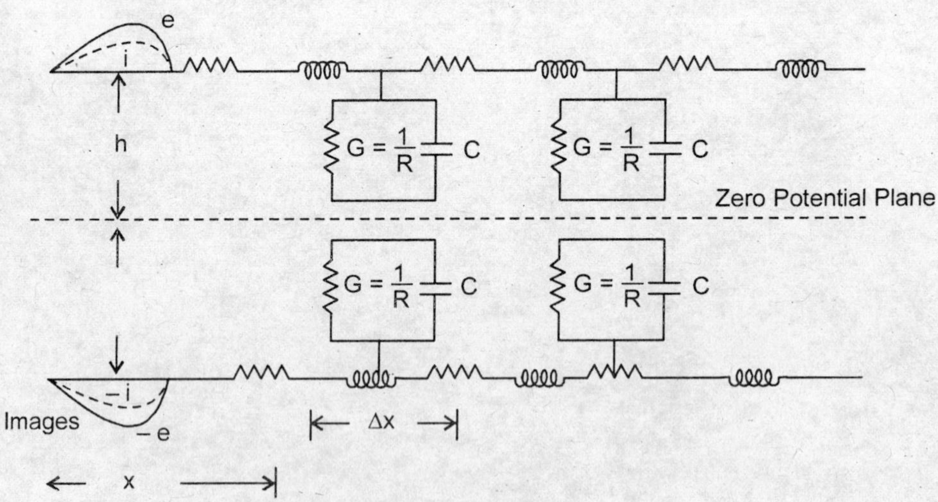

Fig. 5.1 *Voltage and current travelling waves on a line*

Fig. (5.1) Shows a pair of voltage and current travelling waves e and i on a single circuit of two conductors. One of the conductor may be the ground image of the overhead conductor. It is necessary to use Partial differential equations in developing the theory of travelling waves.

Let the line constants of transmission line to neutral (Zero potential plane) are,

$R \rightarrow$ Line resistance per unit length, ohms/meter

$L \rightarrow$ Line inductance per unit length, H/m

$C \rightarrow$ Line capacitance per unit length, F/m

$G \rightarrow$ Leakage conductance per unit length, siemens/m

For an element Δx of the line parameters are $r\,\Delta x$, $L\Delta x$, $C\Delta x$ and $G\Delta x$, The voltage drop in the element is Δx in the positive direction of x is

$$-\Delta e = r\Delta x i + L\Delta x \frac{\partial i}{\partial t} \qquad ...(5.1)$$

If the elements shrinks in length dx, then

$$\frac{\partial e}{\partial x} = -ri - L\frac{\partial i}{\partial t} \qquad ...(5.2)$$

The change in current in the element Δx is

$$-\Delta i = eG\Delta x + C\Delta x \frac{\partial e}{\partial t} \qquad ...(5.3)$$

$$\therefore \quad \frac{\partial i}{\partial x} = -Ge - C\frac{\partial e}{\partial t} \qquad ...(5.4)$$

Applying Laplace transform and substituting the operator p for $\frac{\partial}{\partial t}$, Equation (5.2) and (5.4) become

$$\frac{\partial e}{\partial x} = -(r + Lp)i = -Zi \qquad ...(5.5)$$

$$\frac{\partial i}{\partial x} = -(G + Cp)e = -Ye \qquad ..(5.6)$$

Where $Z = r + Lp$ is the impedance operator and $Y = G + Cp$ is the admittance operator. The negative signs in the above equations arise from the convention being used

➡ (both e and i decreases by increasing x).

Differentiate equation (5.5) with respect to x and substitute for $\frac{\partial i}{\partial x}$ from equation (5.6)

$$\frac{\partial^2 e}{\partial x^2} = -Z\frac{\partial i}{\partial x} = (-Z)(-Ye) = ZYe = \gamma^2 e \qquad ...(5.7)$$

Where,

$\gamma \rightarrow$ Propagation constant and equals $\sqrt{ZY}$

Similarly, $$\frac{\partial^2 i}{\partial x^2} = Y\frac{\partial e}{\partial x} = ZYi = \gamma^2 i \qquad ...(5.8)$$

The solution of equation (5.7) is of the form

$$\varepsilon = \varepsilon^{x\gamma} f_1(t) + \varepsilon^{-x\gamma} f_2(f) \qquad ...(5.9)$$

Where,

$f_1(t)$ and $f_2(t) \rightarrow$ integrating constants with respect to x from equations (5.6) and (5.9)

$$i = -Y\int edx = -\sqrt{\frac{Y}{Z}}\left[\varepsilon^{x\gamma} f_1(t) - \varepsilon^{-x\gamma} f_2(t)\right] \qquad ...(5.10)$$

To simplify the analysis, the line is assumed as lossless *i.e,* $r = O$ and $G = 0$. Then

$$\gamma = \sqrt{(r+Lp)(G+Cp)} = p\sqrt{LC} = \frac{p}{v} \qquad ...(5.11)$$

Where, $v = 1/\sqrt{LC}$

Then. $$\varepsilon = \varepsilon^{xp/v} f_1(t) + \varepsilon^{-xp/v} f_2(t) \qquad ...(5.12)$$

$$i = -\sqrt{\frac{C}{L}}\left[\varepsilon^{xp/v} f_1(t) - \varepsilon^{-xp/v} f_2(t)\right] \qquad ...(5.13)$$

By Taylor's theorem

$$f(t+a) = f(t) + af'(t) + \frac{a^2}{2!}f''(t) + ...$$

$$= \left[1 + ap + \frac{a^2}{2!}p^2 + ...\right] f(t) = \varepsilon^{ap} f(t) \qquad ...(5.14)$$

By comparing with equation (5.14) we can write equation (5.12) and (5.13) as

$$e = f_1\left(t + \frac{x}{v}\right) + f_2\left(t - \frac{x}{v}\right) \qquad ...(5.15)$$

$$i = -\sqrt{\frac{C}{L}}\left[f_1\left(t+\frac{x}{v}\right)-f_2\left(t-\frac{x}{v}\right)\right] \quad ...(5.16)$$

A function of the form $f\left[t \pm \frac{x}{v}\right]$ or $f(x \pm vt)$ represents a travelling wave because for any value of t, a corresponding value of x can be found such that $(x \pm vt)$ has a constant value and, therefore, defines a fixed point on $f(x \pm vt)$. Corresponding values of x and t which define the same points on a wave are given by

$t - x/v = \lambda_1$ for waves travelling in the direction of $+ x$

(forward waves) ...(5.17)

$t + x/v = \lambda_2$ for waves travelling in the direction of $- x$

(back ward waves) ...(5.18)

5.3 VELOCITY OF PROPAGATION

The velocity of propagation can be found by differentiating the equation (5.17) in section (5.2) Keeping in view that λ_1 and λ_2 are constants.

$$\left.\begin{aligned}\frac{dx}{dt} &= v = 1\sqrt{LC} \text{ for forward wave}\\ \frac{dx}{dt} &= -v = -1/\sqrt{LC} \text{ for backward wave}\end{aligned}\right] \quad ...(5.19)$$

From equation (5.15) and (5.16)

$$\left.\begin{aligned}\frac{e}{i} &= 1\sqrt{LC} \text{ for forward wave}\\ \frac{e}{i} &= -\sqrt{LC} \text{ for backward wave}\end{aligned}\right] \quad ...(5.20)$$

The ratio $\sqrt{L/C}$ is known as surge impedance (or) characteristic impedance and is written as Zc. The reciprocal of surge impedance is surge admittance Y_c. Surge impedance has the characteristics of a resistance and has the dimensions of ohm.

The voltage and current waves are replicas of each other. The forward waves are written with *a* subscript f and the backward waves are written with a subscript b. Fig. (5.2) shows the schematic representation of polarities and direction of propogation for forward and backward waves. The forward voltage and current waves (Fig. 5.2 *a* and *c*) propagate in the direction of *x* ad have the same polarity because the ratio *e/i*, for a forward wave is positive as given by equation (5.20). The backward voltae and current waves (Fig. 5.2 *b* and *d*) propagate in the negative *x* direction and have different polarities because the ratio *e/i* for a back ward wave is negative. Thus, if both forward and backward voltage waves are *+ve* then the backward current wave differs from the forward one.

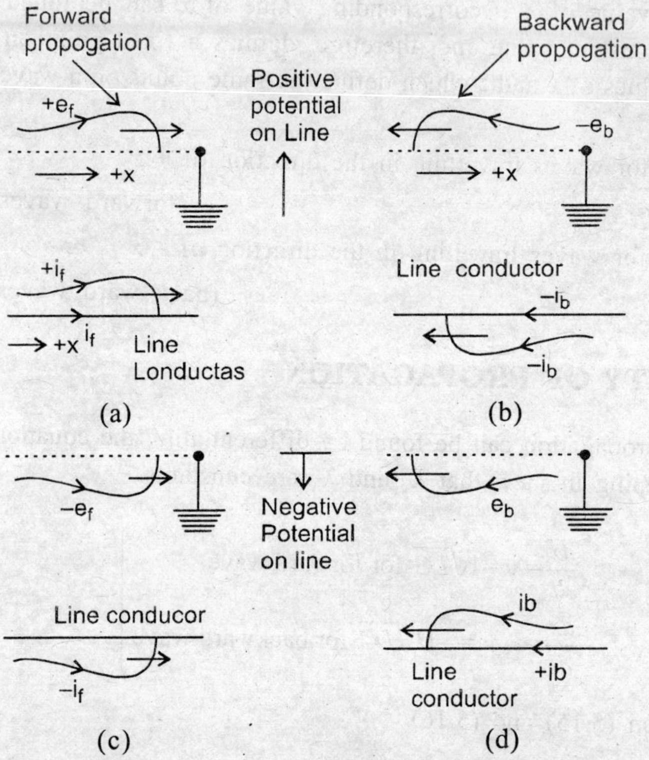

Fig. 5.2 *Schematic represention of polarties and directions of voltage and current waves.*

Let ΔQ is charge on the element Δx

$$\Delta Q = Ce\Delta x \quad ...(5.21)$$

$$i = \frac{\Delta Q}{\Delta t} = Ce\frac{\Delta x}{\Delta t} = Cev \quad ...(5.21)$$

The change in flux $\Delta\phi$ on element Δx is

$$\Delta\phi = L\Delta xi = L\Delta x\ Cev \quad ...(5.22)$$

$$e = \frac{\Delta\phi}{\Delta t} = LCev\frac{\Delta x}{\Delta t} = LCev^2 \quad ...(5.23)$$

$$\therefore \qquad e = LCev^2$$

or
$$v = 1/\sqrt{LC}$$

Where, v = velocity of wave Propagation.

5.4 SURGE IMPEDANCE

We know voltage in inductor of small element Δx is,

$$\frac{\partial e}{\partial k} = -L\frac{\partial i}{\partial t} \qquad ...(5.24)$$

Similarly in capacitor

$$\frac{\partial i}{\partial x} = -C\frac{\partial e}{\partial t} \qquad ...(5.25)$$

integrating equation (5.25) both the sides

$$\int\frac{\partial i}{\partial x}dx = -\int C\frac{\partial e}{\partial t}dx \qquad ...(5.26)$$

$$\int\frac{\partial i}{\partial x}dx = -C(-u)\int\frac{\partial e}{\partial t}\partial x \qquad ...(5.27)$$

$$\therefore \qquad \text{i} = \text{Cue} \qquad ...(5.28)$$

$$\frac{e}{i} = \frac{1}{Cu} = \frac{1}{C}\sqrt{LC} = \sqrt{\frac{L}{C}} \qquad ...(5.29)$$

Where, $\frac{e}{i}$ is impedance i.e), $\frac{e}{i}$ = Z

The quantity $\sqrt{\frac{L}{C}}$ is called the characteristic impedance and it is denoted as Zo.

i.e, $$\text{Zo} = \sqrt{\frac{L}{C}}$$

Surge impedance is the impedance of transmission line when the surge voltage occurs at the transmission lines due to external (or) internal disturbances.

5.5 REFLECTION AND TRANSMISSION OF TRAVELING WAVES UNDER DIFFERENT LOADINGS:

When a wave arrives at a discontinuity in a line, where the characteristic impedance of the line changes, some adjustment must occur if the proportionality is not be violated. This adjustment takes in the form of initiation of two new wave pairs. The reflected voltage wave and its associated current wave travel backward to the line, superimposed on the initial or forward (or) incident wave. The superimposed on the initial or forward (or) incident wave. The refracted or transmitted wave penetrates beyond the discontinuity. The amplitudes of reflected and refracted waves are such that the voltage to current proportionality is preserved for each as demanded by the surge impedances of the lines on which they are travelling, current and voltages at the discontinuity are themselves continuous and energy is conserved.

Let consider a line with surge impedance Z_0 terminated by an impedance Z. Which is shown in Fig. (5.3) and Let,

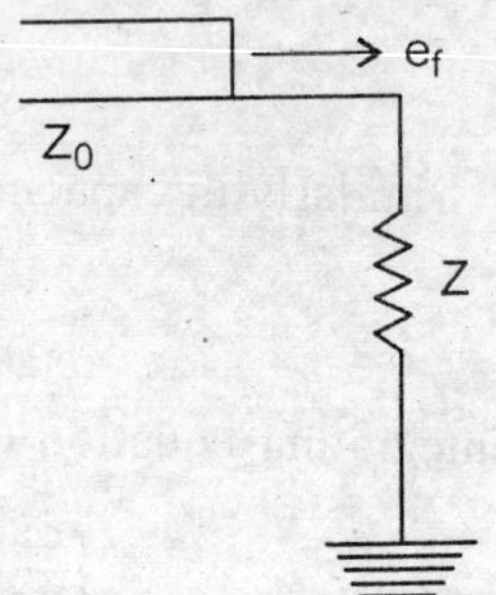

Figure 5.3 *A line terminated by Z*

$e_f \rightarrow$ Forward voltage

$i_f \rightarrow$ Forward current

$e_r \rightarrow$ Reflected voltage

$i_r \rightarrow$ Reflected current

$e_t \rightarrow$ Refracted voltage

$i_t \rightarrow$ Refracted current

Then
$$\frac{e_f}{i_f} = Z_0 \quad \text{...(5.30)}$$

$$\frac{e_r}{i_r} = -Z_0 \quad \text{...(5.31)}$$

$$\frac{e_t}{i_t} = Z \quad \text{...(5.32)}$$

The sign in equation (5.31) is because of the reason that e_r and i_r are travelling in the negative direction of x or backwards on the same line.

Since voltage and current waves are continuous at the junction

$$e_t = e_f + e_r \quad \text{...(5.33)}$$

$$i_t = i_f + i_r \quad \text{...(5.34)}$$

Substitute the values of i_f, i_r and i_t from equation (5.30,), (5.31) and (5.32) in equating (5.34)

$$\frac{e_t}{Z} = \frac{e_t}{Z_0} - \frac{e_r}{Z_0} \quad \text{...(5.35)}$$

From equation (5.33) and (5.35)

$$e_t\left[1+\frac{Z_0}{Z}\right] = 2e_f$$

$$e_t = e_f \frac{2Z}{Z + Z_0} \quad \text{...(5.36)}$$

$$e_r = e_t - e_f = e_f \frac{Z - Z_0}{Z + Z_0} \quad \text{...(5.37)}$$

$$i_r = \frac{e_t}{Z} = e_f \frac{2}{Z + Z_0}$$

$$i_r = \frac{-e_r}{Z_0} = \frac{-e_f}{Z_0} \cdot \frac{Z - Z_0}{Z + Z_0} = i_f \frac{Z_0 - Z}{Z_0 + Z} \quad \text{...(5.38)}$$

The coefficients $\frac{2Z}{Z + Z_0}$ and $\frac{Z - Z_0}{Z + Z_0}$ are called

Refraction and reflection coefficients respectively.

Now we will be under different loaded condition.

(i) Line terminated by a resistance equal to surge impedance.

If we substitute Z_0 for Z in equation (5.35 to 5.38) we find that $e_t = e_f$; $i_t = i_f$; $e_r = 0$ and $i_r = 0$ thus, the incident wave continues as it is and there is **no** reflection.

(ii) Open circuited line

If we substitute $Z = \infty$ in equation (5.35 to 5.38) we find that voltage at **open** end $e_t = 2e_f$

$$i_t = 0$$
$$e_r = e_f$$
$$i_r = -i_f$$

An open circuit at the end of a line demands that the current at that point be Zero at all times. Thus when an incident current wave i_f arrives at the open circuit, a reflected wave equal to $-i_f$ is at once initiated to satisfy this boundary condition.

Conclusion: The resultant voltage at the open end is twice the incident voltage. The wave is shown in (Fig. 5.4)

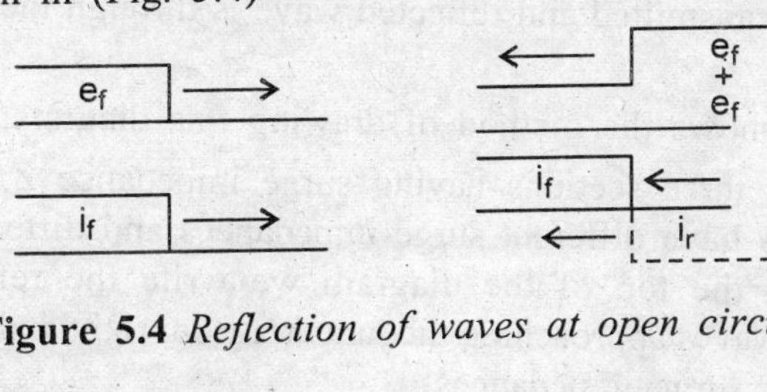

Figure 5.4 *Reflection of waves at open circuit*

(iii) Short circuit line

If we substitute $Z = 0$ in equations (5.35 to 5.38) we get

$$e_t = 0$$
$$i_t = 2i_f$$
$$e_r = -e_f$$
$$i_r = i_f$$

The unique characteristic of the short circuit is that the voltage across them are Zero. When an incident voltage wave e_f reaches a short circuited condition, then the reflected voltage wave must be $-e_f$ to satisfy the condition that two voltage across the short circuit is Zero. The waves are shown in Fig. (5.5).

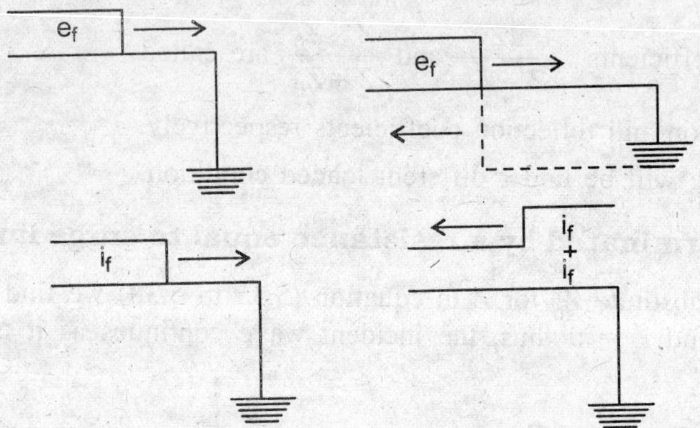

Figure 5.5 *Reflection of waves at short circuit*

Conclusion: Resultant current at shorted end will be twice that of incident currents.

5.6 BEWLAY'S LATFICE DIAGRAM

In large network with many junctions and terminations, the number of transmitted and reflected waves initiated by a single incident wave increases as the wave meets various junctions. It is true that we cannot have more energy than that possessed by the initial surge and this energy will be shared by many waves. At some points may build up due to reinforcing action of the waves. A systematic way of keeping track of all the transmitted and reflected wave is through the use of Bewlay Lattice diagram.

Figure 5.5 shows the method of drawing this diagram.

A line with three sections having surge impedance Z_1, Z_2 and Z_3 is shown. These 3 sections have different surge impedances and different velocities of wave propagation. At the top of the diagram we write the reflection and refraction coefficients for waves approaching the junction from right and left. These coefficients are functions of surge impedances.

Thus, a = Reflection coefficient for waves approaching from left

a' = Reflection coefficient for waves approaching from Right

b = Refraction coefficient for waves approaching from left

b' = Refraction coefficient for waves approaching from right

In drawing the Lattice, the horizontal distances between junctions are laid proportional to the time of passage of wave on these sections. The advantage of doing this is that all diagonals have the same slope and the time scale is applicable to each branch. The vertical time scale can be suitably selected depending on the time for which computations are to be made.

A unit incident wave approaching junction 1 from left gives rise to a refracted wave b_1 and reflected wave a_1. The wave b_1 reaches junction 2 and gives rise to refracted wave $b_1\, b_2$ and reflected wave $b_1\, a_2$. Wave $b_1\, a_2\, a'_1$ and so on. Thus starting with the initial incident wave, the various reflected and refracted waves are determined and indicated on the diagram.

From diagram 5.6, the total potential at any point, at any time, is the superposition of all the waves which have arrived at that point.

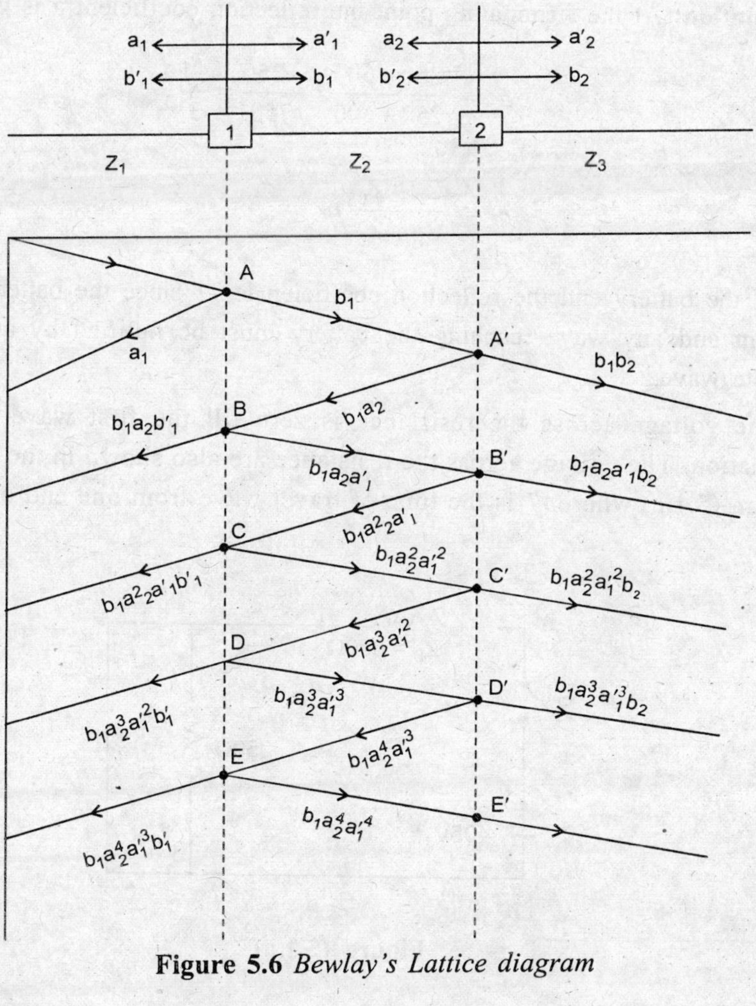

Figure 5.6 *Bewlay's Lattice diagram*

Guidelines to draw Bewlay's Lattice diagram

- At point A ⇒ a_1 and b_1 get reflected.
- At point A' ⇒ b_1 gets divided by two so b_1 is common to both the directions. Then according to directions of reflection and refraction add corresponding element with b_1.
- At point B ⇒ $b_1 a_2$ get divided into two parts so, $b_1 a_2$ is common to both the directions then according to direction of reflected and refraction wave add the corresponding element.

Example 5.1: A line surge impedance 500Ω is charged from a battery of constant voltage 405V. The line is 400m long and is terminated in a resistance of 250Ω. Plot reflection lattice and the voltage across the terminating resistance.

Solution: At the terminating point the reflection coefficient a is given by,

$$a = \frac{250-500}{250+500} = \frac{-250}{750} = \frac{-1}{3}$$

i.e.,

$$a = \frac{Z_1 - Z_0}{Z_1 + Z_0}$$

At the battery end the reflection coefficien is –1 since the battery voltage is constant and any wave reaching the battery must be nulified by an equal and opposite wave.

The voltage across the resistance is zero till the first wave reaches the termination. The voltage across the resistance are also shown in the same lattice diagram (5.1.c) where. T is the time of travel wave from one end of line to the other.

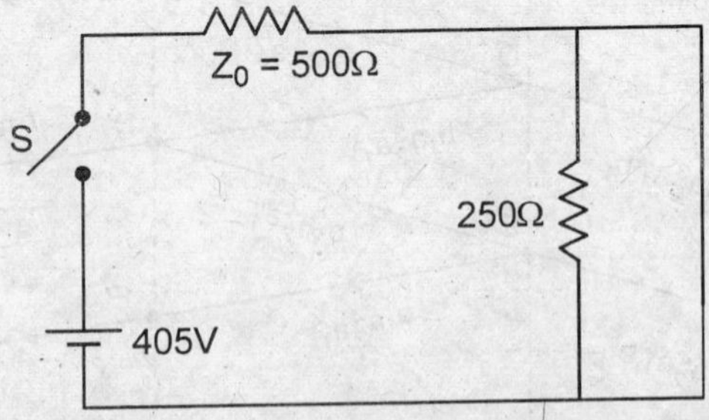

Figure (5.1.a)

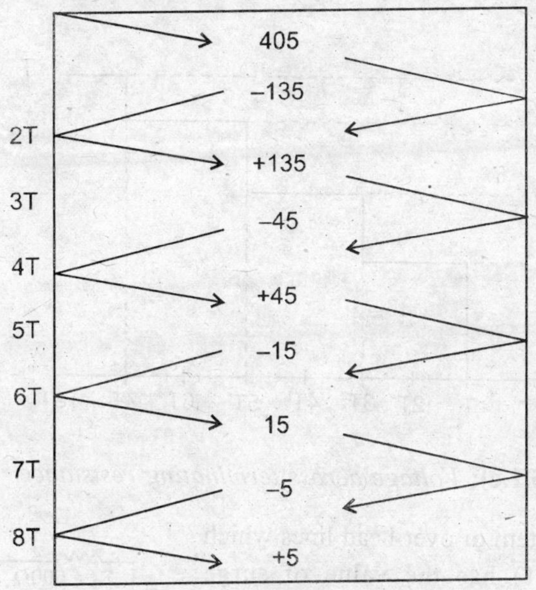

Figure (5.1.b) *Latlice diagram*

➥ **Explantation:** We know $a = \frac{-1}{3}$

∴ The value increased by three times.

First $5 \Rightarrow 5 \times 3 \Rightarrow 15$

$15 \times 3 \Rightarrow 45 \times 3 = 135$

$135 \times 3 \Rightarrow 405$

Then $405 - 5 = 400$

* $405 - 15 = 390$ * $405 - 45 = 360$

* $405 - 135 = 270$

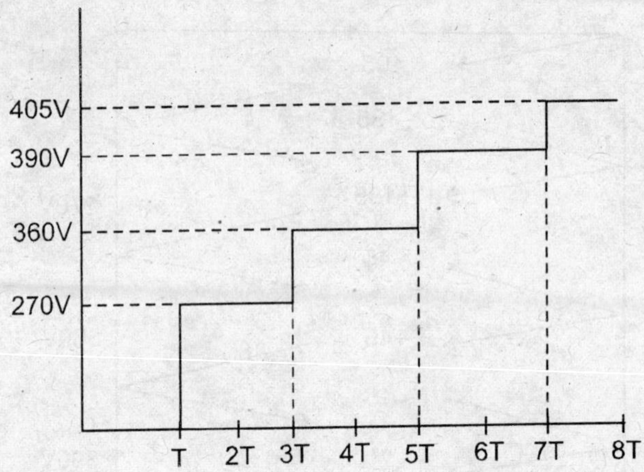

Figure (5.1.c): *Voltage across terminating resistance*

Example 5.2: A system of over head lines which is shown in fig. (5.2.1) has the value of surge impedances and lengths of different sections are shown. The velocities of propogotion are 300 m/μ-sec for head line and 200m/μ-sec for cable. A unit step wave approaches the system from left. Draw reflection lattice diagram and plot voltage waveform till a time of 8 μ-seconds.

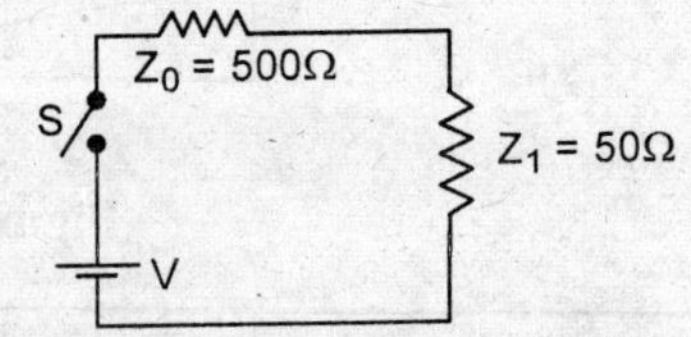

Figure 5.2.1

Solution: Let we find reflection and refraction coefficients $a_1 = -a_2 = a_3 = -$

$$a_4 = a'_4 = a'_3 = a'_2 = a'_1 = \frac{Z_0 - Z_1}{Z_0 + Z_1}$$

$$= \frac{50 - 500}{50 + 500} = -0.818$$

$$b_1 = b_3 = b'_4 = b'_2 = \frac{2Z_0}{Z_1 + Z_0} = 0.182$$

$$b_2 = b_4 = b'_3 = b'_1 = \frac{2Z_1}{Z_1 + Z_0} = 1.82$$

The time of travel of the wave on each section of cable is 1μ-sec. For the 150m over head line the time of travel of the wave is 0.5 μ-sec.

Let see the Lattice diagram.

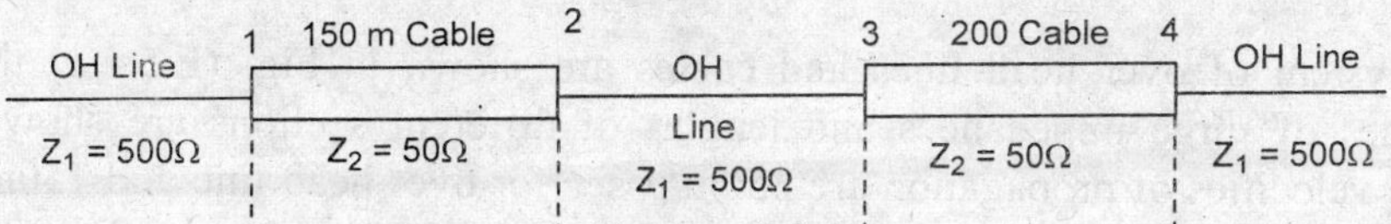

Figure 5.2.1

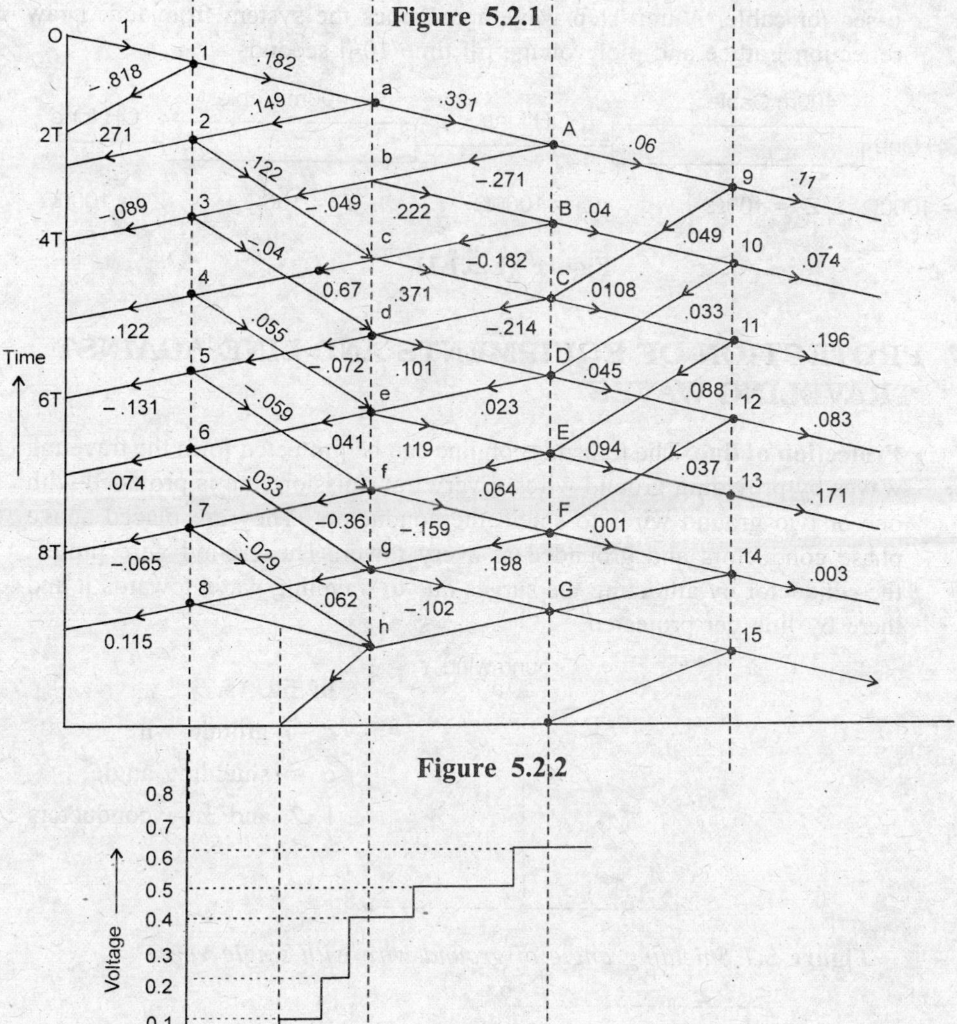

Figure 5.2.2

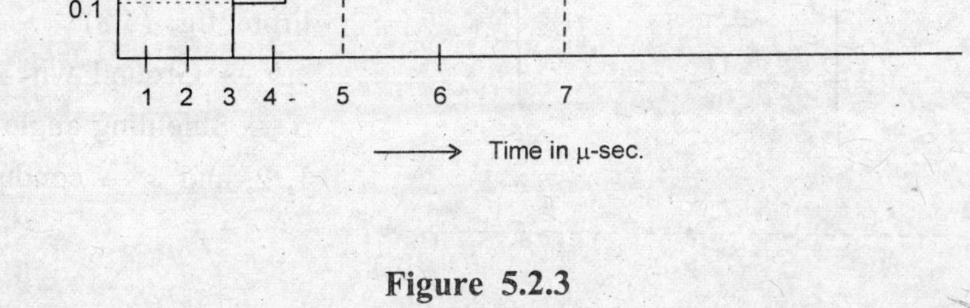

Figure 5.2.3

EXERCISE

E.5.1: A system of over head lines and cables are shown in Fig. (E.5.1.1) the values of surge impedances and lengths of different sections are shown. The velocities of propagation are 350m/μ-sec for over head line and 150m/μ-sec for cable. A unit step wave approaches the system from left. Draw reflection Lattice and plot voltage till time 10/μ-seconds

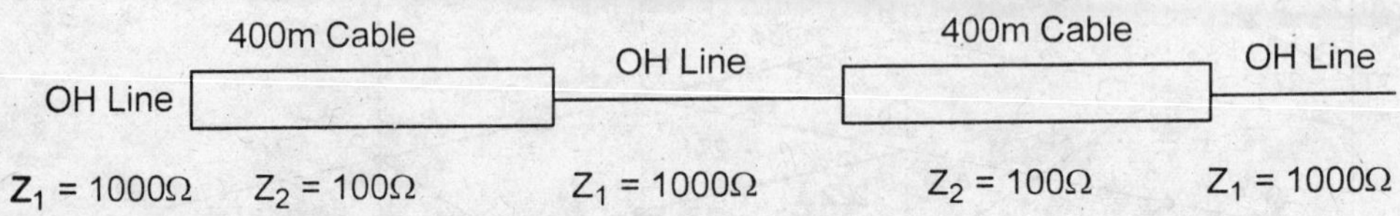

Figure (E.5.1.1)

5.7 PROTECTION OF EQUIPMENTS AND LINE AGAINST TRAVELING WAVES

(*i*) **Protection of line**: The transmission line can be protected form the traveling waves by providing ground wires. Every transmission line is provided with one or two ground wires to shield the conductors. They are placed above phase conductors and grounded at every tower. The ground wire protect the conductor by attracting the surges due to traveling wave towards it and there by line get protected.

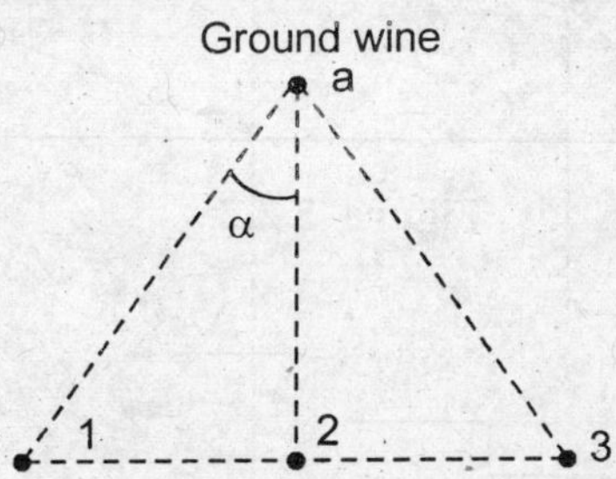

In fig. (5.7)

a → ground wire

α → shielding angle

1, 2, and 3 → conductors

Figure 5.7 *Shielding angle of ground wire with single wire*

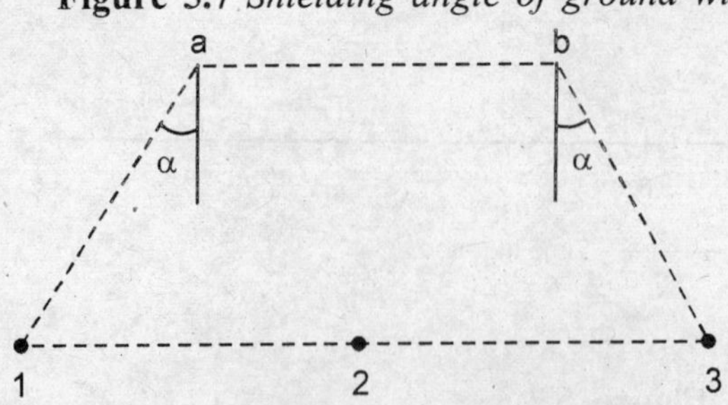

From fig. (5.8)

a,b → Ground wire

α → Shielding angle

1, 2, and 3 → conductors

Figure 5.8

The selection of size, number and arrangement of ground wires is of great importance in line design. The selection of size is based on the consideration of mechanical strength. The number and arrangement depends on the degree of lightning protection needed for the line.

The shielding angle can be defined as the angle between a vertical plane through the ground wire and a slanting plane connecting the ground wire with the outermost phase conductor. Shedding angle of about 30° gives proper (or) required protection. A double ground wires provides better protection as compared to single ground wire. Shielding angle can also be named as protecting angle.

(*ii*) **Protection of equipments:** The equipments of power system can be protected from surges due to traveling wave by surge arrester (or) surge diverter (or) Lightning arrester.

Lightning arrester is a device which is connected between line and earth. i.e., in parallel with the equipment to be protected at the substation. It is a safety valve which limits the magnitude of lightning and switching over voltages at the substation and provides a low resistance path for the surge current to flow to the ground. It is situated (or) installed adjacent to the equipment to be protected.

Requirements of Surge Arrester

- Under normal condition there should not be current flow in surge arrester.
- Spark over Voltage must be higher than the normal voltage.
- This should operate as quickly as possible to any type of surge voltages.
- Immediately after the surge voltage falls below the breakdown value, it must interrupt the following power frequency current instantaneously so that the circuit outage is avoided.

There are many types of surge protecting devices for power system equipments which are

1. Rod gap arrester
2. Horn gap arrester
3. Valve-type surge diverter
4. Expulsion type surge diverter
5. Metal-oxide surge diverter

1. Rod gap arrester

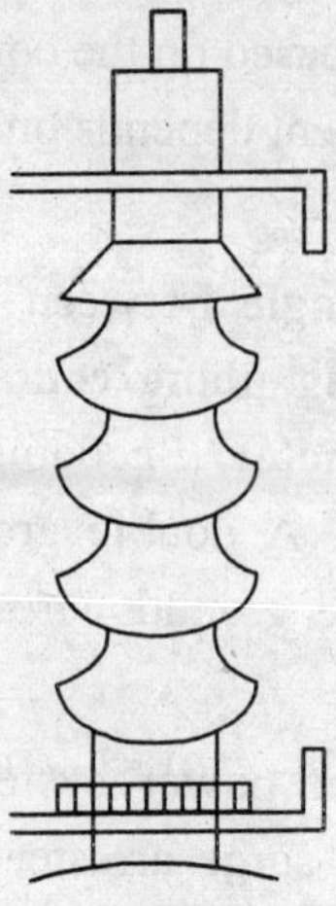

Figure 5.9 *Rod gap on an insulator string*

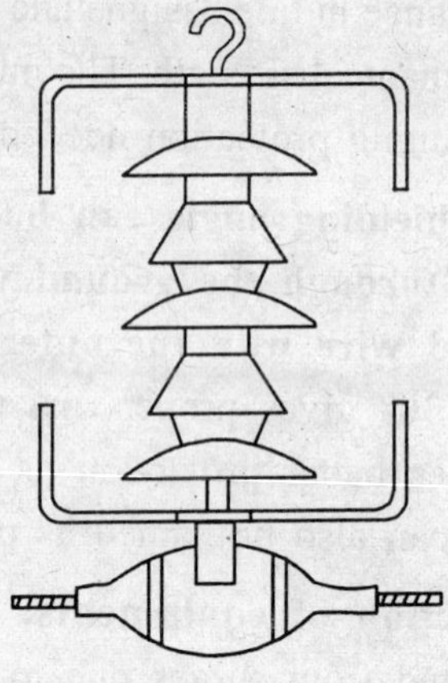

Figure 5.10 *Rod gap on a transformer*

Rod gap means air gap between ends of two rods. When surge occurs in any device (specially in transformer) rod gap should break before the equipment which is to be protected.

Due to limitations of this type of arresters, it is not used much.

Disadvantages

- Due to atmospheric condition such as humidity, temperature and barometric pressure affect the performance of gaps.
- Due to arc produced between two rod gaps there will be wear and tear there by performance gets reduced.

2. Expulsion type surge diverter

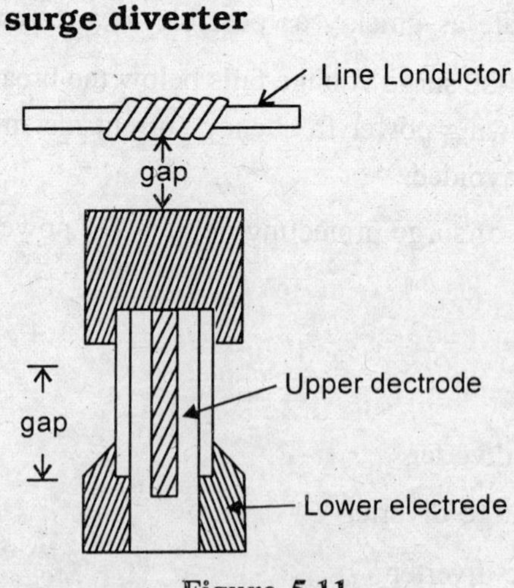

Figure 5.11

The Fig. 5.10 shows the constructional details of surge diverter it has two electrodes namely upper electrode and lower electrode. The upper electrode is kept some distance apart from line conductor in order to have gap. Under normal condition this gap avoids the Leakage currents. The bottom electrode is connected to earth.

When a surge voltage appears then gas produced inside the tube which extinguish the arc between the electrodes.

Arc is interrupted when the current passes through zero point.

It is better than rod gap type.

3. Valve-type surge diverter

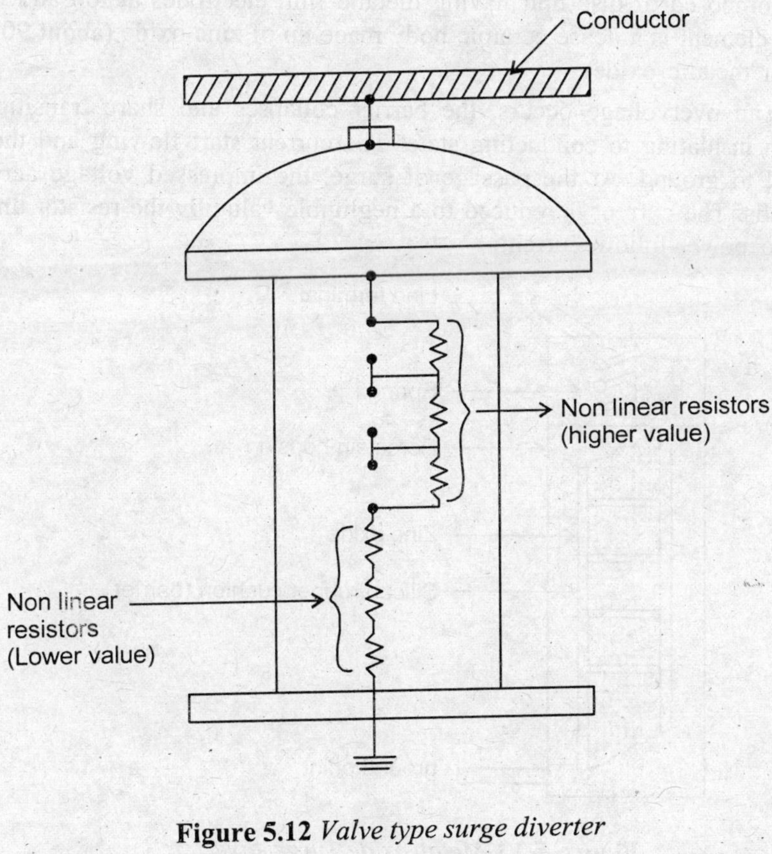

Figure 5.12 *Valve type surge diverter*

This type also called as thyrite surge arrester.

The concept behind this operation is that resistance value decreases as the voltage across it increases.

i.e. $$i = kv^{\alpha}$$

Where, i → crest current

v → crest voltage

k → constant which depends on design exponent α.

In Fig. 5.11 we have shown only one spark gap but practically according to application we can have any number of spark gaps.

When surge voltage occurs then there will be voltage distribution across the gaps which will be non uniform. In order to equalise the surge voltage the gaps are shunted by non linear high resistors. Grading capacitors also can be used to equalise the voltage. This is the way the value type surge diverter diverts the surge voltage to earth.

4. Mutal oxide Gapless surge diverter

It is composed of disc unit having metalic film electrodes at top and bottom. The value element is a dense ceramic body made up of zinc oxide (about 90%) and some other metalic oxides.

When an overvoltage occurs, the barrier collapses and sharp transition take place from insulating to conducting state. The current start flowing and the surge is diverted to ground. At the passage of surge, the impressed voltage across the diverter falls. The current is reduced to a negligible value by the resistor units and there is no power-follow current.

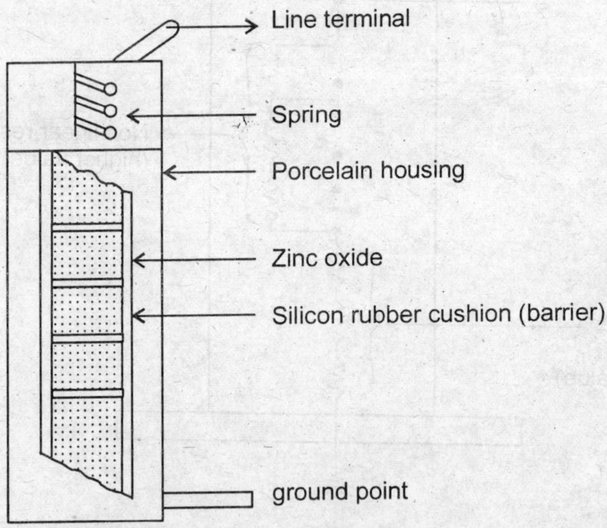

Figure 5.13 *Metal oxide surge diverter*

Advantages

1. Simple construction
2. Superior energy absorption capability

3. Better surge protection
4. More stable protective characteristics

Short Questions and Answers

Q. 1. How traveling waves are produced?

Ans. When disturbances occurs on transmission line due to switching (or) lightning etc., there will be surge voltage produced on transmission line which travels along the transmission line till equipment and which is called as traveling wave.

Q. 2. Write the wave equation with description

Ans.
$$e = f_1\left(t+\frac{x}{v}\right)+f_2\left(t-\frac{x}{v}\right)$$

$$i = -\sqrt{\frac{c}{L}}\left[f_1\left(t+\frac{x}{v}\right)-f_2\left(t-\frac{x}{v}\right)\right]$$

$v \rightarrow$ velocity

$c \rightarrow$ line capacitance

$L \rightarrow$ line inductance

$f_1 \rightarrow$ Arbitray single valued function of $\left(t+\frac{x}{v}\right)$

$f_2 \rightarrow$ Arbitray single valued function of $\left(t-\frac{x}{v}\right)$

Q. 3. What is surge impedance

Ans. It is the value of impedance of transmission line when surge occurs on it. Which gives ratio between surge voltage (*e*) and surge current (*i*)

i.e. $$Z_0 = \frac{e}{i}$$

Q. 4. What is the value of surge impedance?

Ans. Surge impedance $Z_0 = \sqrt{\frac{L}{C}}$

where, $L \rightarrow$ line inductance

$C \rightarrow$ line Capacitance.

Q. 5. Write the formula for reflection and refraction coefficients.

Ans. (*i*) **Refraction coefficient**

$$\frac{2Z}{Z+Z_0}$$

where, $Z \rightarrow$ line impedance

$Z_0 \rightarrow$ Surge impedance

Ans. (*ii*) **Leflction coefficient**

$$\frac{Z - Z_0}{Z + Z_0}$$

Q. 6. What will happen to traveling wave when $Z_0 = Z_1$?

Ans. Incident wave will be continues but there will not be reflection.

Q. 7. What will happen to traveling wave under open circuit condition?

Ans. (*i*) Reflected wave will be equal to $-\ i_f$

(*ii*) Resultant voltage at opened end will be twice the incident voltage.

Q. 8. What will happen to traveling wave under short circuit condition?

Ans. (*i*) The value of reflected voltage wave will be $-\ e_f$

(*ii*) Resultant current at shorted end will be twice the incident current.

Q. 9. What is Bewlay's Lattice diagram?

Ans. It is a technique which is used to track the traveling wave of surge voltage and surge current along transmission line by drawing reflected and refracted wave on sequence.

Q. 10. What are the required data's to draw Lattice diagram?

Ans. (*i*) Surge impedance values

(*ii*) Transmission line Length

(*iii*) Line impedance value.

(*iv*) Reflection and refraction coefficients

> ➡ Reflection and refraction coefficients can be determined by using surge impedance and line impedance.

Q. 11. How to protect transmission line from surges?

Ans. By using ground wires (Ground wires are used shield the conductors they are placed above phase conductors and grounded at every tower.)

Q. 12. What is shielding angle?

Ans. The angle between vertical plane through the ground wire and a slanting plane connecting ground wire with outer most phase conductor.

Q. 13. How to protect the Power system equipments?

Ans. By providing surge diverters (or) surge Arresters.

Q. 14. What is surge divertor?

Ans. It is a protective device which diverts the surge voltage to the ground without affecting the equipment operation.

Q. 15. What are the types of surge diverters?

Ans. (*i*) Rod gap Arrester

(*ii*) Valve-type surge arrester

(*iii*) Expulsion type surge diverter

(*iv*) Metal oxide surge diverter

Choose The Best Answer

1. Surge voltage occurs at transmission line due to

(a) Transients (b) Lightning

(c) both a & b (d) None of the above

2. Wave equation depends on

(a) line inductance (b) line capacitance

(c) both a & b (d) None of the above

3. Velocity of propagation depends on

(a) line inductance (b) line capacitance

(c) both a & b (d) None of the above

4. Which is the correct answer for velocity of propagation?

(a) $\sqrt{\frac{L}{C}}$ (b) $\sqrt{\frac{C}{L}}$

(c) $\frac{1}{\sqrt{LC}}$ (d) None of the above

5. Which is the correct answer for surge impedance?

(a) $\sqrt{\frac{L}{C}}$ (b) $\sqrt{\frac{C}{L}}$

(c) $\frac{1}{\sqrt{LC}}$ (d) None of the above

6. Value of resultant voltage under open circuited condition?

(a) twice the incident voltage

(b) Zero

(c) equal to incident voltage

(d) None of the above

7. What is the value of shielding angle which is provided by single ground wire?
   (a) 30° (b) 60°
   (c) 0° (d) 90°

ANSWERS

1. c 2. c 3. c 4. c 5. a 6. a 7. a

Fill in the Blanks

1. Surges are produced by ________
2. Lattice diagram used to ________ along transmission line.
3. Reflection and refraction coefficients depends on ________ and ________
4. The value of reflected voltage ware under open circuit is ________
5. The value of reflected current wave under short circuit is ________
6. Ground wires are used to protect ________ against ________
7. Power system equipments can be protected by ________

ANSWERS

1. Internal and External disturbances
2. Track the traveling wave
3. Surge impedance, line impedance.
4. $-e_f$
5. $-i_f$
6. Transmission line, surge voltage
7. Surge arrester, (or) surge diverter.

INDEX

NOTES